PULSE OF THE PLANET No.3:

On Peaceful Versus Violent Societies, Nuclear Accidents, and Wilhelm Reich's Orgone Energy for Healing Land and Life

The occasional research report and journal of the
Orgone Biophysical Research Laboratory

Editor: James DeMeo, Ph.D.
Production Assistant: Theirrie Cook, BA

I0120159

CONTENTS: Page

Cover: Rock art depicting a camel-mounted warrior, from the central Sahara desert, c.2000 BCE, a socially turbulent period which developed after the region transformed from lush, wet conditions into dry conditions. See page 3.

www.naturalenergyworks.net   ISBN: 978-0989139069   ISSN: 1041-6773   150706

Editor's Page

More than one year has passed since the last issue of *Pulse of the Planet* was published. During this time, a number of new and exciting research projects have been initiated at the Orgone Biophysical Research Laboratory. These projects include research focused upon the issue of forest death in Europe (usually called "acid rain" in the USA), continued drought abatement work in the USA and overseas, plus considerable new steps regarding the Lab's Desert Greening Project. This work has been the central focus of activity at the Lab, and even the normal schedule of educational workshops has been necessarily reduced. Publication of the *Pulse* has also consequently been delayed. Additionally, given the absence of funds to hire staff members for full-time work on the *Pulse*, we have been forced to eliminate one major section which appeared in prior issues, notably the various weekly "Climate Features" maps, with related geophysical data. At some future date, we hope to re-introduce the maps, and expand them to include social and human factors not contained in prior issues. The decision to eliminate this section was made with much regret, but the maps were the most time-consuming materials to prepare, and were partly the cause for the great delay in publication. The materials contained in this issue remain centrally focused upon the science of orgonomy, however, and given the expanded size with many new and exciting articles, our readers should not be disappointed.

This issue of the *Pulse* contains articles which provide substantial support and new findings regarding the sex-economic and orgone biophysical discoveries of Wilhelm Reich. The articles on "Body Pleasure and the Origins of Violence", and on "Saharasia" provide solid details and evidence on the roots of social and international violence. These findings are constantly being confirmed by daily reports in the newspapers. Regarding the current crisis over petroleum resources, which tragically blossomed into full-fledged war in the Persian/Arabian Gulf, the reader will find insights regarding the underlying behavioral and historical aspects which are not available elsewhere. New findings on the medical applications of orgone energy are also presented, along with a fascinating report on observations of environmental conditions – which include significant *oranur* effects – near the Three Mile Island nuclear power plants during the serious accident there in 1979. We also present a report from Summerhill School in England, along with reports on several orgonomic conferences held during the last few years.

Research reports specifically focused upon the cloudbuster are also provided, including a detailed paper on the breakthrough "OROP Arizona 1989" research project, which confirmed the Desert Greening possibilities of cloudbusting work, as first detailed by Reich in the 1950s. Some of these materials have been published elsewhere; but as brought together in this single volume, they constitute a very powerful confirmation of orgonomic principles.

The *Pulse* will henceforth appear approximately once per year, published as quickly as new materials are gathered and typeset. Thank you for your continued interest and support.

James DeMeo, Ph.D.
Orgone Biophysical Research Lab
El Cerrito, California, USA
Summer 1991

Postscript, to the 2015 Reprint

From 1989 through 1991, the Orgone Biophysical Research Lab (OBRL) published three issues of *Pulse of the Planet*. This was our own in-house journal using older printing and distribution methods that predated the internet, with 500 to 1000 copies of each being gifted or sold. Their content was important material, both for orgonomic science as well as for the history of American and European orgonomy in the post-Reich period.

While the content of the *Pulse* has retained its historical and general natural scientific value, in retrospect I note the change in my thinking on some issues. Previously, as a young university graduate of Environmental Science, I sometimes embraced what I today consider to be scientifically shaky declarations from the environmental groups. As a young professor I would teach the CO2 theory of global warming as one of several competing ideas, and for a few years, considered it to be the best explanation. Today, I do not.

The hidden Marxist agendas of many social activist groups also escaped my notice for a time, but their own extremism increasingly unmasked, and today I am a severe critic of the Marx-Engels totalitarian ideology. For example, see this: www.orgonelab.org/MarxEngelsQuotes.htm

Also the post-911 world of Islamic conquest and terror shattered what small hopes I previously had for self-reform within the Islamic world, an illusion that was unfortunately occasionally expressed in this and prior back issues of *Pulse*.

The long-distance influences of the *Saharasian Desert Belt* on global temperatures and climate became increasingly apparent in my own research and field work. In a parallel manner, the alarming violence of the Saharasian tribes as so aptly detailed in my book *Saharasia*, has also been exported, to wreak havoc in host nations where they have emigrated but refused to abandon misogynistic and violent Sharia Law dictates. But this also was a powerful refutation of Marxist economic determinism, as even massive free wealth from oil revenues did not soften Saharasian character structures, merely providing the means by which to diffuse their totalitarianism around the world.

By contrast, I strongly retain my authentic environmental concerns, notably about nuclear power plants and atomic bomb testing, based upon empirical science. Reich wrote about how atom bomb tests in Nevada could be detected at his laboratory in rural Maine, thousands of miles away. In the early 1990s we also detected such effects, even after the bomb testing had moved underground. This and other issues of *Pulse* helped to detail such long-distance atmospheric, biological and geophysical reactions to the underground atomic explosions, a fact that can only be understood within the context of orgone biophysics. Nuclear power plant accidents pose similar problems.

For those who have followed these issues since the turn of the century, when they review these back issues of *Pulse* from 1989, 1990 and 1991, these differences will become apparent. Nevertheless, the articles and most of the Notes contained in this issue of *Pulse* have stood the test of time, and hold much value for science, medicine, life, health and history.

For all these reasons, the decision to reprint the old *Pulse* issues was not difficult. This issue of *Pulse* has not been changed in the reprint, other than to correct or strike out a few inaccurate postal addresses.

James DeMeo, Ph.D.
OBRL, Ashland, Oregon, USA
Summer 2015

The Origins and Diffusion of Patrism in Saharasia, c.4000 BCE: Evidence for a Worldwide, Climate-Linked Geographical Pattern in Human Behavior*

James DeMeo, Ph.D.**

ABSTRACT

Global geographical patterns of repressive, painful, trau-matic, and violent, armored, patrist behaviors and social insti-tutions, which thwart maternal-infant and male-female bonds, were correlated and developed through a systematic analysis of anthropological data on 1170 subsistence-level cultures. When the behavior data were mapped, the hyperarid desert belt encompassing North Africa, the Near East, and Central Asia, which I call Saharasia, was found to possess the greatest areal extent of the most extreme patrist behaviors and social institutions on Earth. Regions farthest removed from Sahara-sia, in Oceania and the New World, were found to possess the most gentle, unarmored, matrist behaviors, which support and protect maternal-infant and male-female bonds. A systematic review of archaeological and historical materials suggests that patrism first developed in Saharasia after c.4000 BCE, the time of a major ecological transition from relatively wet grassland-forest conditions to arid desert conditions. Settlement and migration patterns of patrist peoples were traced, from their earliest homelands in Saharasia, to explain the later appear-ance of patrism in regions outside of Saharasia. Prior to the onset of dry conditions in Saharasia, evidence for matrism is widespread, but evidence for patrism is generally nonexistent. It is argued that matrism constitutes the earliest, original, and innate form of human behavior and social organization, while patrism, perpetuated by trauma-inducing social institutions, first developed among Homo Sapiens in Saharasia, under the pressures of severe desertification, famine, and forced migra-tions. The psychological insights of Wilhelm Reich provide an understanding of the mechanism by which patrist (armored, violent) behaviors become established and continue long after the initial trauma has passed.

* Previously published in: *Kyoto Review* 23: 19-38, Spring 1990 (Japan) ; *Emotion* 10, 1991 (Germany); and *World Futures: The Journal of General Evolution*, 30: 247-271, 1991. A more comprehensive account of Dr. DeMeo's work on this subject has appeared in the *Journal of Orgonomy*, serialized in each issue since 1987. A book providing more details on this subject is planned for late 1991.

** Director of Research, Orgone Biophysical Research Lab., Ashland, Oregon, USA demeo@mind.net

INTRODUCTION

The present paper summarizes the evidence and conclu-sions of my own seven-year geographical study on the world-wide, regional variation in human behavior, and related socio-environmental factors, a study which constituted my doctoral dissertation (DeMeo 1985, 1986, 1987). In this research, I specifically focused upon a major complex of traumatic and repressive attitudes, behaviors, social customs and institutions which are correlated with violence and warfare. My study proceeded from clinical and cross-cultural observations on the biological needs of infants, children, and adolescents, the repressive and damaging effects that certain social institutions and classes of harsh natural environment have upon those needs, and the behavioral consequences of such repression and damage.

The geographical approach to the origins of human behav-ior, as presented here, has allowed the reconstruction of a much clearer global picture of our most ancient cultural history than has heretofore been possible. The causal relationship between traumatic and repressive social institutions to de-structive aggression and warfare has been verified and strength-ened in my approach, which has confirmed the existence of an ancient, worldwide period of relatively peaceful social condi-tions, where warfare, male domination, and destructive ag-gression were either absent, or at extremely minimal levels. Moreover, it has become possible to pinpoint both the exact times and places on Earth where human culture first trans-formed from peaceful, democratic, egalitarian conditions, to violent, warlike, despotic conditions.

These findings were made possible only by virtue of recent paleoclimatic and archaeological field studies (which revealed previously hidden social and environmental conditions), and by the development of large, global anthropological data bases composed of cultural data from hundreds to thousands of different cultures from around the world. The microcomputer, also a recent innovation, allowed easy access to such data, and the preparation within a few years of global behavior maps which otherwise would have taken a lifetime to prepare. My approach to these questions also constituted one of the first systematically derived, global geographical reviews of human behavior and social institutions, uncovering a previously unob-served, but clear-cut global pattern in human behavior. Before presenting the maps, which display in spatial form the core of my findings, some discussion of the variables of interest, and the theory behind the maps, is in order.

Matrist Versus Patrist Culture: The Roots of Violence in Childhood Trauma and Sex-Repression

My research was initially aimed at developing a global geographical analysis of social factors related to early childhood trauma and sexual repression, as a test of the sex-economic theory of Wilhelm Reich (1935, 1942, 1945, 1947, 1949, 1953, 1967, 1983). Reich's theory, which developed and diverged from psychoanalysis, labeled the destructive aggression and sadistic violence of *Homo sapiens* a completely abnormal condition, resultant from the traumatically-induced chronic inhibition of respiration, emotional expression, and pleasure-directed impulses. According to this viewpoint, inhibition is made chronic within the individual by virtue of specific painful and pleasure-censoring rituals and social institutions, which consciously or unconsciously interfere with maternal-infant and male-female bonds. These rituals and institutions exist among both subsistence-level "primitives" and technologically developed "civilized" societies. Some examples are: unconscious or rationalized infliction of pain upon newborn infants and children through various means; separation and isolation of the infant from its mother; indifference towards the crying, upset infant; immobilizing, round-the-clock swaddling; denial of the breast to, and premature weaning of the infant; cutting of the child's flesh, usually the genitals; traumatic toilet training; and demands to be quiet, uncurious, and obedient, enforced by physical punishment or threats. Other social institutions aim to control or crush the child's budding sexual interests, such as the female virginity taboo, demanded by every culture worshiping a patriarchal high god, and the punishment- and guilt-enforced arranged or compulsive marriage. Most of these ritual punishments and restraints fall more painfully upon the female, though males are also greatly affected. Demands for pain endurance, emotion-suppression, and uncritical obedience to elder (usually male) authority figures regarding basic life decisions are integral aspects of such social institutions, which extend to control adult behavior as well. These repressive institutions are supported and defended by the average individual within a given society, irrespective of their painful, pleasure-reducing, or life-threatening consequences, and are uncritically viewed as being "good", "character building" experiences, a part of "tradition". Nevertheless, from such a complex of painful and repressive social institutions, it is argued, comes the neurotic, psychotic, self-destructive and sadistic components of human behavior, which are expressed in a plethora of either disguised and unconscious, or blatantly clear and obvious ways.

According to Reich's sex-economic viewpoint, a chronic characterological and muscular *armor* is set up in the growing human according to the type and severity of painful trauma it experiences. The biophysical processes which normally lead to full and complete respiration, emotional expression, and sexual discharge during orgasm are chronically blocked by the armor, to a greater or lesser extent, leading to the accumulation of pent-up, undischarged emotional and sexual (bioenergetic) tension. The dammed-up reservoir of internal tension drives the organism to behave in a generally unconscious, distorted, self-destructive, and/or sadistic manner (Reich 1942, 1949). The above processes occur whenever, and only whenever, attempts are made to irrationally deflect or mold human primary biological needs or urges according to the demands of "culture". The denial of the breast to an infant, the beating of a child for defecation or sexual expression, or the forced marriage of young girls to old men ("child betrothal", "bride price"), are examples.

Pain-inflicting and pleasure-censoring rituals and social institutions have been present in most, but by no means all, historical and contemporary cultures. There are, for instance, some cultures (a minority, to be sure) which neither inflict pain upon infants and children, consciously or otherwise, nor repress the sexual interests of children or adults. Of great interest is the fact that these are also nonviolent societies, with stable monogamous family bonds, and congenial, friendly social relations.

Malinowski (1927, 1932) first pointed to such cultures as a rebuttal to Freud's assertion of a biological, pan-cultural nature for childhood sexual latency and the Oedipal conflict. Reich (1935) argued that conditions within Trobriand society proved the correctness of his clinical and social findings relating sexual repression to pathological behavior. Other ethnographic descriptions of similar cultures have been made (Elwin 1947, 1968; Hallet & Relle 1973; Turnbull 1961). Prescott's (1975) and my own (DeMeo 1986, pp.114-120) global cross-cultural studies have confirmed these findings: Societies which heap trauma and pain upon their infants and children, and which subsequently repress the emotional expressiveness and sexual interests of their adolescents, invariably exhibit a spectrum of neurotic, self-destructive, and violent behaviors. Contrawise, societies which treat infants and children with great physical affection and gentle tenderness, and which view emotional expressiveness and adolescent sexuality in a positive light, are by contrast psychically healthy and nonviolent. Indeed, cross-cultural research has demonstrated the difficulty, perhaps the impossibility, of locating any disturbed, violent society which does not also traumatize its young and/or sexually repress them.

A systematic survey of global historical literature independently confirmed the above correlations, between childhood traumas, sex-repression, male-dominance, and family violence, in the descriptions of various warlike, authoritarian and despotic central states (DeMeo 1985, Chapters 6 & 7 of 1986) (1). From similar historical data, Taylor (1953) developed a dichotomous schema of human behavior in various societies. Using Taylor's terminology, and expanding upon his schema according to sex-economic findings, such violent, repressive societies are called *patrist*, and they differ in almost every respect from *matrist* cultures, whose social institutions are designed to protect and enhance the pleasurable maternal-infant and male-female bonds. (2) Table 1 gives a contrast between extreme forms of patrist (armored) and matrist (unarmored) culture.

Many aspects of patrism interfere with the biology of the infant and child in a manner generally unseen elsewhere in the animal world, and some clearly increase infant and maternal mortality and morbidity. Besides the painful or pleasure-reducing rites given in Table 1, it is important to note that most patrist societies possessed, at some time in their recent or distant past, severe psychopathological social disorders designed for the socially-approved, organized discharge of mur-

---

1. My survey involved over 100 seperate sources, to include a number of classical sexological works: Brandt 1974; Bullough 1976; Gage 1980; Hodin 1937; Kiefer 1951; Levy 1971; Lewinsohn 1958; Mantegazza 1935; May 1930; Stone 1976; Tannahill 1980; Taylor 1953; Van Gulik 1961.

2. Some time after my dissertation had been completed, I learned of Riane Eisler's (1987a) study *Chalice and the Blade*, which indentified *dominator* and *partnership* types of social organization. These are nearly identical in concept to the respective patrist and matrist forms of social organization as defined here.

> **" Patrism... must have had *specific times and places of origins* among some, but not all, of the earliest human societies."**

derous rage towards children and women (ie., ritual murder of children, widows, "witches", "prostitutes", &c.), with a complement deification of the most aggressive and sadistically cruel males (totalitarianism, divine kingship). A few contemporary cultures express such conditions in a fully-blown form, or exhibit residues of such conditions, and these are facts which have distinct geographical implications.

For example, given that clinical, cross-cultural, and historical evidence indicates that adult violence is rooted in early childhood trauma and sex-repression, and does not exist where maternal-infant and male-female bonds are protected and nurtured by matrist social institutions, a question naturally arises as to how the cultural gestalt of trauma, repression and violence (patrism) could have gotten started in the first instance. Patrism, with its great outpouring of violence toward infants, children, and women, which is passed from one generation to the next through painful and life-threatening social institutions, must have had *specific times and places of origins* among some, but not all of the earliest human societies. The assumed absence of an innate character to patrism, which derives from the chronic blocking, inhibition, and damming-up of biological urges, demands that this be so. Matrism, however, which springs from freely-expressed, unimpeded biological impulse, and which therefore is innate, would have been global in nature, ubiquitous among all of humankind at the earliest times. Indeed, natural selection would have favored matrism, given the fact that it does not generate the sadistic urges which lead to deadly violence toward women and children, nor does it disturb the emotional bonds between mothers and infants, which impart distinct psycho-physiological survival advantages (Klaus & Kennell 1976; LeBoyer 1975; Montagu 1971; Stewart & Stewart 1978a, 1978b, Reich 1942, 1949).

Confirmation and support for the above assumptions and inferences exists in the geographical aspects of the global anthropological and archaeological data, and it was a central focus of my research to examine the spatial aspects of the facts and observations gathered by different field researchers. (3) For example, certain aspects of matrism and peaceful social conditions had previously been identified in the deepest archaeological layers of some regions, with demonstrated transitions toward more violent, male-dominated conditions in later years. While some researchers have either been unaware of these newer findings, have tended to ignore them, or have objected to their implications, a growing number of studies

---

3. The structure of the argument here demands that we make a sharp distinction between facts, and theories about facts. All behavior science theories attempt to explain a variety of observed clinical and social facts. A few even make the attempt to incorporate into theory the facts of anthropology, that is, behavior in other cultures. However, most of such theories fail to be either global or geographical in nature. That is, they do not attempt to simultaneously explain human behavior among a significant number of the better-studied cultures within each world region. Most behavioral theories, if they address the anthropological literature at all, focus only upon patrist cultures, and fail to pass the test of being both systematically-derived and global. Cross-cultural studies are a great step forward in these matters, but the combined global geographical and cross-cultural approach is an additional, necessary refinement, which will force all behavioral theories to henceforth address the specific facts of history, migration, culture-contact, and natural environment.

have demonstrated major social transitions in ancient times, from peaceful, democratic and egalitarian conditions, to violent, male-dominated, warlike conditions (Bell 1971; Eisler 1987a, 1987b; Huntington 1907, 1911; Gimbutas 1965, 1977, 1982; Stone 1976; Velikovsky 1950, 1984). The geographical aspects of these findings are most telling.

A systematic and global review of such evidence (DeMeo 1985, Chapters 6 & 7 of 1986) revealed distinct global patterns in these archaeological transitions, wherein entire regions were transformed from matrism to patrism within the same general time periods, or where the transition to patrism swept across major portions of a continent, from one end to the other, over a period of centuries. Of major significance was the finding that the earliest of these cultural transformations occurred in specific Old World regions (notably in North Africa, the Near East, and Central Asia, around 4000-3500 BCE), *in concert with major environmental transformations, from relatively wet to arid conditions in those regions.* Later transformations generally occurred in regions outside of the new-formed deserts, associated with the abandonment of the new arid zones, and subsequent invasion of moister borderland territories. The existence of these timed environmental and cultural transitions was most important, given other evidence which suggested that severe drought and desertification had the potential to traumatically disrupt maternal-infant and male-female bonds, just as certainly as any harsh and painful patrist social institution.

Social Devastation in Regions of Drought, Desertification and Famine

Other lines of evidence lead to the conclusion that severe and repeated drought and desertification, which promotes famine, starvation, and mass migrations among subsistence-level cultures, must have been a crucial factor which would have gradually, or even rapidly, pushed early matrist cultures towards patrism. For example:

1) Recent eyewitness reports of culture-change occurring during famine and starvation conditions indicate a resultant breakdown of social and family bonds. Turnbull's (1972) heartbreaking account of the Ik peoples of East Africa is most clear on this point, but other, similar observations have been made (Cahill 1982; Garcia 1981; Garcia & Escudero 1982; Sorokin 1975). Under the most severe famine conditions, husbands often leave their wives and children in search of food; they may or may not return. Starving children and elderly family members are eventually abandoned to struggle on their own, or to die. Children may form roving bands dedicated to stealing food, and the remaining social fabric may be utterly torn apart. The maternal-infant bond appears to endure the longest, but eventually starving mothers will also abandon their young.

2) Clinical research on the effects of severe protein-calorie malnutrition of infants and children indicates that starvation is a trauma of the most severe proportions. A child suffering from marasmus or kwashiorkor will exhibit symptoms of contactlessness and immobility, with, in the most extreme cases, a cessation of body and brain growth. If the starvation has lasted long enough, recuperation to full potential may not occur after food supply is restored, and mild to severe physical and

Table 1: DICHOTOMOUS BEHAVIORS, ATTITUDES, AND SOCIAL INSTITUTIONS

Trait	Patrist (armored)	Matrist (unarmored)
Infants, Children, & Adolescents:	Less indulgence Less physical affection Infants traumatized Painful initiations Dominated by family Sex-segregated houses or military	More indulgence More physical affection Infants not traumatized Absence of pain in initiations Children's democracies Mixed sex children's houses or age villages
Sexuality:	Restrictive attitude Genital mutilations Female virginity taboo Adolescent lovemaking severely censured Homosexual tendency plus severe taboo Incest tendency plus severe taboo Concubinage/prostitution may exist	Permissive attitude No genital mutilations No female virginity taboo Adolescent lovemaking freely permitted Absence of homosexual tendency or strong taboo Absence of strong incest tendency or strong taboo Absence of concubinage or prostitution
Women:	LImits on freedom Inferior status Vaginal blood taboo (hymenal, menstrual & childbirth blood) Cannot choose own mate Cannot divorce at will Males control fertility	More freedom Equal status No vaginal blood taboo Can choose own mate Can divorce at will Females control fertility
Cultural & Family Structure:	Authoritarian Hierarchical Patrilineal Patrilocal Compulsive lifelong monogamy Often polygamous Military structure Violent, sadistic	Democratic Egalitarian Matrilineal Matrilocal Noncompulsive monogamy Rarely polygamous No full time military Nonviolent
Religion & Beliefs	Male/father oriented Asceticism, avoidance of pleasure Inhibition, fear of nature Full time religious specialists Male shamans Strict behavior codes	Female/mother oriented Pleasure welcomed and institutionalized Spontaneity, nature worshiped No full time religious specialists Male or female shaman Absence of strict codes.

emotional retardation may occur. Other effects of famine and starvation upon children and adults have been noted, to include reductions in general emotional vitality and sexual energy, some effects of which may persist even after food supply is restored. Importantly, the infant biophysically and emotionally withdraws and contracts under conditions of famine and starvation in a manner nearly identical to the equally traumatic effects of maternal deprivation and isolation. Both sets of experiences have clear, lifelong effects which disturb the ability of adults to emotionally bond with both mate and offspring. (Aykroyd 1974; Garcia & Escudero 1982; Prescott, Read & Coursin 1975).

3) A number of other traumatic factors specifically related to the hard life in deserts and droughty regions were identified. One major example was the use of the restraining, head-molding, back-pack cradle by migratory peoples of Central Asia, which appears to have inadvertently led to the dual traumas of infant cranial deformation and swaddling. Infant cranial deformation as a social institution died out around the turn of the century, but swaddling today appears to persist in the same general regions. Normally, an infant subjected to painful restraint struggles to free itself and will cry loudly, quickly attracting the help of alert caretakers. Not so, I speculate, among famished infants strapped into a body-restraining (and oftentimes head-squashing) back-pack cradle for a long march during a parching drought. Under extreme drought and famine conditions, caretakers would become less attentive, contactless, and less willing to constantly stop and quiet a child hurting in the cranial-deforming restraints of a back-pack cradle. As desertification progressed in Central Asia, migration from region to region became a relatively permanent way of life. The archaeological record suggests that cranial deformations and swaddling subsequently became institutionalized parts of child-rearing tradition in those same areas (DeMeo 1986, pp.142-152; Dingwall 1931; Gorer & Rickman 1962). Indeed, painful cranial deformations and swaddling became an identifying mark and cherished social institution of such peoples, to persist even after they gave up the nomadic existence for a settled lifestyle. Other major social institutions, such as male and female genital mutilations (circumcision, infibulation), were found to be geographically centered on, and have their earliest origins within the great Old World desert belt, though for reasons that are less clear.

In the process of making the above determinations, it became increasingly apparent to me that early matrist social bonds might have first been shattered among subsistence-level cultures which had survived the devastating effects of severe, sequential droughts, desertification, and prolonged famine. With the progressive, generation-after-generation disruption of maternal-infant and male-female social bonds by hyperaridity, famine, starvation, and forced migrations, there would be a consequent development and intensification of patrist attitudes, behaviors, and social institutions. And these would gradually replace the older matrist ones. Patrism would have become fixed into the character structure just as hyper-arid, desert conditions became fixed into the landscape. And once so fixed, patrism would remain with the afflicted people, irrespective of subsequent climate or food supply, given the

behavior-affecting, self-duplicating character of social institutions. Patrism would thereafter appear in the moister regions of plenty by virtue of irruptions of migrating, warlike peoples from adjacent desert regions.

From the above considerations, a very clear geographical test was thereby suggested. If a mapped, worldwide spatial correlation existed between harsh desert environments and extreme patrist culture, then a clear mechanism for initiating the first trauma and repression among ancient human cultures would be identified. This would also directly corroborate Reich's sex-economic theory, which necessitated some ancient mechanism of trauma to explain the genesis of armoring. The spatial correlations which emerged from this approach were startling.

Swaddled infant drawing by Deborah Carrino, based on a photo by Dean Conger. Deformed crania plates from Dingwall (1931).

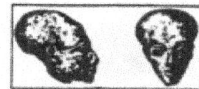

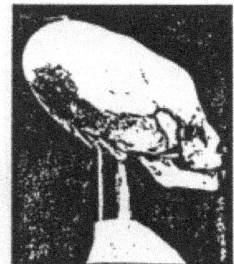

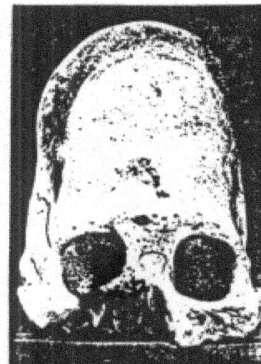

Swaddling and Artifically Deformed Crania appear as complementary practices, as first developed in Central Asia with use of the back-pack cradle by migrating peoples. Infant cranial deformation has died out, but swaddling, a remnant practice, persists in most regions influenced by such peoples.

Male Genital Mutilations

■ Extremely Severe Forms:
(Flaying, Circumcision,
Subincision)

○ Forms of Lesser Severity:
○○ (Incision)

From the data of Murdock (1967)
and Montagu (1945, 1946)

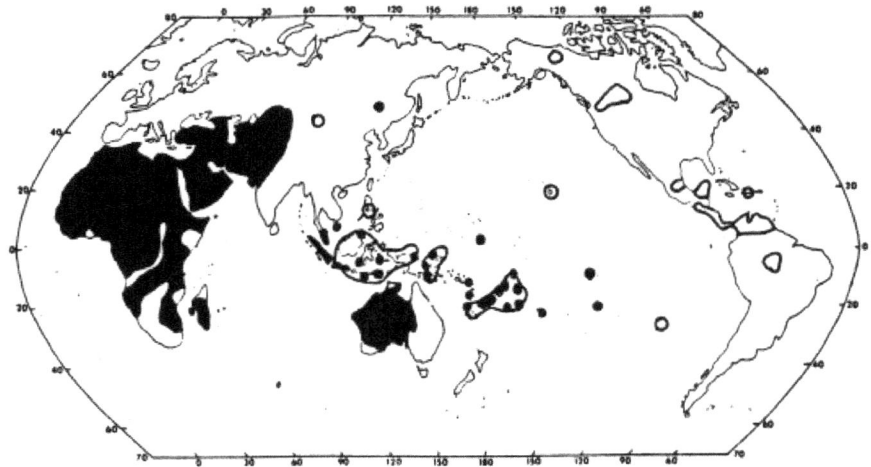

Female Genital Mutilations

■ Extremely Severe Forms:
(Infibulation)

▨ Severe Forms:
(Clitoridectomy, Excision)

○○ Present, but Type Unclear

From the data of Hosken (1979)
and Montagu (1945, 1946)

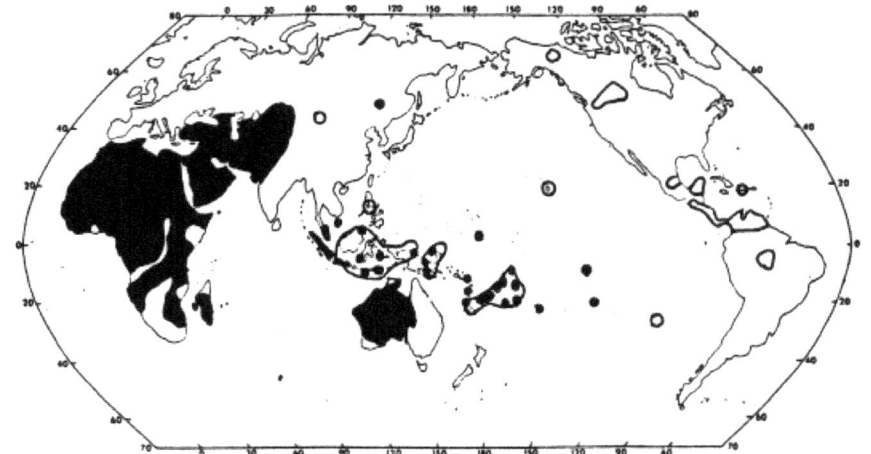

Infant Cranial Deformation
and Swaddling

▨ Present; Arrows Mark
Diffusion Patterns

From the data of Dingwall (1931)

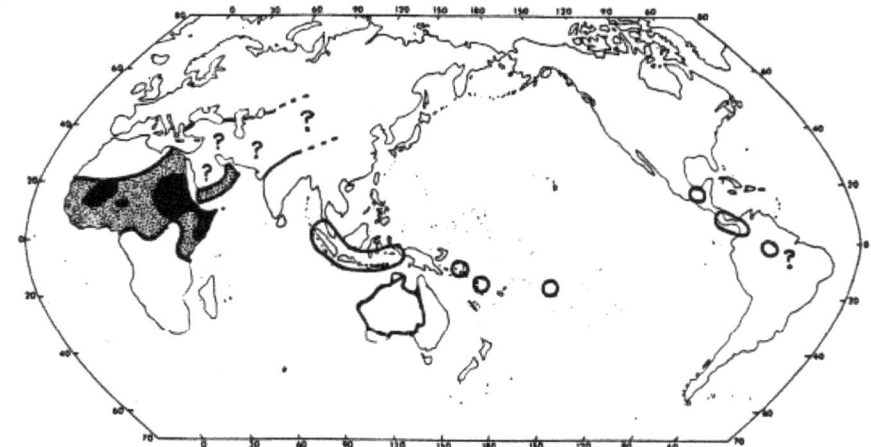

NOTE: All maps are composed of data from native, aboriginal, subsistence-level peoples.
In the Americas and Oceania, these data reflect conditions generally prior to the arrival of European settlers.

Normal Versus Marasmatic Infants

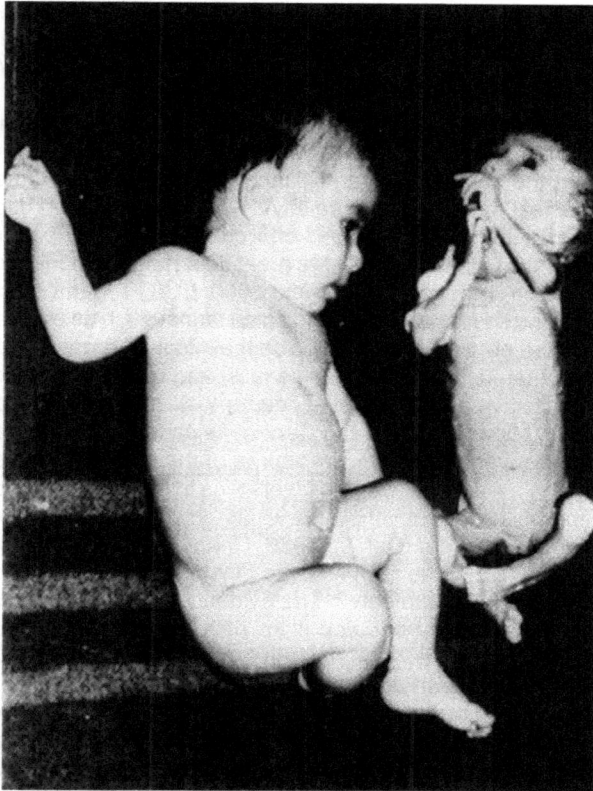

Left infant
5 months old
healthy

Right infant
7 months old
marasmatic.

Reproduced Courtesy of F. Monckeberg (in Prescott, et al. 1975)

THE GEOGRAPHICAL ASPECTS OF ANTHROPOLOGY AND CLIMATOLOGY

My preliminary review of behavior and social institutions in a sample of 400 different subsistence-level, aboriginal cultures from around the world indicated that the most extreme of patrist peoples lived in desert environments (DeMeo 1980), though not exclusively so. A more systematic and definitive global analysis derived from 1170 different cultures later confirmed the desert-patrist connection, but demonstrated that the generality was *not* valid for all semiarid lands or even hyperarid deserts of limited geographical size, where food and water supplies could be obtained by making a short journey. Moreover, wetland regions adjacent to the largest, most hyperarid deserts were likewise found to be patrist in character, a fact which was later explained in the demonstrated migrations of peoples (DeMeo 1986, 1987). Cultural data used for this later analysis were taken from Murdock's *Ethnographic Atlas* (1967), which did not contain any maps, and was composed almost exclusively of descriptive tabular data on aboriginal peoples living in their native regions. Data for North and South America, and Oceania, in large measure, reflected native, pre-European conditions. Murdock's data was gathered from hundreds of reliable sources published roughly between 1750 to 1960; his data has been constructively reviewed by other scholars, and is widely used for cross-cultural theory testing. Each of the 1170 individual cultures was separately evaluated (by computer) according 15 different variables which approximated the matrist-patrist schema previously given.(4) Cultures exhibiting a high percentage of patrist characteristics received an appropriately high score, while cultures with a low percentage of

---

4. The 15 variables were: Female Premarital Sex Taboos, Segregation of Adolescent Boys, Male Genital Mutilations, Bride Price, Family Organization, Marital Residence, Post-Partum Sex Taboo, Cognatic Kin Groups, Descent, Land Inheritance, Movable Property Inheritance, High God, Class Stratification, Caste Stratification, and Slavery.

Transillumination of the Skulls of Normal (left), Malnourished (center), and Marasmatic (right) Infants.

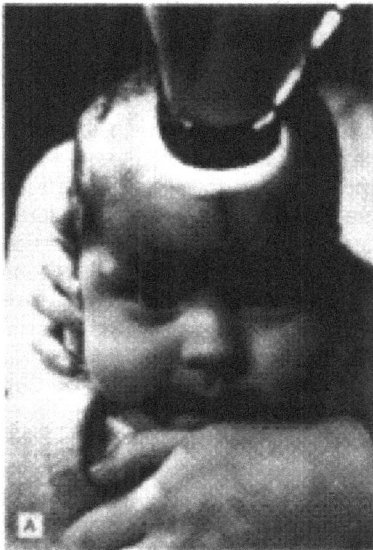

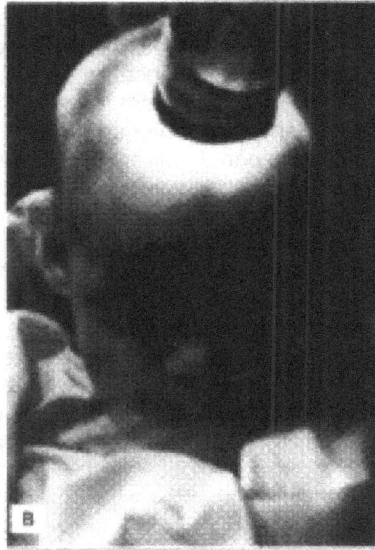

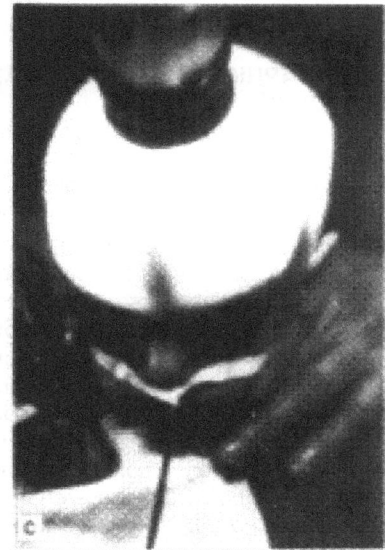

The skull is illuminated in proportion to the amount of fluid-filled space between the brain and skull. A well-fed infant has a well-developed brain with little space and fluid between the brain and skull. Not so, the malnourished or starving child. Reproduced courtesy of F. Monckeberg (in Prescott, et al. 1975)

patrist characteristics (with a high degree of matrism) received an appropriately low score. Latitudes and longitudes were obtained for each culture, and a regional percent-patrist average was extracted for each 5° by 5° block of latitude and longitude. Figure 1, the World Behavior Map, emerged from this procedure (DeMeo 1986, Chapter 4).

The patterns on the World Behavior Map were independently supported by separate maps of each of the 15 variables used in its construction, and by maps of other related variables (genital mutilations, infant cranial deformation, swaddling) given in the original dissertation (DeMeo 1986, Chapter 5). The World Behavior Map clearly demonstrates that patrism was neither ubiquitous nor random in its worldwide distribution. Old World cultures were clearly more patrist than those in either Oceania or the New World. Furthermore, the area of most extreme patrism in the Old World is found in one large, contiguous swath, stretching across North Africa, the Near (Middle) East, and into Central Asia. Of major significance is the fact that *this same geographical territory encompasses what is today the most intense, widespread, and hyperarid of desert environments found on Earth.*

Maps of environmental factors related to desert conditions demonstrate distributions very similar to that of extreme patrism on the World Behavior Map. Figure 2 is, for instance, a map identifying the most hyperarid of desert environments as determined from the Budyko-Lettau dryness ratio (Budyko 1958; Hare 1977). This ratio contrasts the amount of evaporative energy available in a given environment relative to the amount of precipitation. It is a more sensitive indicator of stress in arid environments than those used in more standard climate classification systems, which may mislead one into thinking that all "desert" environments are similar in nature. Maps identifying other stressful environmental extremes, such as greatest precipitation variability, highest mean monthly maximum temperatures, vegetation-barren regions, regions of lowest carrying capacity, regions of desert soils, and uninhabited regions show very similar distributions of their most intense, widespread aspects within this same extreme desert-patrist territory (DeMeo 1986, Chapter 2; DeMeo 1987). I have given the name *Saharasia* to this broad expanse of correlated extreme climate and culture.

THE GEOGRAPHICAL ASPECTS OF ARCHAEOLOGY AND HISTORY

The highly structured distributions on the World Behavior Map suggested that patrism developed within Saharasia, perhaps only in ancient historical times, after which it was carried outward by migrating peoples to affect surrounding moister regions. The testing of this hypothesis regarding behavior, migrations, and climate in ancient times necessitated the creation of a new data base composed of information on ancient climatic conditions, the migrations of peoples, past social factors relevant to the treatment of infants, children, and women, and tendencies towards male dominance, despotism, sadistic violence, and warfare. A new data base containing over 10,000 individual time- and location-specific notecards was developed and assembled chronologically; each card contained information from the archaeological or historical literature identifying artifacts and/or ecological conditions for specific field sites or regions at specific times. Over 100 separate authoritative sources were consulted and outlined to compose this new data base, which allowed identification and comparison of ancient conditions across broad geographical

regions for similar time periods. Times and places of widespread ecological and cultural transition, as well as the migrations and settlement patterns of peoples, were thereby identified. My predominant focus was on Saharasia and its moister Afro-Euro-Asian borderlands, but a significant amount of data was also collected for Oceania and the New World (DeMeo 1985, Ch. 6 & 7 of 1986).

From the patterns observed in this data base, I was able to confirm that patrism developed first and earliest in Saharasia, at the same time that the landscape underwent a major ecological transition, from relatively wet to arid, desert conditions. Evidence from dozens of archaeological and paleoclimatic studies indicates that the great desert belt of modern day Saharasia was, prior to c.4000-3000 BCE, a semiforested grassland savanna. Large and small fauna, such as elephant, giraffe, rhino, and gazelle, lived on the highland grasses, while hippopotamus, crocodile, fish, snails, and mollusks thrived in streams, rivers and lakes. Today, most of this same North African, Middle-Eastern and Central Asian terrain is hyperarid and often vegetation-barren. Some of the now-dry basins of Saharasia were then filled to levels tens to hundreds of meters deep, while the canyons and wadis flowed with permanent streams and rivers (DeMeo 1986, Chapter 6).

But what of the peoples who inhabited Saharasia during the wetter times of plenty? The evidence is also clear on this point: *These early peoples were peaceful, unarmored, and matrist in character.* Indeed, *I have concluded that there does not exist any clear, compelling or unambiguous evidence for the existence of patrism anywhere on Earth significantly prior to c.4000 BCE.* However, strong evidence exists for early matrist social conditions. These inferences are made partly from the *presence* of certain artifacts from those earliest times, which include: the sensitive and careful burial of the dead, irrespective of sex, with a relatively uniform grave wealth; sexually realistic female statues; and naturalistic, sensitive artwork on rock walls and pottery which emphasized women, children, music, the dance, animals, and the hunt. In later centuries, some of these same peaceful matrist peoples would progress technologically, and develop large, unfortified agrarian and/or trading states, notably in Crete, the Indus Valley, and Soviet Central Asia. The inference of matrism in these early times is also made from the *absence* of archaeological evidence for chaos, warfare, sadism, and brutality, which becomes quite evident in more recent strata, after Saharasia dried up. This latter archaeological evidence includes: weapons of war; destruction layers in settlements; massive fortifications, temples, and tombs devoted to big-man rulers; infant cranial deformation; ritual murder of females in the tombs or graves of generally older men; ritual foundation sacrifices of children; mass or unkept graves with mutilated bodies thrown in helter-skelter; and caste stratification, slavery, extreme social hierarchy, polygamy and concubinage, as determined from architecture, grave goods and other mortuary arrangements. Artwork style and subject matter of the later, dry periods also changes, to emphasize mounted warriors, horses, chariots, battles, and camels. Scenes of women, children, and daily life vanish. Naturalistic female statues and artwork simultaneously become abstract, unrealistic, or even fierce, losing their former gentle, nurturing, or erotic qualities; or they disappear entirely, to be replaced by statues of male gods or god-kings. Artwork quality as well as architectural styles decline for Old World sites at such times, to be followed in later years by monumental, warrior, and phallic motifs (DeMeo 1986, Chap-

Figure 1. The World Behavior Map: For the period roughly between 1840 and 1960, as reconstructed from aboriginal cultural data given in Murdock's *Ethnographic Atlas* (1967), with minimal historical interpretation.

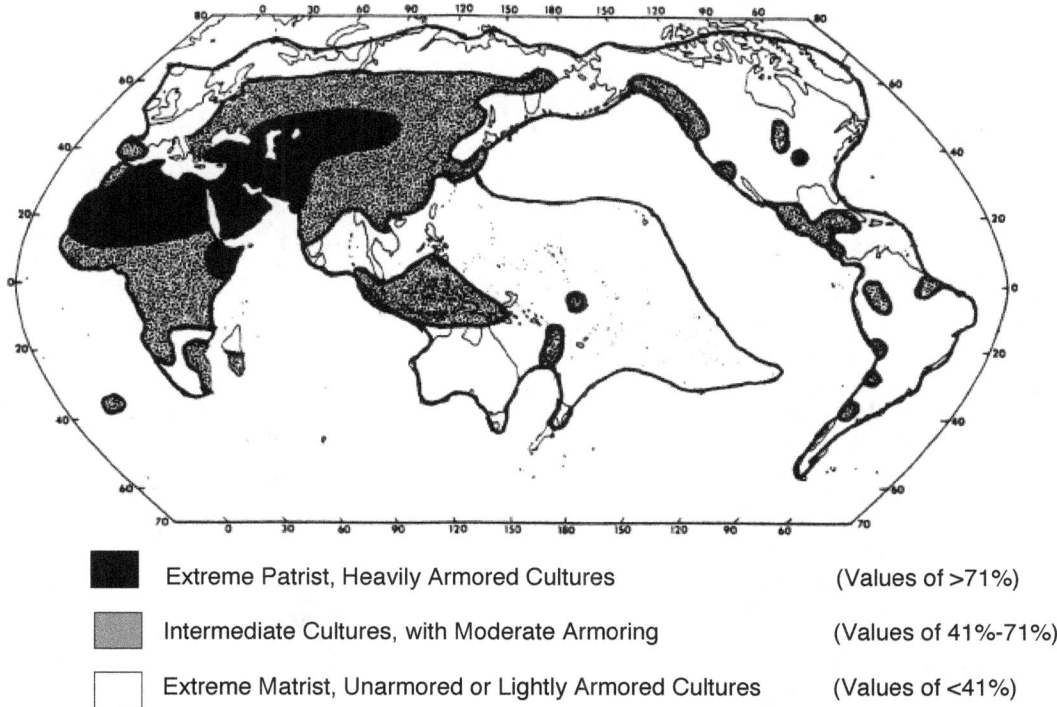

	Extreme Patrist, Heavily Armored Cultures	(Values of >71%)
	Intermediate Cultures, with Moderate Armoring	(Values of 41%-71%)
	Extreme Matrist, Unarmored or Lightly Armored Cultures	(Values of <41%)

Figure 2. Budyko-Lettau Dryness Ratio: Contrasting the relative dryness of different arid lands around the world. Values reflect the ratio between precipitation and evaporative energy; values of 2 receive twice as much evaporative solar heat as moisture from precipitation, while values of 10 receive ten times as much.

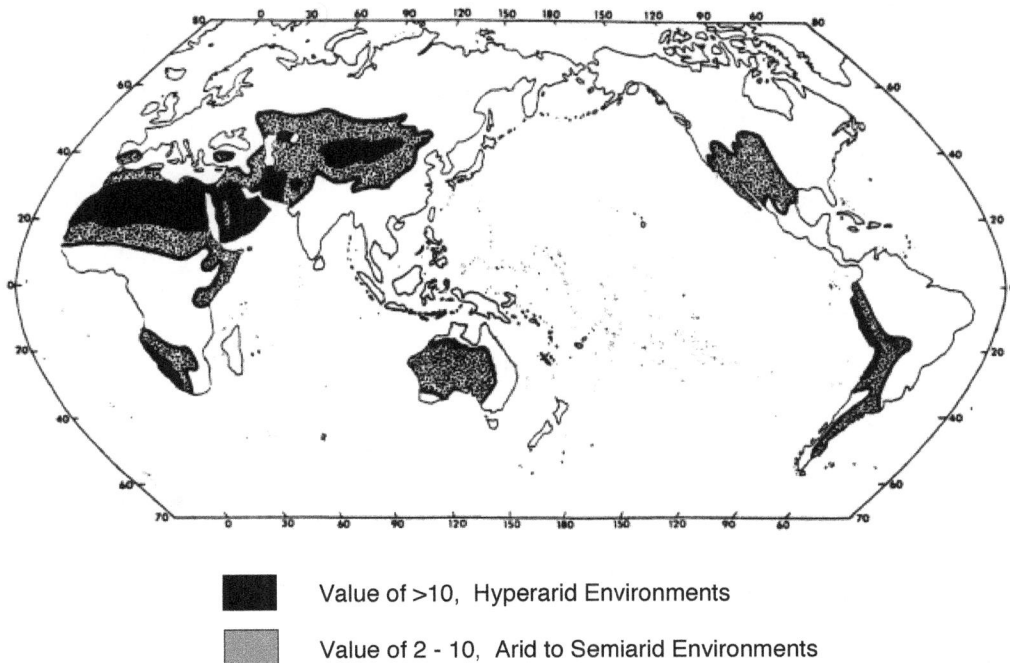

| | Value of >10, Hyperarid Environments |
| | Value of 2 - 10, Arid to Semiarid Environments |

ters 6 & 7). I was not the first to note the existence of cultural transitions in the archaeological and historical record, or to note the powerful effects of environmental change upon culture, to be sure.(5) However, my work was the first to simultaneously be global in scope, systematically derived, and both time- and location-specific.

With a few special exceptions, the first and earliest evidence for chaotic social conditions and patrism on Earth can be found in those parts of Saharasia which began to dry up first, namely within, or very close to Arabia and Central Asia. Those special exceptions are sites in Anatolia and the Levant, which contain some fleeting evidence suggesting that a very limited patrism may have existed as early as 5000 BCE; but this evidence exists alongside other evidence suggesting an early arid subphase in those same regions, with a complement shift towards migration and nomadic pastoralism. As such, they appear to be exceptions which prove the rule: Severe desertification and famine trauma greatly disturbed the original matrist social fabric, and promoted the development of patrist behaviors and social institutions; patrism was, in turn, compounded and intensified by widespread land-abandonment, migratory adjustments, and competition over scarce water resources.

The Genesis of Patrism in Saharasia

After c.4000-3500 BCE, radical social transformations are apparent in the ruins of previously peaceful, matrist settlements along river valleys in Central Asia, Mesopotamia, and North Africa. In each case, evidence for increasing aridity and land abandonment coincides with migratory pressures upon settlements with secure water supplies, such as those at oases, or on exotic rivers. Central Asia also experienced a shifting in lake levels and river beds coincidental to climatic instability and aridity, stimulating abandonment of large lakeshore or irrigation agricultural communities.

Settlements on the Nile and Tigris-Euphrates, as well as in the moister highland portions of the Levant, Anatolia, and Iran, were invaded and conquered by peoples abandoning Arabia and/or Central Asia, which continued to dry out. New despotic central states emerged thereafter. Tomb, temple, and fortification architecture, with evidence for ritual widow murder (eg., mother murder, when performed by the eldest son), cranial deformations, emphasis on the horse and camel, and growth of the military occurs following such invasions in almost every case I have studied. As these new despotic central states grew in power, they expanded their territories, sometimes to conquer the nomadic pastoral tribes still present on the desiccating steppe. Some of these despotic states periodically invaded into the wetlands adjacent to Saharasia to expand their territories. They either conquered local peoples in the wetlands or, failing to do so, stimulated defensive reactions among them, which can be seen in the subsequent appearance of fortifications, weapons technology, and an intermediate level of patrism in those wetlands. Other despotic Saharasian states eventually vanished from the history books as aridity intensified and dried up their subsistence (DeMeo 1985, Chapter 6 of 1986).

---

5. My study was possible only by the grace of the prior good works of many other scholars. Besides the work of Reich, my ideas on environmental and cultural transformations drew in large measure from the prior works of Bell (1971), Gimbutas (1965), Huntington (1907, 1911), Stone (1976) and Velikovsky (1950, 1984), though I take full responsibility for the conclusions and maps presented here.

North African Rock Art

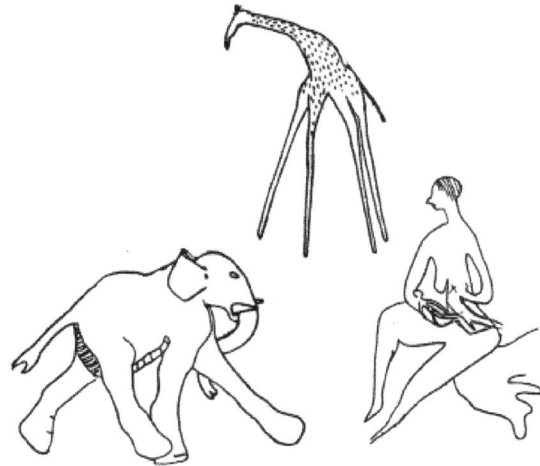

Moist Neolithic Hunter-Gatherer Period, c.7000 BCE

Moist Neolithic Pastoralist Period, c.5000 BCE

Dry Bronze Age, Warrior, Horse, Chariot, Camel Period, c.2000-500 BCE

The Diffusion of Patrism into the Saharasian Borderlands

Patrism appeared in the wetter Saharasian borderlands after, and only after, it first developed within the desiccating Saharasian core. As aridity gripped Saharasia, and as the armored, patrist response increasingly gripped Saharasian peoples, migrations out of the dry regions increasingly put such peoples into contact with the more peaceful peoples of the moister Saharasian borderlands. Increasingly, the migrations out of Saharasia took place in the form of massive invasions of the more fertile border territories. In these borderlands, patrism took root not by virtue of desertification or famine trauma, but by the killing off and replacement of the original matrist populations by the invader patrist groups, or by the forced adoption of new patrist social institutions introduced by the invading, conquering peoples. For example, Europe was sequentially invaded after c.4000 BCE by Battle-Axe peoples, Kurgans, Scythians, Sarmatians, Huns, Arabs, Mongols, and Turks. Each took a turn at warring, conquering, looting, and generally transforming Europe towards an increasingly patrist character. European social institutions progressively turned away from matrism towards patrism, with the far western parts of Europe, notably Britain and Scandinavia, developing patrist

conditions much later and in a more dilute form, than either Mediterranean or Eastern Europe, which were more profoundly influenced by Saharasian peoples.

Across the Old World, in the moister parts of China, peaceful matrist conditions likewise prevailed until the coming of the first extreme patrist Central Asian invaders, the Shang and Chou, after c.2000 BCE. Subsequent invasions by the Huns, Mongols and others would reinforce patrism in wetland China. Japanese culture remained matrist a bit longer, given the isolating influence of the China Sea and Korean Strait, until the coming of the first invading patrist groups from the Asian mainland, such as the Yayoi, around c.1000 BCE. In South Asia, the peaceful, largely matrist settlements and trading states of the Indus River valley collapsed after c.1800 BCE, under the combined pressures of aridity and patrist warrior-nomad invaders from arid Central Asian lands. Patrism spread thereafter into India, and was intensified in later centuries by Hunnish, Arab, and Mongol invasions, which also came from Central Asia. Matrism similarly predominated in Southeast Asia until the onset of progressive patrist migrations and invasions, by both land and sea, from the patrist kingly states of China, India, Africa, and Islamic regions. In sub-Saharan Africa, available evidence suggests that patrism first appeared

Figure 3. Areas Influenced or Occupied by Arab Armies Since 632 AD.
(after Jordan & Rowntree, 1979)

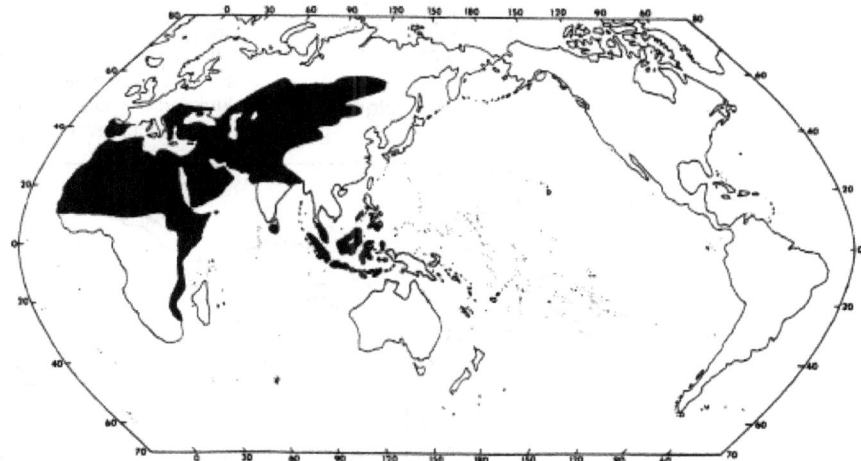

Figure 4. Areas Influenced or Occupied by Turkish Armies Since 540 AD
(after Pitcher, 1972)

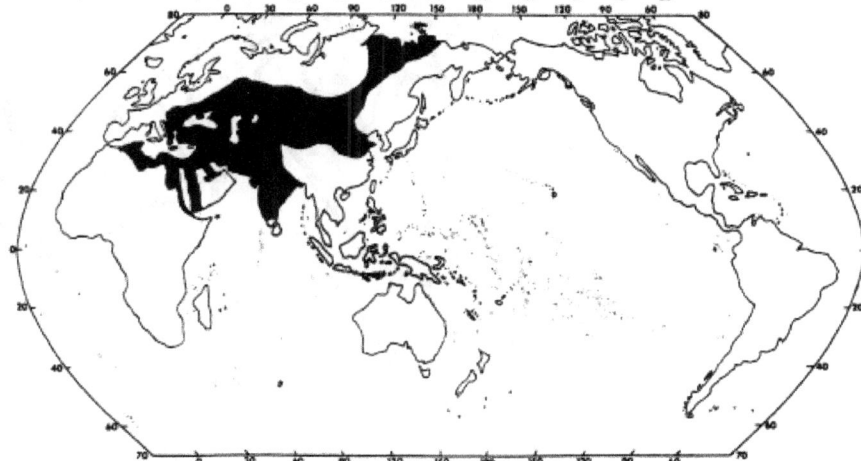

with the arrival of various southward-migrating peoples, around the time that North Africa dried up and was abandoned. Pharaonic Egyptian, Carthaginian, Greek, Roman, Byzantine, Bantu, Arab, Turkish, and Colonial European influences also increased African patrism in later years (DeMeo 1985, Chapter 6 of 1986).

The geographical patterns in these migrations, invasions, and settlement patterns are most striking. Two major patrist core zones appear in the data after c.4000 BCE, one in Arabia and the other in Central Asia, the respective homelands from which Semitic and Indoaryan peoples would migrate. These were also the first parts of Saharasia to start desiccating, though other portions of Saharasia would begin to dry up and convert to patrism within a few centuries. Another historical aspect of these irruptions of desert warrior nomads can be seen in Figures 3 and 4, which map the territories occupied at one time or another by the Arabs and Turks, respectively (Jordan & Rowntree 1979; Pitcher 1972). The territories of these two groups, who were the last of a series of invaders coming from Arabia and Central Asia, encompass fully 100% of desert Saharasia, spilling outward into its moister borderlands.

These facts of geography explain why matrism was preserved to a greater extent in those regions most far removed from Saharasia. Regions at the periphery of Saharasia (particularly islands), such as England, Crete, Scandinavia, the Asian Arctic, Southern Africa, Southern India, Southeast Asia, and Island Asia, demonstrate a later historical acquaintance with or adoption of patrism, and a consequent dilution of patrism with pre-existing native matrist social institutions. From the various sources used to construct my data base, Figure 5 was developed, suggesting patterns of diffusion of patrism within the Old World. The vectors are only a first approximation, but are in agreement with prior studies on the migrations and diffusion of peoples. These geographical patterns, taken from the literature of archaeology and history, are independently supported by a very similar spatial pattern in the more recent anthropological data, as previously given in Figure 1, the World Behavior Map.

The Diffusion of Patrism into Oceania and the New World

These observations regarding the migrations of patrist peoples may be extended to include the trans-oceanic diffusion of patrism from the Old World, through Oceania, and possibly even into the New World. A map of these suggested pathways is given in Figure 6, which assumes no source region for patrism other than Saharasia. This last map was derived from the various maps presented above, including the World Behavior Map, and from other sources given in my dissertation. Additional research will clearly be needed to confirm or clarify these suggested pathways.

It is significant that patrism in the Americas was identified on the World Behavior Map primarily among peoples who lived along the coasts or among peoples whose ancestors developed their earliest patrist communities on coastal regions. Furthermore, it is significant that the early patrist peoples of the Americas were the very same cultures for whom others have argued, on the basis of material culture, artwork, or linguistics, a pre-Columbian connection with the ocean-navigating patrist states of the Old World.(6) Nevertheless, a more limited patrism may have developed independently in Oceania and the

---

6. This finding directly challenges the assertion that all Pre-Columbian peoples of the New World arrived by migrating across the Bering Strait during glacial times predating c.10,000 BCE. If patrism had been carried into the New World at that time, it would have been more homogenously distributed. The quantity and quality of data supporting the idea of Pre-Columbian contacts has grown tremendously in recent years. For a sumary of such evidence, see Chapter 7 of DeMeo, 1986.

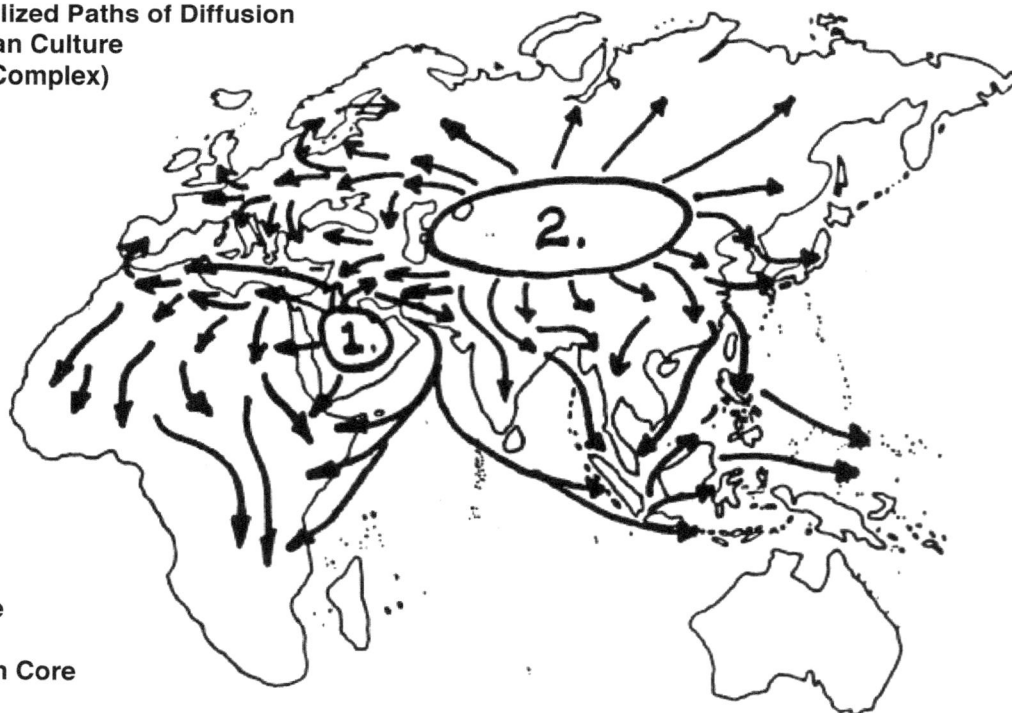

Figure 5: Generalized Paths of Diffusion of Armored Human Culture (Patrist Cultural Complex) in the Old World.

1. Arabian Core

2. Central Asian Core

New World through a desert-famine-migration mechanism similar to that argued for Saharasia, possibly within the Australian Desert, in the arid Great Basin of North America, and/or in the Atacama Desert (DeMeo 1986, Chapter 7).

CONCLUSIONS

The theory of the Saharasian origins of armored patrism was developed from a systematic geographical review of archaeological, historical, and anthropological data. The mapping of the various data was undertaken in an attempt to better understand the genesis of patrism, and to test the predictive power of the basic starting assumptions. This was accomplished through examination of the geographical dimensions of specific social institutions which either thwart basic biological maternal-infant and male-female bonding impulses, or which indicate a high level of male dominance, social hierarchy, and destructive aggression. As such, the basic starting assumptions of the study, namely the sex-economic theory of human behavior, the matrist-patrist schema, and the causal links between desertification and patrism, have been further verified and strengthened.

These findings strongly suggest that the innate portions of behavior are limited to the pleasure-directed aspects of life and social living, which impart distinct survival and health advantages to the growing child, and work to preserve the social unit. These are the matrist behaviors and social institutions, which support and protect the bonding functions between newborn babies and their mothers, which nurture the child through its various developmental stages, and which encourage and protect the bonds of love and pleasurable excitation which spontaneously develop between the young male and female. From these pleasure-directed biological impulses come other socially cooperative tendencies, and life-protecting, life-enhancing social institutions. Such impulses and behaviors, which are prochild, profemale, sex-positive and pleasure-oriented, have been demonstrated to exist in more recent times predominantly outside the bounds of the Saharasian desert belt. However, they once were the dominant forms of behavior and social organization everywhere on the planet, before the great Old World desiccation occurred. Given the new evidence presented here, patrism, to include its child-abusive, female-subordinating, sex-repressive, and destructively aggressive components, is best and most simply explained as a contractive emotional and cultural response to the traumatic famine conditions that first developed when Saharasia dried up after c.4000 BCE, a response which subsequently spread out of the desert through the diffusion of traumatized and affected peoples, and their altered social institutions.

REFERENCES

Aykroyd, W. 1974. *The Conquest of Famine*. London: Chatto & Windus.

Bell, B. 1971. "The Dark Ages in Ancient History, 1: The Firs Dark Age in Egypt". *American J. Archaeology*. 75:1-26.

Budyko, M.I. 1958. *The Heat Balance of the Earth's Surface*. N.A.Stepanova, trs. Washington, DC: US Dept. of Commerce.

Brandt, P. 1974. *Sexual Life in Ancient Greece*. NY: AMS Press.

Bullough, V. 1976. *Sexual Variance in Society and History*. NY: J. Wiley.

Cahill, K. 1982. *Famine*. Maryknoll, NY: Orbis Books.

DeMeo, J. 1980. "Cross Cultural Studies as a Tool in Geographic Research". *AAG Program Abstracts, Louisville, 1980*. Washington, DC Association of America Geographers. Annual Meeting. p.167.

DeMeo, J. 1985. "Archaeological/Historical Reconstruction of Late Quaternary Environmental and Cultural Changes in Saharasia." Unpublished Monograph.

Figure 6. Suggested Patterns of Diffusion of Patrism Around the World.

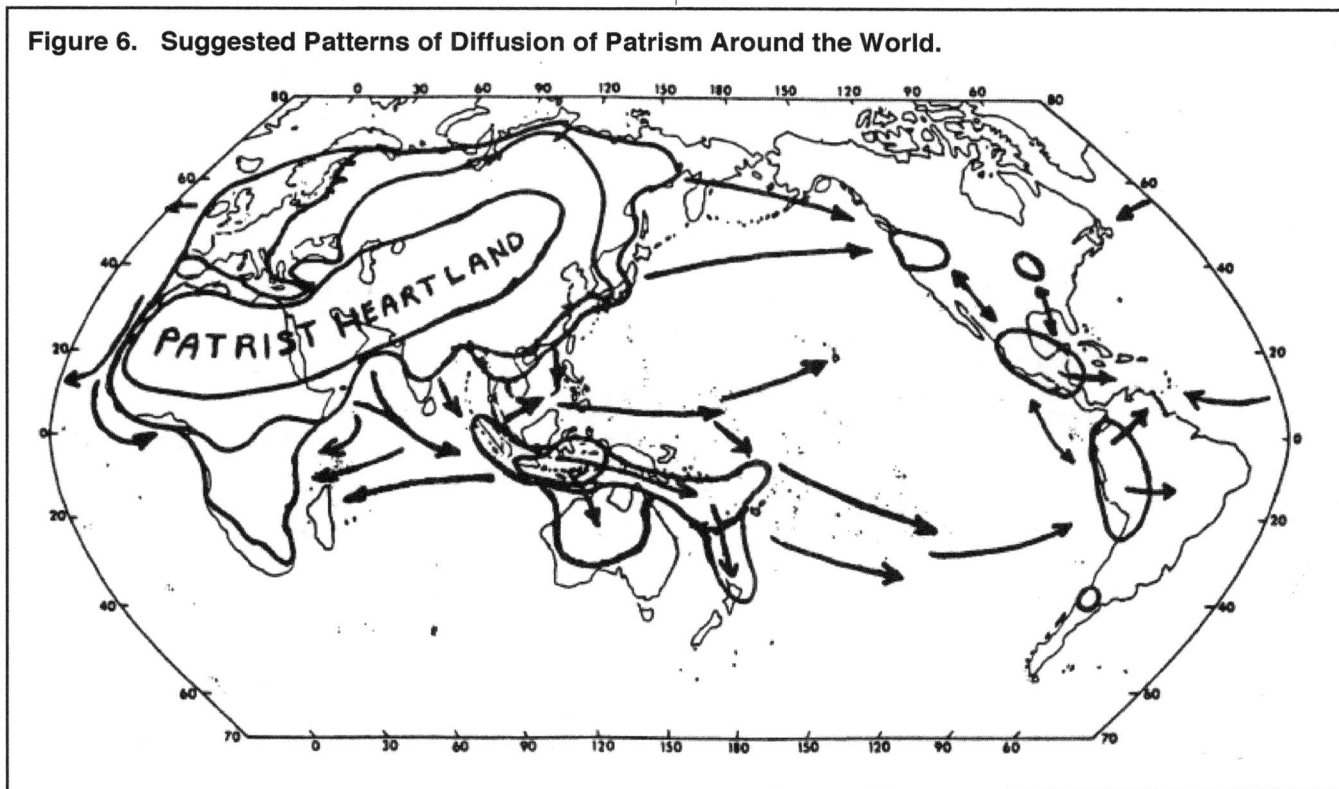

DeMeo, J. 1986. *On the Origins and Diffusion of Patrism: The Saharasian Connection.* Dissertation. University of Kansas Geography Department. Xerox available from Natural Energy Works, PO Box 864, El Cerrito, CA 94530.

DeMeo, J. 1987. "Desertification and the Origins of Armoring, Part 1", *Journal of Orgonomy.* 21(2):185-213.

DeMeo, J. 1988. "Desertification and the Origins of Armoring, Parts 2 & 3", *Journal of Orgonomy.* 22(1):185-213 & 22(2):268-289.

Dingwall, E.J. 1931. *Artificial Cranial Deformation.* London: J. Bale, Sons, & Danielson, Ltd.

Eisler, R. 1987a. *The Chalice and the Blade.* San Francisco: Harper & Row.

Eisler, R. 1987b. "Woman, Man, and the Evolution of Social Structure." *World Futures.* 23(1):79-92.

Elwin, V. 1947. *The Muria and their Ghotul.* Calcutta: Oxford U. Press.

Elwin, V. 1968. *The Kingdom of the Young.* Bombay: Oxford U. Press.

Fisher, H. 1982. *The Sex Contract: The Evolution of Human Behavior.* NY: William Morrow.

Gage, M. 1980. J. *Woman, Church & State.* Watertown, MA: Persephone Press.

Garcia, R. 1981. *Nature Pleads Not Guilty*, Vol. 1 of the Drought and Man series. IFIAS Project. NY: Pergamon Press.

Garcia, R. & Escudero, J. 1982. *The Constant Catastrophe: Malnutrition, Famines, and Drought,* Vol. 2 of the *Drought and Man* series. IFIAS Project. NY: Pergamon Press.

Gimbutas, M. 1965. *Bronze Age Cultures in Central and Eastern Europe.* The Hague: Mouton.

Gimbutas, M. 1977. "The First Wave of Eurasian Steppe Pastoalists into Copper Age Europe". *Journal of Indo-European Studies,* 5(4), Winter.

Gimbutas, M. 1982. *The Goddesses and Gods of Old Europe.* Berkeley: U. of California Press.

Gorer, G. & Rickman, J. 1962. *The People of Great Russia: A Psychological Study.* NY: W.W. Norton.

Hallet, J.P. & Relle, A. 1973. *Pygmy Kitabu.* NY: Random House.

Hare, K. 1977. "Connections Between Climate and Desertification". *Environmental Conservation.* 4(2):81-90.

Hodin, M. 1937. *A History of Modern Morals.* NY: AMS Press.

Hosken, F. 1979. Hosken Report on Genital and Sexual Mutilation of Females, 2nd Ed., Lexington, MA: Women's International Network News.

Huntington, E. 1907. *The Pulse of Asia.* NY: Houghton-Mifflin.

Huntington, E. 1911. *Palestine and its Transformation.* NY: Houghton-Mifflin.

Jordan, T & Rowntree, L. 1979. *The Human Mosaic.* NY: Harper & Row. p.187.

Kiefer, O. 1951. *Sexual life in Ancient Rome.* NY: Barnes & Nobel.

Klaus, M.H. & Kennell, J.H. 1976. *Maternal-Infant Bonding: The Impact of Early Separation or Loss on Family Development.* St. Louis: C.V. Mosby.

LeBoyer, F. 1975. *Birth Without Violence.* NY: Alfred Knopf.

Levy, H.S. 1971. *Sex, Love, and the Japanese.* Washington, DC: Warm-Soft Village Press.

Lewinsohn, R. 1958. A *History of Sexual Customs.* NY: Harper Bros.

Malinowski, B. 1927. *Sex and Repression in Savage Society.* London: Humanities Press.

Malinowski, B. 1932. *The Sexual Life of Savages.* London: Routledge & Keegan Paul.

Mantegazza, P. 1935. T*he Sexual Relations of Mankind.* NY: Eugenics Press.

May, G. 1930. *Social Control of Sex Expression.* London: George Allen & Unwin.

Montagu, A. 1945. "Infibulation and Defibulation in the Old and New Worlds". *Am. Anthropologist,* 47:464-7.

Montagu, A. "Ritual Mutilation Among Primitive Peoples". *Ciba Symposium.* pp.421-36, October.

Montagu, A. 1971. *Touching: The Human Significance of the Skin.* NY: Columbia U. Press.

Murdock, G.P. 1967. *Ethnographic Atlas.* U. Pittsburgh Press.

Pitcher, D.E. 1972. *An Historical Geography of the Ottoman Empire.* Leiden: E.J. Brill. Map V.

Prescott, J. 1975. "Body Pleasure and the Origins of Violence". *Bulletin of Atomic Scientists.* November, pp.10-20.

Prescott, J., Read, M. & Coursin, D. 1975. *Brain Function and Malnutrition.* NY: J. Wiley & Sons.

Reich, W. 1935. *The Invasion of Compulsory Sex-Morality.* 3rd Edition. NY: Farrar, Straus & Giroux edition. 1971.

Reich, W. 1942. *Function of the Orgasm.* NY: Farrar, Straus & Giroux edition. 1973.

Reich, W. 1945. *The Sexual Revolution.* 3rd Edition. NY: Octagon Books edition. 1973.

Reich, W. 1947. *The Mass Psychology of Fascism.* 3rd Edition. NY: Farrar, Straus & Giroux edition. 1970.

Reich, W. 1949. *Character Analysis.* 3rd Edition. NY: Farrar, Straus & Giroux edition. 1971.

Reich, W. 1953. *People in Trouble.* NY: Farrar, Straus & Giroux edition. 1976.

Reich, W. 1967. *Reich Speaks of Freud.* NY: Farrar, Straus & Giroux.

Reich, W. 1983. *Children of the Future.* NY: Farrar, Straus & Giroux.

Stewart, D. & Stewart, L. 1978a. *Safe Alternatives in Childbirth.* Chapel Hill, NC: NAPSAC.

Stewart, D. & Stewart, L. 1978b. *21st Century Obstetrics Now!* Vols. 1 & 2. Chapel Hill, NC: NAPSAC.

Stone, M. 1976. *When God Was a Woman.* NY: Dial.

Sorokin, P. 1975. *Hunger as a Factor in Human Affairs.* Gainesville: Univ. Florida Press.

Tannahill, R. 1980. *Sex in History.* NY: Stein & Day.

Taylor, G.R. 1953. *Sex in History.* London: Thames & Hudson.

Turnbull, C. 1961. *The Forest People.* NY: Simon & Schuster.

Turnbull, C. 1972. *The Mountain People.* NY: Simon & Schuster.

Van Gulik, R. 1961. *Sexual Life in Ancient China.* Leiden: E.J. Brill.

Velikovsky, I. 1950. *Worlds in Collision.* NY: Macmillan.

Velikovsky, I. 1984. *Mankind in Amnesia.* NY: Doubleday.

Body Pleasure and the Origins of Violence*

James W. Prescott, Ph.D.**

Human violence is fast becoming a global epidemic. All over the world, police face angry mobs, terrorists disrupt the Olympics, hijackers seize airplanes, and bombs wreck buildings. During the past year (1975), wars raged in the Mideast, Cyprus, and Southeast Asia, and guerilla fighting continued to escalate in Ireland. Meanwhile, crime in the United States grew even faster than inflation. Figures from the Federal Bureau of Investigation show that serious crimes rose 16% in the first six months of 1974—one of the largest increases since FBI record-keeping began.

Unless the causes of violence are isolated and treated, we will continue to live in the world of fear and apprehension. Unfortunately, violence is often offered as a solution to violence. Many law enforcement officials advocate "get tough" policies as the best method to reduce crime. Imprisoning people, our usual way of dealing with crime, will not solve the problem because the causes of violence lie in our basic values and the way in which we bring up our children and youth. Physical punishment and violent films and TV programs teach our children that physical violence is normal. But these early life experiences are not the only or even the main source of violent behavior. Recent research supports the point of view that the deprivation of physical pleasure is a major ingredient in the expression of physical violence. The common association of sex with violence provides a clue to understanding physical violence in terms of deprivation of physical pleasure.

Unlike violence, pleasure seems to be something the world can't get enough of. People are constantly in search of new forms of pleasure, yet most of our "pleasure" activities appear to be substitutes for the natural sensory pleasures of touching. We touch for pleasure or for pain or we don't touch at all. Although physical pleasure and physical violence seem worlds apart, there seems to be a subtle and intimate connection between the two. Until the relationship between pleasure and violence is understood, violence will continue to escalate.

As a developmental neuropsychologist, I have de-voted a great deal of study to the peculiar relationship between violence and pleasure. I am convinced that the deprivation of physical sensory pleasure is the principal root cause of violence. Laboratory experiments with animals show that pleasure and violence have a reciprocal relationship, that is, the presence of one inhibits the other. A raging, violent animal will abruptly calm down when electrodes stimulate the pleasure centers of its brain. Likewise, stimulating the violence centers in the brain can terminate the animals's sensual pleasure and peaceful behavior. When the brain's pleasure circuits are "on," the violence circuits are "off," and vice-versa. Among human beings, a pleasure-prone personality rarely displays violence or aggressive behaviors, and a violent personality has little ability to tolerate, experience, or enjoy sensuously pleasing activities. As either violence or pleasure goes up, the other goes down.

The reciprocal relationship of pleasure and violence is highly significant, because certain sensory experiences during the formative periods of development will create a neuropsychological predisposition for either violence-seeking or pleasure-seeking behaviors later in life. I am convinced that various abnormal social and emotional behaviors resulting from what psychologists call "maternal-social" deprivation, that is, a lack of tender, loving care, are caused by a unique type of sensory deprivation, somatosensory deprivation. Derived from the Greek word for "body," the term refers to the sensations of touch and body movement which differ from the senses of sight, hearing, smell, and taste. I believe that the deprivation of body touch, contact, and movement are the basic causes of a number of emotional disturbances which include depressive and autistic behaviors, hyperactivity, sexual aberration, drug abuse, violence, and aggression.

These insights were derived chiefly from the controlled laboratory studies of Harry F. and Margaret K. Harlow at the University of Wisconsin. The Harlow's and their students separated infant monkeys from their mothers at birth. The monkeys were raised in single cages in an animal colony room, where they could develop social relationships with the other animals through seeing, hearing, and smelling, but not through touching or movement. These and other studies indicate that it is the deprivation of body contact and body movement—not deprivation of the other senses—that produces the wide variety of abnormal emotional behaviors in these isolation-reared animals. It is well known that human infants and children who are hospitalized or institutionalized for extended periods with little physical touching and holding develop almost identical abnormal behaviors, such as rocking and head banging.

* Reprinted with permission from The Futurist, April 1975, published by the World Future Society, 4916 Saint Elmo Ave., Bethesda, MD 20814.

** Formerly a developmental neuropsychologist and Health Scientist Administrator for the National Institute of Child Health and Human Development, National Institutes of Health; Presently, Director of the Intitute of Humanistic Science, Lewes, Deleware, jprescott34@comcast.net

Although the pathological violence observed in isolation-reared monkeys is well documented, the linking of early somatosensory deprivation with physical violence in humans is less well established. Numerous studies of juvenile delinquents and adult criminals have shown a family background of broken homes and/or physically abusive parents. These studies have rarely mentioned, let alone measured, the degree of deprivation of physical affection, although this is often inferred from the degree of neglect and abuse. One exceptional study in this respect is that of Brandt F. Steele and C. B. Pollock, psychiatrists at the University of Colorado, who studied child abuse in three generations of families who physically abused their children. They found that parents who abused their children were invariably deprived of physical affection themselves during childhood and that their adult sex life was extremely poor. Steele noted that almost without exception the women who abused their children had never experienced orgasm. The degree of sexual pleasure experienced by the men who abused their children was not ascertained, but their sex life, in general, was unsatisfactory. The hypothesis that physical pleasure actively inhibits physical violence can be appreciated from our own sexual experiences. How many of us feel like assaulting someone after we have just experienced orgasm?

The contributions of Freud to the effects of early experiences upon later behaviors and the consequences of repressed sexuality have been well established. Unfortunately time and space does not permit a discussion of his differences with Wilhelm Reich and this writer concerning his *Beyond the Pleasure Principle*.

The hypothesis that deprivation of physical pleasure results in physical violence requires a formal systematic evaluation. We can test this hypothesis by examining cross-cultural studies of child-rearing practices, sexual behaviors, and physical violence. We would expect to find that human societies which provide their infants and children with a great deal of physical affection (touching, holding, carrying) would be less physically violent than human societies which give very little physical affection to their infants and children. Similarly, human societies which tolerate and accept premarital and extramarital sex would be less physical violent that societies which prohibit and punish premarital and extramarital sex.

Cross-Cultural Studies of Physical Violence

Cultural anthropologists have gathered exactly that data required to examine this hypothesis for human societies—and their findings are conveniently arranged in R. B. Textor's *A Cross-Cultural Summary* (HRAF Press, 1967). Textor's book is basically a research tool for cross-cultural statistical inquiry. The survey provides some 20,000 statistically significant correlations from 400 culture samples of primitive societies.

Infant Neglect and Adult Violence

Certain variables which reflect physical affection (such as fondling, caressing, and playing with infants) were related to other variables which measure crime and violence (frequency of theft, killing, etc.). The important relationships are displayed in the tables accompanying this article. The percent figures reflect the relationships among the variables, e.g. high affection/low violence plus low affection/high violence. This procedure is followed for all tables.

Societies ranking high or low on the Infant Physical Affection Scale were examined for degree of violence. The results (Table 1) clearly indicated that those societies which give their infants the greatest amount of physical affection were characterized by low theft, low infant physical pain, low religious activity, and negligible or absent killing, mutilating, or torturing of the enemy. These data directly confirm that the deprivation of body pleasure during infancy is significantly linked to a high rate of crime and violence.

Some societies physically punish their infants as a

The Long-Term Consequences of Infant Pleasure and Pain

Human societies differ greatly in their treatment of infants. In some cultures, parents lavish physical affection on infants, while in others, the parents physically punish their infants. Pleasure-receiving infants grow into peaceable adults while the pleasure-deprived infants are likely to become violent when they reach adulthood.

Prescott's study of anthropological data found that those societies which give their infants the greatest amount of physical affection have less theft and violence among adults. Deprivation of bodily pleasure during infancy is significantly linked to a high rate of crime and violence.

The Tables below show how the physical affection -- or punishment -- given infants correlates with other variables. For example, cultures which inflict pain on infants appear to be more likely to practice slavery, polygyny, etc. In the tables, N refers to the number of cultures in the comparison while P is the probability that the observed relationship could occur by chance, as calculated by the Fisher Exact Probability Test.

TABLE 1: *High Infant Physical Affection*

Adult Behaviors	Percent %	Number N	Probability P
Invidious display of wealth is low	66	50	.06
Incidence of theft is low	72	36	.02
Overall infant indulgence is high	80	66	.0000
Infant physical pain is low	65	63	.03
Negligible killing, torturing or mutilating the enemy	73	49	.004
Low religious activity	81	27	.003

The coded scales on infancy were developed by cultural anthropologists Barry, Bacon and Child; on sexual behavior by Westbrook, Ford and Beach, and on physical violence by Slater.

TABLE 2: *Pain Inflicted on Infant by Parent or Nuturing Agent*			
Adult Behaviors	**Percent**	**Number**	**Probality**
	%	**N**	**P**
Slavery is Present	64	66	.03
Polygyny (multiple wives) practiced	79	34	.001
Women status is inferior	78	14	.03
Low infant physical affection	65	63	.03
Low overall infant indulgence	77	66	.000
Developing nurturant behavior in child is low	67	45	.05
Supernaturals (gods) are aggressive	64	36	.01

matter of discipline, while others do not. We can determine whether this punishment reflects a general concern for the infant's welfare by matching it against child nurturant care. The results (Table 2) indicate that societies which inflict pain and discomfort upon their infants tend to neglect them as well.

Adult physical violence was accurately predicted in 36 of 49 cultures (73%) from the infant physical affection variable. The probability that a 73% rate of accuracy could occur by chance is only four times out of a thousand.

Sexual Repression and Adult Violence

Thirteen of the 49 societies studied seemed to be exceptions to the theory that a lack of somatosensory pleasure makes people physically violent. It was expected that cultures which placed a high value upon physical pleasure during infancy and childhood would maintain such values into adulthood. This is not the case. Child rearing practices do not predict patterns of later sexual behavior. This initial surprise and presumed discrepancy, however, becomes advantageous for further prediction. Two variables that are highly correlated are not as useful for

predicting a third variable as two variables that are uncorrelated. Consequently, it is meaningful to examine the sexual behaviors of the 13 cultures whose adult violence was not predictable from physical pleasure during infancy. Apparently, the social customs which influence and determine the behaviors of sexual affection are different from those which underlie the expression of physical affection toward infants.

When the six societies characterized by both high infant affection and high violence are compared in terms of their premarital sexual behavior, it is surprising to find that five of them exhibit premarital sexual repression where virginity is a high value of these cultures. It appears that *the beneficial effects of infant physical affection can be negated by the repression of physical pleasure (premarital sex) later in life.*

The seven societies characterized by both low infant physical affection and low adult physical violence were all found to be characterized by permissive premarital sexual behaviors. Thus, *the detrimental effects of infant physical affectional deprivation seem to be compensated for later in life by sexual body pleasure experience during adolescence.* These findings have led to a revision of the somatosensory pleasure deprivation theory from a one-stage to a two-stage developmental theory where the physical violence in 48 of the 49 cultures could be accurately classified. In short, violence may stem from deprivation of somatosensory pleasure either in infancy or in adolescence. The only true exception in this culture sample is the headhunting Jivaro tribe of South America. Clearly, this society

requires detailed study to determine the causes of its violence. The Jivaro belief system may play an important role, for as anthropologist Michael Harner notes in *Jivaro Souls*, these Indians have a "deep-seated belief that killing leads to the acquisition of souls which provide a super-natural power conferring immunity from death."

The strength of the two-stage deprivation theory of violence is most vividly illustrated when we contrast the societies showing high rates of physical affection during infancy *and* adolescence against those societies which are consistently low in physical affection for both developmental periods. The statistics associated with this relationship are extraordinary: The percent likelihood of a society being physically violent if it is physically affectionate toward its infants *and* tolerant of premarital sexual behavior is 2% (48/49). The probability of this relationship occurring by chance is 125,000 to one. I am not aware of any other developmental variable that has such a high degree of predictive validity. Thus, we seem to have a firmly based principle: *Physically affectionate human societies are highly unlikely to be physically violent.*

Accordingly, when physical affection and pleasure during adolescence as well as infancy are related to measures of violence, we find direct evidence of a significant relationship between the punishment of premarital sex behaviors and various measures of crime and violence. As Table 4 shows, additional clusters of relationships link the punishment and repression of premarital sex to large community size, high social complexity and class stratification, small extended families, purchase of wives, practice of slavery, and a high god present in human morality. The relationship between small extended families and punitive premarital sex attitudes deserves emphasis, for it suggests that the nuclear family structure in contemporary Western cultures may be a contributing factor to our repressive attitudes toward sexual expression. The same can be suggested on community size, social complexity, and class stratification.

Not surprisingly, when high self-needs are combined with the deprivation of physical affection, the result is self-interest and high rates of narcissism. Likewise, exhibitionistic dancing and pornography may be interpreted as a substitute for normal sexual expression. Some nations which are most repressive of female sexuality have rich pornographic art forms.

Acceptance of Premarital and Extramarital Sex May Reduce Violence

I also examined the influence of premarital and extramarital sex taboos upon crime and violence. The data clearly indicate the punitive-repressive attitudes toward sex are linked

> **"Physically affectionate human societies are highly unlikely to be physically violent. (There is a) connection between rejection of physical pleasure with (the) expression of physical violence."**

with physical violence, personal crime, and practices of domination, such as slavery. Societies which value compulsive monogamy emphasize military glory and worship aggressive gods.

These findings overwhelmingly support the thesis that deprivation of body pleasure throughout life—but particularly during the formative periods of infancy, childhood, and adolescence—are very closely related to the amount of warfare and interpersonal violence. These insights should be applied to large and complicated industrial and post-industrial societies.

Sexual Pleasure vs. Sexual Violence

Crime and physical violence have substantially increased over the past decade in the United States. According to FBI statistics, both murder and aggravated assault increased 53% between 1967 and 1972, while forcible rape rose 70%.

These figures again raise the question of the special relationship between sexuality and violence. In addition to our rape statistics, there is other evidence that points to America's preference for sexual violence over sexual pleasure. This is reflected in our acceptance of sexually explicit films that involve violence and rape, and our rejection of sexually explicit films for

Societies that provide infants with a great deal of physical affection ("tender loving care") are later characterized by relatively non-violent adults, Prescott says. In 36 of the 49 cultures he studied, a high degree of infant physical affection was associated with a low degree of adult physical violence—and vice versa. When Prescott investigated the 13 exceptions he found that all but one (the Jivaro tribe of South America) provide adolescents with a great deal of sexual freedom, thus allowing them to compensate for the lack of physical affection that they experienced in infancy.

TABLE 3:
Relationship of Infant Physical Affectional Deprivation to Adult Physical Violence

High Infant Physical Affection	Low Infant Physical Affection	High Infant Physical Affection	Low Infant Physical Affection
Low Adult Physical Violence	High Adult Physical Violence	High Adult Physical Violence	Low Adult Physical Violence
Andamanese	Alorese	Cheyenne	Ainu
Arapesh	Aranda	Chir-Apache	Ganda
Balinese	Araucanians	Crow	Kwakiutl
Chagga	Ashanti	Jivaro	Lepcha
Chenchu	Aymara	Kurtatchi	Pukapuka
Chuckchee	Azande	Zuni	Samoans
Cuna	Comanche		Tanala
Hano	Fon		
Lau	Kaska		
Lesu	Marquesans		
Maori	Masai		
Murngin	Navaho		
Nuer	Ojibwa		
Papago	Thonga		
Siriono			
TallensiTikopia			
Timbira			
Trobriand			
Wogego			
Woleaians			
Yahgan			

Derived from: R.B. Textor, A Cross Cultural Summary, HRAF Press, New Haven Connecticut, 1967. Infant ratings from: Barry, Bacon, & Child. Adult violence ratings from: P.E. Slater.

Premarital sexual freedom for young people can help reduce violence in a society. Prescott's study indicates that the physical pleasure that youth obtain from sex can offset a lack of physical affection during infancy. Other research also indicates that societies which punish premarital sex are likely to engage in wife purchasing, to worship a high god in human morality, and to practice slavery. Other results are shown in the table below:

TABLE 4:
Premarital Sex is Strongly Punished

Adult Behaviors	Percent %	Number N	Probability P
Community size is larger	73	80	.0003
Slavery is present	59	176	.005
Societal complexity is high	87	15	.01
Personal crime is high	71	28	.05
Class stratification is high	60	111	.01
High incidence of theft	68	31	.07
Small extended family	70	63	.008
Extramarital sex is punished	71	58	.005
Wives are purchased	54	114	.02
Castration anxiety is high	65	37	.009
Longer post-partum sex taboo	62	50	.03
Bellicosity is extreme	68	37	.04
Sex disability is high	83	23	.004
Killing, torturing and mutilating the enemy is high	69	35	.07
Narcissism is high	66	38	.04
Exhibitionistic dancing is emphasized	65	66	.04
High god in human morality	81	27	.01

pleasure only. Apparently, sex with pleasure is immoral and unacceptable, but sex with violence and pain is moral and acceptable.

A questionnaire I developed to explore this question was administered to 96 college students whose average age was 19 years. The results of the questionnaire support the connection between rejection of physical pleasure (and particularly of premarital and extramarital sex) with expression of physical violence. Respondents who reject abortion, responsible premarital sex, and nudity within the family were likely to approve of harsh physical punishment for children and to believe that pain helps build strong moral character. These respondents were likely to find alcohol and drugs more satisfying than sex. The data obtained from the questionnaire provide strong statistical support for the basic inverse relationship between physical violence and physical pleasure. If violence is high, pleasure is low, and conversely, if pleasure is high, violence is low. The questionnaire bears out the theory that the pleasure-violence relationship found in primitive cultures also holds true for a modern industrial nation.

Violence and Pleasure: The Attitudes of American College Students

The reciprocal relationship of violence and pleasure holds true in modern industrial nations as well as in primitive societies. Prescott tested this viewpoint by means of a questionnaire given to American college students (average age: 19). The results showed that students who have relatively negative attitudes toward sexual pleasure tend to favor harsh punishment for children and to believe that violence is necessary to solve problems. In Prescott's questionnaire, students rated a series of statements on a scale of 1 to 6, where 1 indicated strong agreement and 6 strong disagreement. Through a statistical technique (factor analysis), he developed a personality profile of the violent person.

Table 5 shows the degree of relationship among the various statements which reflect social and moral values. The figures, know as "loadings", are treated like correlation coefficients. They indicate the strength with which each variable contributes to the overall personality description of the respondent as defined by this specific profile.

Table 6 shows how student's attitudes toward premarital and extramarital sex correlated with their attitudes toward alcohol and drugs. The number in the r (correlation coefficient) column reflects the strength of the relationship between attitudes toward drugs and attitudes toward premarital and extramarital sex. A perfect relationship would be 1.0, indicating that one variable can be perfectly predicted from the other. Correlations in the .50 to .70 range are regarded as highly significant. The results indicate that people who reject premarital sex are likely to prefer alcohol and drugs to sex, and are also likely to become hostile and aggressive when they drink alcohol.

Drugs, Sex, and Violence: The Unholy Trinity

Another way of looking at the reciprocal relationship between violence and pleasure is to examine a society's choice of drugs. A society will support behaviors that are consistent with its values and social mores. U.S. society is a competitive, aggressive, and violent society. Consequently, it supports drugs that facilitate competitive, aggressive, and violent behaviors and opposes drugs that counteract such behaviors. Alcohol is well known to facilitate the expression of violent behaviors, and, although addicting and very harmful to chronic users, is acceptable to American society.

The data from my questionnaire support this view. As Table 5 shows, very high correlations between alcohol use and parental punishment indicate that people who received little affection from their mothers and had physically punitive fathers are likely to become hostile and aggressive when they drink. Such people find alcohol more satisfying than sex. There is an even stronger relationship between parental physical punishment and drug usage. Respondents who were physically punished as children showed alcohol-induced hostility and aggression and were likely to find alcohol and drugs more satisfying than sex. The questionnaire also reveals high correlations between sexual repression and drug usage. Those who de-

TABLE 5:
Somatosensory Index of Human Affection

Violence Approved
.85	Hard physical punishment is good for children who disobey a lot.
.81	Physical punishment and pain help build a strong moral character.
.76	Capital punishment should be permitted by society.
.75	Violence is necessary to really solve our problems.
.74	Physical punishment should be allowed in the schools.
.69	I enjoy sadistic pornography.
.54	I often feel like hitting someone.
.43	I can tolerate pain very well.

Physical Pleasure Condemned
.84	Prostitution should be punished by society.
.80	Abortion should be punished by society.
.80	Responsible premarital sex is not agreeable to me.
.78	Nudity within the family has a harmful influence upon children.
.73	Sexual pleasures help build a weak moral character.
.72	Society should interfere with private sexual behavior between adults.
.69	Responsible extramural sex is not agreeable to me.
.61	Natural fresh body odors are often offensive.
.47	I do not enjoy affectional pornography.

Alcohol and Drugs Rated Higher than Sex
.70	Alcohol is more satisfying than sex.
.65	Drugs are more satisfying than sex.
.60	I get hostile and aggressive when I drink alcohol.
.49	I would rather drink alcohol than smoke marijuana.
.45	I drink alcohol more often than I experience orgasm.

Other
.82	I tend to be conservative in my political points of view.
.77	Age (Older).
.51	I often dream of either floating, flying, falling, or climbing.
.45	My mother is often indifferent towards me.
.42	I often get "uptight" about being touched.
.40	I remember when my father physically punished me a lot.

scribe premarital sex as "not agreeable" are likely to become aggressive when drinking and to prefer drugs and alcohol to sexual pleasures. This is additional evidence for the hypothesis that drug "pleasures" are a substitute for somato-sensory pleasures.

> "...deprivation of body pleasure throughout life -- but particularly during the formative periods of infancy, childhood, and adolescence -- are very closely related to the amount of warfare and interpersonal violence."

New Values for a Peaceful World

It is clear that the world has only limited time to change its custom of resolving conflicts violently. It is uncertain whether we have the time to undo the damage done by countless previous generation, nor do we know how many future generations it will take to transform our psychobiology of violence into one of peace.

If we accept the theory that the lack of sufficient somatosensory pleasure is a principal cause of violence, we can work toward promoting pleasure and encouraging affectionate interpersonal relationships as a means of combatting aggression. We should give high priority to body pleasure in the context of meaningful human relationships. Such body pleasure is very different from promiscuity, which reflects a basic inability to experience pleasure. If a sexual relationship is not pleasurable, the individual looks for another partner. A continuing failure to find sexual satisfaction leads to a continuing search for new partners, that is, to promiscuous behavior. Affectionately shared physical pleasure, on the other hand, tends to stabilize a relationship and eliminate the search. However, a variety of sexual experiences seems to be normal in cultures which permit its expression, and this may be important for optimizing pleasure and affection in sexual relationships.

Available data clearly indicate that the rigid value of monogamy, chastity, and virginity help produce physical violence. The denial of female sexuality must give way to an acceptance and respect for it, and men must share with women the responsibility for giving affection and care to infants and children. As the father assumes a more equal role with the mother in child-rearing and becomes more affectionate toward his children, certain changes must follow in our socioeconomic system. A corporate structure which tends to separate either parent from the family by travel, extended meetings, or overtime work weakens the parent-child relationship and harms family stability. To develop a peaceful society, we must put more emphasis on human relationships.

Family planning is essential. Children must be properly spaced so that each can receive optimal affection and care. The needs of the infant should be immediately met. Cross-cultural evidence does not support the view that such practices will "spoil" the infant. It is harmful for a baby to cry itself to sleep. By not answering an infant's needs immediately and consistently, we not only teach a child distrust at a very basic emotional level, but also establish patterns of neglect which harm the child's social and emotional health. The discouragement of breast feeding in favor of bottle feeding and the separation of healthy newborns from their mothers in our "modern" hospitals are other examples of harmful child rearing practices.

About 25% of marriages in the U.S. now (1975) end in divorce, and an even higher percentage of couples have experienced extramarital affairs. This suggests that something is basically wrong with the traditional concept of compulsive monogamy. When viewed in connection with the cross-cultural evidence of the physical deprivations, violence, and warfare associated with compulsive monogamy, the need to create a more pluralistic system of marriages becomes clear. Contemporary experiments with communal living and group marriage are attempting to meet basic needs that remain unfulfilled in the isolation of a nuclear marriage. We must seriously consider new options, such as extended families comprised of two or three couples who share values and lifestyles. By sharing the benefits and responsibilities of child rearing, such families could provide an affectionate and varied environment for children as well as adults, and thereby reduce the incidence of child abuse and runaways.

The communal family -- like the extended family group -- can provide a more stimulating and supportive environment for both children and adults than can the average nuclear family. Communal living should not, of course, be equated with group sex, which is not a sharing, but more often an escape from intimacy and emotional vulnerability.

Openness About The Body

No matter what type of family structure is chosen, it will be important to encourage openness about the body and its functions. From this standpoint, we would benefit from redesigning our homes along the Japanese format, separating the toilet from the bathing facilities. The family bath should be used

Table 6:
Drugs, Sex, and Violence

Drug Relationships	Mean	Premarital Sex Not Agreeable (5.3)	Extramarital Sex Not Agreeable (3.5)
I use and experiment with drugs quite often.	4.8	.52	.18
I smoke marijuana quite often.	4.0	.27	.05
I drink alcoholic beverages quite often.	3.8	.32	.11
I get hostile and aggressive when I drink alcohol.	5.1	.68	.30
I would rather drink alcohol than smoke marijuana.	3.4	.35	.32
Alcohol is more satisfying than sex.	5.3	.70	.34
Drugs are more satisfying than sex.	5.4	.73	.34
I take drugs more often than I experience orgasm.	4.8	.47	.13
I drink alcohol more often than I experience orgasm.	4.8	.44	.17

> **"By not answering an infant's needs immediately and consistently, we not only teach a child distrust at a very basic emotional level, but also establish patterns of neglect which harm the child's social and emotional health."**

for socialization and relaxation, and should provide a natural situation for children to learn about male-female differences. Nudity, like sex, can be misused and abused, and this fear often prevents us from accepting the honesty of our own bodies.

The beneficial stimulation of whirlpool baths should not be limited to hospitals or health club spas, but brought into the home. the family bath should be large enough to accommodate parents and children, and be equipped with a whirlpool to maximize relaxation and pleasure. Nudity, openness, and affection within the family can teach children and adults that the body is not shameful and inferior, but rather is a source of beauty and sensuality through which we emotionally relate to one another. Physical affection involving touching, holding, and caressing should not be equated with sexual stimulation, which is a special type of physical affection.

Teaching Children to Love, not Compete

The competitive ethic, which teaches children that they must advance at the expense of others, should be replaced by values of cooperation and a pursuit of excellence for its own sake. We must raise children to be emotionally capable of giving love and affection, rather than to exploit others. We should recognize that sexuality in teenagers is not only natural, but desirable, and accept premarital sexuality as a positive moral good. Parents should help teenagers realize their own sexual selfhood by allowing them to use the family home for sexual fulfillment. Such honesty would encourage a more mature attitude toward sexual relationships and provide a private supportive environment that is far better for their development than the back seat of a car or other undesirable locations outside the home. Early sexual experiences are too often an attempt to prove one's adulthood and maleness or femaleness rather than a joyful sharing of affection and pleasure.

Sexual Equality of Women

Above all, male sexuality must recognize the equality of female sexuality. The traditional right of men to multiple sexual relationships must be extended to women. The great barrier between man and woman is man's fear of the depth and intensity of female sensuality. Because power and aggression are neutralized through sensual pleasure, man's primary defense against a loss of dominance has been the historic denial, repression, and control of the sensual pleasure of women. The use of sex to provide mere release from physiological tension (apparent pleasure) should not be confused with a state of sensual pleasure which is incompatible with dominance, power, aggression, violence, and pain. It is through the mutual sharing of sensual pleasure that sexual equality between women and men will be realized.

The Psychobiology of Moral Behavior

The sensory environment in which an individual grows up has a major influence upon the development and functional organization of the brain. Sensory stimulation is a nutrient that the brain must have to develop and function normally. How the brain functions determines how a person behaves. At birth a human brain is extremely immature and new brain cells develop up to the age of two years. The complexity of brain cell development continues up to about 16 years of age. Herman Epstein of Brandeis University has evidence that growth spurts in the human brain occur at approximately three, seven, 11, and 15 years of age. How early deprivations affect these growth spurts has yet to be determined; however, some data suggest that the final growth spurt may be abolished by early deprivations.

W. T. Greenough, a psychologist at the University of Illinois, has demonstrated that an enriched sensory environment produces a more complex brain cell in rats than an ordinary or impoverished sensory environment. His studies show that extreme sensory deprivation is not necessary to induce structural changes in the developing brain. Many other investigators have shown that rearing rats in isolation after they are weaned induces significant changes in the biochemistry of their brain cell functioning. Other investigators have shown abnormal electrical activity of brain cell functioning in monkeys reared in isolation. I have suggested that the cerebellum, a brain structure involved in the regulation of many brain processes, is rendered dysfunctional when an animal is reared in isolation and is implicated in violent-aggressive behaviors due to somatosensory deprivation. It has been shown that cerebellar neurosurgery can change the aggressive behaviors of isolation-reared monkeys to peaceful behavior. Preda-

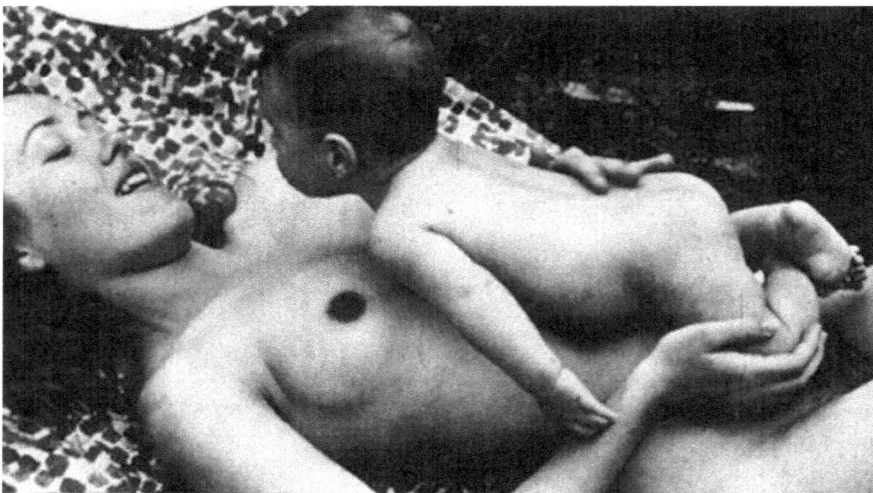

Fondling, breast-feeding and caressing will help this infant to grow into a non-violent adult. Denial of such affection in infancy can have the opposite effect.

tory killing behavior in ordinary house cats can be provoked by stimulating the cerebellar fastigial nucleus, one of the deep brain nuclei of the cerebellum.

Abnormally low levels of platelet serotonin have been found in monkeys reared in isolation and also in institutionalized, highly aggressive children. These findings suggest that somatosensory deprivation during the formative periods of development significantly alters an important biochemical system in the body associated with highly aggressive behaviors. A number of other investigators have documented abnormalities in the adrenal cortical response system in rodents who were isolation-reared and who developed hyperactive, hyperreactive, and hyperaggressive behavior. Thus another important biochemical system associated with aggressiveness is known to be altered by somatosensory deprivation early in life.

Clearly, if we consider violent and aggressive behaviors undesirable then we must provide an enriched somatosensory environment so that the brain can develop and function in a way that results in pleasurable and peaceful behaviors. The solution to physical violence is physical pleasure that is experienced within the context of meaningful human relationships.

Changing the Patterns of Deprivation and Violence

For many people, a fundamental moral principle is the rejection of creeds, policies, and behaviors that inflict pain, suffering and deprivation upon our fellow humans. This principle needs to be extended: We should seek not just an absence of pain and suffering, but also the enhancement of pleasure, the promotion of affectionate human relationships, and the enrichment of human experience.

If we strive to increase the pleasure in our lives this will also affect the ways we express aggression and hostility. The reciprocal relationship between pleasure and violence is such that one inhibits the other; when physical pleasure is high, physical violence is low. When violence is high, pleasure is low. This basic premise of the somatosensory pleasure deprivation theory provides us with the tools necessary to fashion a world of peaceful, affectionate, cooperative individuals.

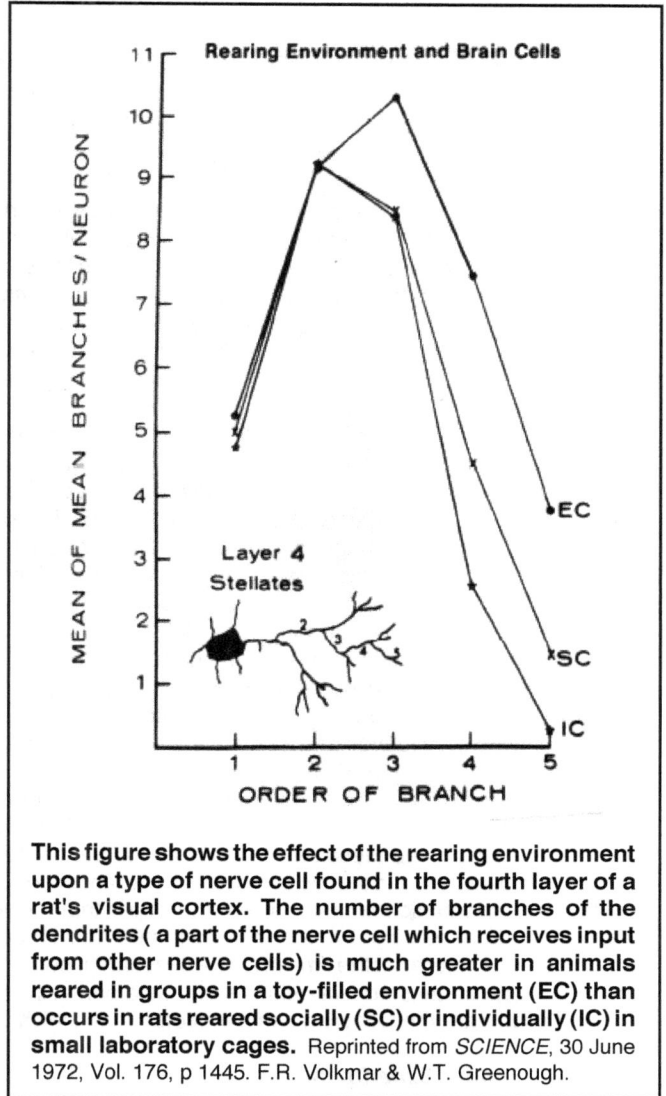

This figure shows the effect of the rearing environment upon a type of nerve cell found in the fourth layer of a rat's visual cortex. The number of branches of the dendrites (a part of the nerve cell which receives input from other nerve cells) is much greater in animals reared in groups in a toy-filled environment (EC) than occurs in rats reared socially (SC) or individually (IC) in small laboratory cages. Reprinted from *SCIENCE*, 30 June 1972, Vol. 176, p 1445. F.R. Volkmar & W.T. Greenough.

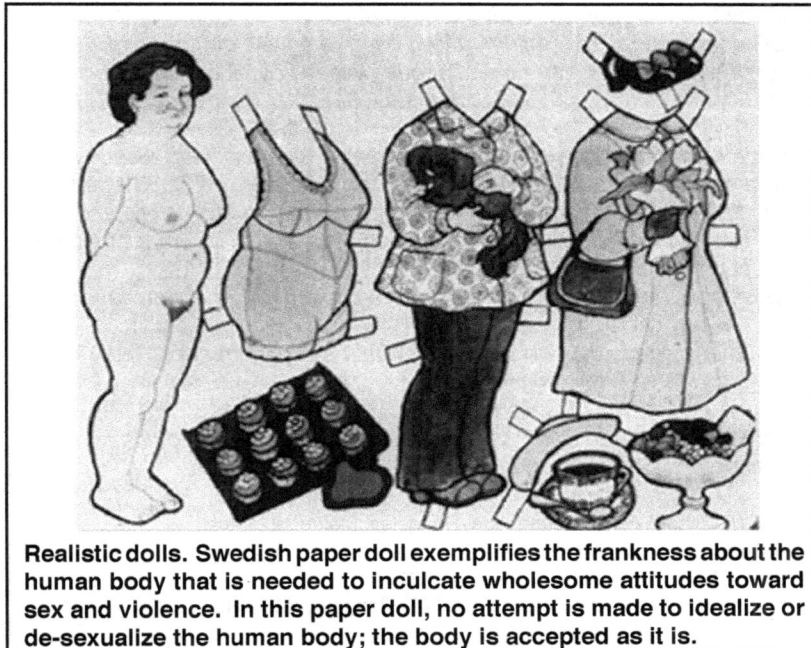

Realistic dolls. Swedish paper doll exemplifies the frankness about the human body that is needed to inculcate wholesome attitudes toward sex and violence. In this paper doll, no attempt is made to idealize or de-sexualize the human body; the body is accepted as it is.

An updated presentation of Dr. Prescott's findings will appear in the chapter "Affectional Bonding for the Prevention of Violent Behavior", in Violent Behavior: Vol.1, Assessment and Intervention, *L.J. Hertzberg, et al. Eds., P.M.A. Publishers, NY, 1991.*

Editor's Note: Oranur Effects from the Three Mile Island Nuclear Power Plant Accident?

In 1951, Dr. Wilhelm Reich published a report on his experiments combining high concentrations of orgone energy with nuclear energy. Relatively small sources of radioactive material (20 millicuries of Phosphorus 32, and 2.26 millicuries of Cobalt 60) were brought into contact with a very high charge of orgone (life) energy in Reich's laboratory. The high charge of orgone had been developed in Reich's lab from many different powerful orgone energy accumulators, including a large room-sized accumulator. This combination of OR energy (as Reich called the orgone) and nuclear energy created a powerful chain reaction which spread beyond the laboratory walls, into the surrounding environment. Almost instantly, workers at the lab were made ill, while laboratory mice being used for experimental purposes died in large numbers. Specific subjective bodily reactions were noted by almost everyone present during the crisis, and many new observations were recorded regarding the behavior of the atmospheric orgone during episodes of assault by nuclear energy. The phenomena observed were new, and could not be explained by classical radiation theory. Reich called this energetic phenomenon *ORANUR*, for the ORgone Anti NUclear Radiation effect.

Reich and his co-workers observed that the air in the orgone-charged, oranur-influenced rooms became heavy and difficult to breathe after the nuclear irritants were introduced. Nausea, cramps and various bodily pains occurred, as well as strange tastes in the mouth. The air in the room appeared stagnated, and took on a hazy characteristic that could be seen, even from outside the building, through the windows. The oranur phenomenon persisted over time, even after the nuclear irritants producing the effects had been removed; it could not be shielded through ordinary means, such as lead plate. Laboratory workers and visitors acquired symptoms of illness from exposure to the irritated energy field of the lab. These included symptoms of nausea, weakness, conjunctivitis (eye irritation), headache, sinus pressure, edemas, dizziness, hot flashes and cold shivers, mottling of the skin, fainting, and loss of equilibrium. In later months, the laboratory had to be abandoned, even though the nuclear materials had long since been removed from the building and property. Weather patterns over the laboratory were observably changed, with a shift towards droughty conditions. Clouds developed a ragged appearance, and a dull, greyish haze dominated the landscape. Trees were badly affected, dying in the areas surrounding the lab, with a loss of leaves and needles at the tops and ends of branches. Unusual blackish and whiteish substances were observed precipitating out of the air onto exposed surfaces. Exposed rock surfaces on the Laboratory building developed a blackish discoloration, and began to crumble away. Several years passed before the oranur effects diminished.

With the exception of Reich's co-workers, and those who have followed in his footsteps in recent decades, the scientific community has ignored these new findings about oranur. Unusual biological and atmospheric reactions to nuclear energy, nuclear power plants, and atomic bomb tests have tended to be dismissed, even when reported by persons with no connections to Reich or orgonomic research. From the paper that follows, author Katagiri recorded many observations of oranur-like phenomena from persons living and working near the Three Mile Island nuclear power plant in 1979, shortly after a major reactor accident there. At the time of his field research, neither Katagiri nor the people he interviewed knew about Reich's work on oranur, and so these observations stand as a partial independent corroboration of Reich's observations. The reader is invited to compare the following sample quotes from Reich to the materials gathered by Katagiri.

From *The Oranur Experiment:* †

"...every one of us could feel the heaviness of the air, the oppression, the pulling pains here and there in the body, headaches and nausea... ventilation did not seem to remove the oppressive air from the laboratory building. After one hour ventilation, it was still impossible to enter the OR energy room, the radium having been removed long ago." (p.281)

"A penetrating salty taste, turning slightly bitter or sour on the outstretched tongue, was felt by all present everywhere within the building and even outside...as far as 50 feet." ... "THE OR ENERGY ITSELF SEEMED TO HAVE CHANGED INTO A DANGEROUS DEADLY POWER." (p..282)

"Three experimental observers remained outside the laboratory within about 100 yards. One assistant rushed the experimental piece of radium into the OR energy room and into the 20x charger... A few minutes later, we could clearly see through the large windows that the atmosphere in the laboratory had become 'clouded'; it was moving visibly, and shined blue to purple through the glass... I felt severe nausea, a slight sensation of fainting, loss of equilibrium, clouding of consciousness, and had to make an effort to keep erect on my feet." (p.283)

"The OR energy seemed to have run amok, possibly even to the extent of a chainlike reaction in the atmosphere, far outside the building." (p.285)

"I put my head inside [the accumulator] for a moment, and felt suddenly as if hit with a sledgehammer..." (p.307)

From *The Blackening Rocks: Melanor:* ‡

" ...after a while, it felt like sitting under an ultra-violet sunlamp. Our faces were burned up, flushed and hot. On the whole, the physical and emotional situation under Oranur felt like a chronic, mild SUN-STROKE". (p.56)

---

† W. Reich, *The Oranur Experiment, First Report (1947-1951)*, Wilhelm Reich Foundation, Rangeley, Maine 1951.
   Partially reprinted in W. Reich, *Selected Writings*, Farrar Straus & Giroux, NY, 1973.
‡ W. Reich, "The Blackening Rocks: Melanor", *Orgone Energy Bulletin*, 5(1-2):28-59, March 1953.
(Xerox copies of both citations available from the *Wilhelm Reich Museum Bookstore*, PO Box 687, Rangeley, ME 04970 USA).

Three Mile Island: The Language of Science Versus The People's Reality*

Mitsuru Katagiri (Nakao Hajime) **

"In my throat I had the feeling that my breath was cut off. It was in my throat, not my chest, but in my throat. I couldn't breathe. It happened on Wednesday, Thursday, and Friday. I felt like something was draining and I had to keep spitting but I couldn't get anything. When I got to Lewistown, it stopped. I was up there for ten days and I didn't have it at all. When I came back here, I had it again for a while." (statement by Joan Fisher, who evacuated the area of the Three Mile Island nuclear power plants plants with her eight-year old son, David, and his grandmother on Friday, March 30th, two days after the accident.)

Joan's husband, Jeremiah, stayed to take care of the cows. He got a burning sensation in the back of his neck and his face down below the cheekbones. His eyes burned badly.

Jane Lee describes Jeremiah's symptoms. *"The whites of the eyes get inflamed for one thing. He could see all right but I don't know why... but I was able to see that the inside of the eye here, you know, where the lid's supposed to be up against, and it was almost like it fell away from the eye. And I figured how did his eyes get so red all the way around? He looked horrible. Now, a lot of people complained about this, burning of the eyes."*

Those apparent symptoms disappeared quickly and haven't been taken seriously in the media nor by the experts. Today, not only is it said that the accident at Three Mile Island was blown up excessively by the news media, but it is also thought that what happened there was not really very serious. Moreover, this very claim is being spread through that same news media. In such circumstances, how can we, who in our daily lives are utterly dependent on the news networks for information, be sure that the nuclear accident actually even happened?

Are we going to give up our ability to speak and act beyond this closed circuit of manufactured information? Of course not. Efforts are still being made to restore the right of lay people to speak up on such seemingly specialized issues as nuclear accidents. First of all, by the victims.

* Translated from the Japanese by Sara Acherman and Rebbeca Jennison. Parts I and 2 are reprinted from *Kyoto Review* 12: 1-21, 1980 and 13: 36-53, 1981. The author traveled to the Three Mile Island area after the nuclear reactor accident of March 28, 1979.

** Dean of Academic Affairs, Kyoto Seika University, 137 Kinoty, Iwakura, Sakyo-ku, Kyoto 606, Japan.

People living near Three Mile Island (TMI) whose well being was and still is being threatened are fighting back; several citizens' groups have been cooperating and have set up an office called the TMI Public Interest Resource Center in Harrisburg. Following are the observations of those people.

Flowing beyond a hillock, the Susquehanna river bends, turning in a southerly direction, and grows in width from two-thirds of a mile to one and a half miles. One of the two long, low islands lying there side by side is Three Mile Island.

It is said that when the river flooded in 1972, the islands were completely submerged. At that time, Three Mile Island Nuclear Station #1 must have been under construction. In 1975 another flood forced local residents to evacuate, but that time, the island was only partially covered with water. It seems only natural then to wonder why anyone would undertake to build a nuclear power plant at a location where such potentially unfavorable conditions exist.

Three Mile Island had already come into the possession of the power company or its predecessor by around 1900. At that time, looking at both Pittsburgh's industrial belt in the Ohio River basin of western Pennsylvania and the Philadelphia Industrial zone along the Delaware River basin in the east, it would probably have appeared that those areas had already been saturated with industrial development. Perhaps this saturation was anticipated by the original speculators who were planning to develop the Susquehanna River basin, making use of its plentiful water supply. It is also possible that in around 1900 it was a nationwide trend for the major industrialists to procure the land along the major river basins. Today, there are indeed a great number of industrial facilities lining the banks of the Susquehanna River. Even if events may not have unfolded exactly as industrial speculators envisioned, those in possession of property continued to profit.

Undoubtedly, then, it was the time and trouble of land acquisition, compared to the lesser cost of potentially damaging floods that might or might not strike, that led to the choice of this site for the nuclear power plant.

On a sunny day in August we set out by car from Lancaster, passing through the lush, green countryside to visit the town of Goldsboro which is situated on the western bank of the Susquehanna River adjacent to the two big islands. When we arrived, we stood looking at the power plant's four cooling towers over the trees on the shore of Shelley Island, which lay between us and Three Mile Island. We were a party of a few Japanese visitors and three local residents.

"It's useless," I had been told by a friend in Japan.

> **"This deafening roar occurred at least six times before the accident; it shook the houses of people living nearby, and when it sounded late at night, they lept out of their beds in surprise."**

"Even if you go there you won't get anything out of it. Do you think you can learn anything from just looking at a power plant? All you'll see is a bunch of buildings."

Yes, that's generally how it is. No matter which plant you go to, whether it's at Tsuruga, Mihama, Takahama or Oi, all you can do is look at the bare, white concrete buildings without any idea about how to really begin to approach them -- the only thing one can do is just have a look and go home. It's not easy to imagine the laborers beyond the wall of the building on their hands and knees scrubbing the floors to mop up the invisible radioactive dust. It is generally true, we can only see things that are reflected in our eyes, but even then we fail to really see many of those very things that are reflected.

From this bank there is nothing in the landscape either to indicate that anything serious has happened. I can even see fishing boats and people water skiing. Wouldn't it be a good idea to take a souvenir photograph with the power plant that has achieved world fame overnight there, standing in the background? Yes, that's about the best we can do.

There is almost no way even to begin to feel interested. But of course, there is no doubt that the nuclear power technocrats are well aware of this unexpressed sense of helplessness and alienation which afflicts the average person. (Wise administrators also undoubtedly realize that when an alienated person has some destructive impulse he/she can, ironically, become an active participant). Doesn't this invisible wall represent the tacit understanding prevailing in the technological society that it is prudent for people who are not specialists in a certain field to stay out of it? And doesn't it remind them that there is no need for them to understand? . . . By now you should have learned not to just plunge into territory where you're not even sure whether or not your efforts will amount to anything. Don't you know that whatever the new territory is, you will just be faced with an endless number of extra chores? Why would you want to add to your life new and complex problems which will just rob you over and over again of your time and energy? Aren't you already losing control over your life? . . .

Articles we read by special correspondents, whether they have been written on the banks of the Susquehanna or in a room in a New York building, do not convey the chaos that actually existed at the scene; rather they are well-organized packages of information, proof of the correspondant's ability. For the most part, we accept what is written in them as representing the world as it really is. The reporter endeavors to do his/her best as a reporter and the reader increasingly devotes him/herself to becoming a better reader. Thanks to this, our modern world has become easy to understand, and there seems to be no necessity to harbor any doubts.

The woman acting as our guide, Jane Lee, is a rapid speaker. She says Mr. Whittock, who lives on the shore of Goldsboro near where we are standing, heard a series of deafening roars which sounded like a Boeing jet flying toward him, skimming close to the surface of the Earth on the morning of the accident. Jane Lee also heard this kind of roaring noise. Gesturing as she speaks, she tells us about how she was terrified and worried -- what on Earth could have happened? We have heard nothing about this in Japan. A deafening roar? What was it? How should I know? I'm a layman. But

what could this noise have been? If it was such an impressive roar, and if it occurred at the time of such a major accident, its importance surely cannot be questioned. But we really don't know. Fujita-san and the other lawyers from Osaka also have puzzled looks on their faces; even these knowledgeable lawyers who have been preparing themselves for several years to work on the Ikata nuclear plant lawsuit, and who have outshone the government scientists serving as witnesses for the state, making them lose face, these same lawyers also shake their heads in dismay.

An engineer specializing in the mechanics of power plants would most likely answer that it was simply a sound that occurs when the acqueous vapor circulating in the secondary cooling system under some tens of atmospheric pressure units is discharged into the environment. To Jane Lee, however, someone who lives near the plant, it was an extremely alarming experience, something beyond comprehension. And listening to her, I can't even begin to think about it -- I must simply take in what I can and try to make some sense out of it later. To the confusion of innumerable unclear and unsolved problems, as well as to the overwhelming quantity of half-baked, superficial interpretations and explanations, yet another question mark has been added.

When I visited the area again in October, I heard another account about this roaring sound -- a footnote to this question mark. This deafening roar occurred at least six times before the accident; it shook the houses of people living nearby, and when it sounded late at night, they lept out of their beds in surprise. Why is it that this kind of simple, straightforward and frightening experience which has been directly confirmed by sensory perception is of no value to the press?

One more footnote: this roaring blast was also heard on the morning of March 30 when the tension about evacuation was continuing to mount. In the Daily Local News of the liberal town of Westchester it was noted that UPI reporter Emil Sveilis had heard the blast when he was at the Observation Center located right on the eastern bank across from the island. At 8:42 a.m. a hold-up tank whose pressure had risen to a dangerously high level emitted radioactive gas into the atmosphere. There must have been many people from the press there, but the DLN was the only paper that reported it. No, to be accurate, other than this report, I have not personally come across any references to this incident in any of the piles of newspapers I have gone through. Of course it didn't appear in the Japanese newspapers, not in the New York Times, nor the Washington Post. No doubt the specialists will tell us that it is a matter of course, to be expected, perfectly natural...There is nothing to be frightened of; what's important is the number of curies of radiation, the number of microcuries, how many roentgens the dose is, how many rads, how many rems, how many millirems and so on. Whatever their minds are really like, to the representatives of the utilities company, the reactor's condition improved somewhat, and the stage where the maximum disaster could be diverted has been reached. (Of course the accident didn't just disappear in a vacuum; and however ironically absurd their circumstances, the operators and engineers on the site deserve no end of respect.)

And indeed, by noon of Friday, March 30, when the state governor had advised that all pregnant women and pre-school aged children should evacuate, radiation which by far exceeded even the politically determined "permissible" levels set by the government and the industrial sector had been released and had undeniably already done irretrievable damage to human life and health.

And practically speaking, it would not have been at all odd if the people of that area had fled every time they heard that deafening roar which sounded repeatedly before the accident. And those hundreds of news reporters who are supposed to be our surrogate eyes and ears, to become our source of knowledge, daring to risk their lives -- they should have followed the shifting wind currents with Geiger counters in hand through the towns, the fields and the woods. This is of course a mean joke, a bit too cruel. Yet we must ask what should have been the function, the responsibility of each sector of society when the top officials of the NRC were forced to admit that they, too, were as helpless as the blind. AP knew enough to send its reporters respirators and protective clothing to shield them from the radiation.

II

Jane Lee lives on a farm owned by the Fishers located about three miles away from the island to the west. She has appealed to public offices, specialists, to anyone who will lend an ear, about the abnormal births, underdevelopment and numerous other strange symptoms of illness that have become increasingly conspicuous on farms in the area since about two years after Three Mile Island Nuclear Station #1 began operation. But as is always the case, up to this very day, nobody has taken any action. Even now, after the accident, no one does a thing. She talks on and on about the cats, the ducks, the cows, and the fact that when the wind blows from the east where the power plant stands, the weather grows bad, or that whatever is in the air is carried down into the gently sloping valley where the farms are. Jerry Fisher had been wearing a hat and the back of his neck felt as thought it has been sunburned after the accident; his eyes were also badly affected...and so the unending stream of facts and concerns continues to pour forth.

While listening to Jane Lee's voice rattle on through the haze of the foreign language which we cannot adequately understand, suddenly, though belatedly, I begin to understand what is happening. We haven't come here to learn what the people of this country have found out nor to hear the conclusions they have drawn, but to be thrown into this unmanageable confusion. Although as amateurs, we have a fair amount of knowledge about the effects of leaked radioactivity from power plants on the environment and human beings, we find that we cannot provide any answers to dispel her anxieties. The only thing I can do is try and keep up with her and do my best not to let her words slip past me and not to forget them.

At Charles Conley's farm which we visited earlier, some kind of white powder-like substance had fallen long before the accident, covering the fields with a thin, white coating. He said that when he tried to take the material out that had settled at the bottom of the trough, the water just turned milky white. We were shown apple trees that had lost their leaves even though it was summer. The farm was located at the top of a hill that sloped down to the river bank; from there we could see very clearly the power plant located about two miles away. The smaller, younger trees had hardly any leaves on them at all, and the larger trees were also losing their leaves from the tips of the branches; on taking a closer look, we could see that they had unusually small fruits on them. The tops of the pine trees lined up in back of the house were bent over, and many had strangely twisting branches and brown needles; those that were not protected by the house were more badly damaged than those which were sheltered. Yes, viewed from the power plant, the less damaged trees were hidden behind the house. A large portion of the foliage of a dome-shaped tree, probably some kind of maple tree standing at the southern corner of the house, had been stripped away leaving a wide, bare section above the middle; the leaves remaining at the top appeared to be new ones, but at that height and against the bright sky, it was difficult to tell.

Here we are just a few minutes later, looking at the familiar shape of the white cylinders of the cooling towers with their spread-out bases. Almost all of the trees on the riverbank near here have lost their peripheral leaves as well as the leaves from their smaller branches... It's like looking into the first layer of hell. You might think us crazy for saying so, but a few months ago this place really was a hell. Nurses left their patients behind and teachers fled, leaving their students in the classrooms. The siren went off, and many people ran from their homes thinking they would never return to them.

Listening to the gentle hum of the motorboats, we looked up at the trees in Mr. Shittock's garden as Jane Lee continues speaking...This big spruce tree didn't used to be this ragged. I remember because two years ago when we let balloons loose here it was perfectly healthy. And she repeats again and again that the air was "steel blue" and that they had a "metallic taste" in their mouths. Disconcerted, we listen. All we can do is soak her words up like a sponge. There is no way to answer her. Everything she says merges into one large question mark. To avoid misunderstanding, let me say that this question mark refers to our own inability. It is not that we had any doubt concerning the existence of the events themselves.

We are going to have to set foot into the quagmire, for we have already heard so much that we must begin a dialogue. Who, when, where? Who saw this steel blue atmosphere? When did they see it? Where was it? What kind of metallic taste? When did the trees begin to lose their leaves? Where? I have to throw these questions back at her. But she hasn't finished talking yet. We have to listen, for she is our hostess.

The time for departure draws near. We, the guests, must hurry so we will be on time for tonight's meeting of the Susquehanna Valley Alliance. We are their guests also. The Susquehanna Valley Alliance, located in Lancaster which is about twenty miles from Three Mile Island, is a citizens group suing to prohibit the dumping of the 3,000 tons of radioactive, contaminated water which leaked out of the reactor vessel. The people living in Lancaster drink the water from the Susquehanna.

Feeling just like a person who on a TV talk show has been told, "Sorry, time's up," and is about to be left in emptiness, we climb into the car, and Jane Lee returns to her home. How many times since the accident this spring has she explained the same things to visitors like us? And how many times have visitors like us hurriedly run away? The accident has affected a wide geographic area, or more accurately, its effects extend limitlessly. Pressed for time, we have to run to listen to the people of Lancaster. However, we have undoubtedly left behind accounts of many incidents of equal importance which must sometime be hauled up from the depths and clarified.

In the car, I leaf through copies of the records of abnormal animals, records which Jane Lee has been collecting since 1977 from neighboring farms and which are signed by the farmers. I shudder anew as I think that I, like her -- no, even more than her -- am an amateur. I am neither a veterinarian nor a plant pathologist. I have never milked a cow and I don't even know the names of the trees. By what fate have I gotten into this kind of thing?

There are about twenty sheets of neatly typed records from local farms about diseased and stillborn animals. Invariably the farmer's signature can be found at the bottom of the page. She explains that this is done so that their credibility won't be questioned. Even Jane Lee, a private citizen who holds no public position, has been compelled to conform to the quasibureaucratic procedure when dealing with public officials. I wonder how many times they refused to deal with her before she adopted this method. One sheet is headed with the name of Charles Conley:

Vance is Jerry's brother. This area is filled with people who are closely related to each other. There are many Fishers and also many Conleys who have probably been here since the land was settled. And this is the kind of thing that is recorded. But what does it all mean? Jane Lee told us about the cat whose back weakened and who couldn't stand up and the butcher who claimed he had never seen cows with such soft bones; but is there some relationship between these things and the veterinarian's prescription for mineral deficiencies? Does it have anything to do with the power plant? And the white, powder-like substance? As far as they know it fell about the beginning of this year (1979). But what about last year, and 1977? ...

Nakao-san, what is this white powder? Why ask me? ... You mean nobody's investigating it? That's right, nobody's investigating it. Isn't that a bit backward? Don't they have any universities in this society where there are nuclear power plants operating all over the place? And don't they even know what the white, powder-like substance that has fallen on the pastures is? Aren't there any specialists or intellectuals?

Since the moment I arrived on the West Coast, I have been asking everyone I meet the same few questions: Has there been any investigation made since the accident of the effects on the environment? What about investigations on the health effects of the people living in the area? Or the accumulation of radioactivity in plant and animal life in the area? There are almost no answers. In coming here, at last, that has been made clear to me. In any case, what do all these records mean?

Charles Conley, Etters, PA (within 1 mile of TMI.)
Mr. Conley has farmed all his life.

1978:   Sept.   Steer Died - was down unable to get up.

              Dec.   Steer Died - was down and unable to get up.
                            One steer down and drags itself about.
                            One steer appears to be normal.

These animals were treated by the vet with shots of vitamins and minerals. Mr. Vance Fisher followed through and administered the shots every day for a week. These animals were purchased sometime in April 1978.

Mr. Conely is puzzled about the white powdery substance that he finds in the watering trough every morning and also on the grass. When he sweeps the substance away, the water turns milky white.

Animals also suffer from diarrhea and weakness. Also, walked in a humped fashion, vet bills over $100.00. The vet informed them that they will have to have 2 1/2 times the required amount of minerals for the animal to survive. Sixty dollars was spent in two months for minerals. The vet said if the minerals loss in the ground continues, the farmer will be unable to grow hay. Last summer 10 kittens from two separate litters all died.

        1978 - Pigs for this year did not develop as rapidly as in the past.
        1979 - Kittens - litter of three - one unable to turn its head -died.

_____
                             Charles Conley

Date:_____

Diarrhea. Weakness. A cow that drags itself about. Is this cow the same one we were shown a little while ago? The record reads that the animals were brought in spring and died in September of that year. Now I get it. They buy the calves, raise them, and then sell them. Where did they buy them? What happened to the other calves that were being sold at the same place?

How about all the other farms? Yes, there are others. I don't know what it means, but cases of stillbirth, premature birth and deformities have all been recorded. Nothing but things that I can't really understand. But isn't it reasonable to say that it is the high concentration of abnormalities that have appeared in this area that has made Jane Lee keep all these records? She isn't just compiling data so that she can make a display of the percent of abnormalities that might be expected to occur anywhere; and there is no doubt that backing her up are the voices of the local farm people who live within this several square mile area in this valley. This has been going on since *before* the accident last spring which brought Three Mile Island to the world's attention.

On the day after Joseph Conley, Charles' brother, learned of the accident, it is said that he gave up his farm land because he thought he couldn't continue farming here anymore, and that he himself was dragging his feet when he left. It is meaningless to say, then, that because Jane Lee's records are not epidemiologically or statistically valid, that they have no meaning.

I would like to comment upon one thing that occurs to me every time I refer to the information that Jane Lee has compiled. For the most part, looking at such records, people quickly dismiss them as being inadequate or as having no value. I am not referring to the evasive words of the officials whose careless attitude we can take for granted; I have heard this reaction from the mouths of people who are afraid of how the accident at Three Mile Island is going to affect their children's future. And I have even heard it from the mouths of people in the movement who gather at meetings. It's an understandable response. Of course Jane Lee's data is not, epidemiologically speaking, solid. Anywhere you go there are abnormal births, and her records don't show how many cases appeared in relation to the total. But can't you people understand the nature of things? Is there no significance in the grievances of people who have, since their childhood, been observing the animals in their pastures? I know you may want proof that can be used quickly at a court trial. Neat, instant, "scientific" proof. You don't understand the meaning of science: someone must make the effort, must struggle to find the answer if the necessary proof is to appear. Are you going to tell the farmers to do epidemiological surveys? Or tell Jane Lee that she had better become an epidemiologist? Why should even you mimic the tone of the bureaucrats?

III

The world we had suddenly found ourselves hurled into seemed very unsettled not only to us, but to the people who were living there as well. It is no wonder that they were worried, knowing about the high incidence of abnormalities in farm animals and the leaves that had been falling in such a strange way during the summer after the accident—and all of this going on right around them. And those symptoms that at the time had felt like sunburn, inflammation of the eyes and so on were undoubtedly not to be forgotten. But for the present, at least on the surface, all of them seemed to be continuing their lives as though nothing were wrong. Even though somewhere deep within there lurked this undeniable anxiety, people had to go on working, had to continue their daily lives; they were enduring everything, unclear as it was.

In spite of lengthy discussions of the issue of atomic power, and in spite of the reality of an accident of such magnitude, there seemed to be almost no signs that the experiences of the people living in the area would be taken up in the expected manner. In my view, they were unable to believe for a second that this was reality—that no one was going to do anything. During that week, or those weeks, or that month of fear and anxiety, those people who had done what they believed to be their best, waited for the concerned specialists to arrive on the scene. Surely, they thought, the experts would be anxious to help them answer their questions.

We are supposed to be social animal who, through communicating with each other, maintain the society in which we live; but in this case we were not awakened by the crisis and did not take action. One might say this society of ours has come to be like a row of signboards which professes to provide solutions to all our problems; or isn't it just like a bunch of partitioned, rigid departments, filled with personnel, that make up the kind of public office one would find in a swelling municipal administration? If you go there with a problem, you are given the run-around by each office you approach as they tell you that they are not in charge of that particular thing. Moreover, narrowly specialized work can only be dealt with by specialists; those outside the field who intervene are generally a nuisance. And then again there is very little communication about practical problems between departments at the bottom.

But even in the case of an incident like this, where it was made so clear that such central administrative agencies as the NBC were incapable of dealing with the situation, there were extremeley few independent individuals who ventured to start inquiries on their own. Was there anyone who had the feeling that he himself should go? It seemed that the experience of anxiety known to those people who had been thrown into that uncertain situation, would only sink further into oblivion.

... The day after we visited Goldsboro, we came to York to talk with a group of citizens called "ANGRY"—Anti-Nuclear Group Representing York. We were in a small restaurant called *Sprouty* which is run entirely by women. One of them is called Cindy Braus. As with everyone else I met, I completely forgot to ask her age; but I would say she is under thirty. She is a small, vigorous-voiced person.

Three Mile Island Nuclear Power Plants, as seen from Goldsboro bank.

> **"...cases of stillbirth, premature birth and deformities have all been recorded.
> ...This has all been going on since *before* the accident last spring ..."**

On the fourth day after the accident began, Saturday, March 31st -- to be exact, at 2:00 o'clock in the morning -- together with the other five members of her family and their three dogs, Cindy ran away from the area, crossing the Appalachian Mountains toward Pittsburg. She said that she had wanted to have mountains between themselves and the nuclear power plant (and good or bad luck aside, we have only our own judgment and physical actions to resort to in such situations). They went along the old Route 74 in a north-northwesterly direction; after crossing the first of the high ridges, they began to head west and then turned onto the highway that leads to Pittsburg.

Most of her family were feeling nauseous and they had to stop at every rest area along the way. The dogs seemed to be in the same condition. Moreover, of the three dogs, two later suffered such symptoms as scabs around the eyes.

Sproutry, the shop where Cindy works, is in York City, about twenty kilometers south of Three Mile Island, and her house is on the south side of the city. According to records from the weather tower located on the north end of Three Mile Island, northerly winds were blowing into this area from the direction of the plant for about two hours during the morning and again from around 3:00 p.m. to 11:00 p.m. on March 29, the second day of the crisis. If they had been exposed to something harmful coming from the plant, would it have been during those times? If the nausea had really been caused by such exposure, then it is probable that these symptoms would have appeared about a day later, isn't it?

Or was it that the direction she chose for their evacuation was an unfortunate one? From the afternoon until late in the evening of Friday, the 30th, the wind had continued to blow in a north-northwesterly direction. That is, until three or four hours before they left the house, the wind had been blowing at several meters per second from the direction of Three Mile Island toward the area into which they were about to travel. If something had in fact been released from the power plant, it is as though they chose the old Route 74 running in a north-northwesterly direction in order to meet whatever that something was as it was being carried along in the wind.

But who can say with certainty that what they felt were the acute symptoms caused by exposure to radiation? Who can decide whether those were symptoms occurring after such and such number of hours? What I have heard and what I am writing here is, if left just as it is, an account of just one isolated incident. What if it had only been car-sickness caused by the tension of their late night exodus? And the dogs' nausea? Could it have been that they just appeared to be in that condition? I didn't make certain whether the dogs had actually vomitted or not -- what on Earth was I doing? Was there really any value in checking the time, direction and speed of the wind? And there is of course no way I can prove that they didn't just make these things up; nor can I prove that they were related to the accident. And all of these experiences are scattered and isolated so every person involved feels uncertain. Now it is January, 1980. Soon a year will have gone by, but I believe no one has even begun to solve the problem. ...

Cindy says she doesn't remember reading any ac-counts of significant statistics or findings and that those things she does remember she has heard in conversations with and among friends or acquaintances as they were talking of their experiences.

"But that kind of thing is very important."

"Yeah, but it's also obviously important that they didn't do anything. They didn't record that. They didn't make a note of it."

"So, you can record things."

"Yeah... We put a notice in the newspaper saying that if anybody had had certain symptoms such as diarrhea, nausea, skin rashes and so on to please call in, and I took... and I have a list of about 23, no, 25 names of the people, with how long the symptoms lasted and so on. And I want to get back to them again just to see how they're getting along. But some of them called up and they did have problems with the skin rashes around their eyes, particularly, and said that their doctors had noticed an unusual number of such cases, and they the doctors said it was unusual. They hadn't really seen anything like it."

Having come this far, you might think we have only to take one more step to be able to get hold of something by the tail. I felt the same way, but that single step is an incredible distance away.

There is daily life, inertia and there are things which we have to do as a matter of course. And these things, combined with the complexity of our narrowly specialized society, its extreme degree of centralization, and the chronic state of crisis we find ourselves in because of our technology, seem to be molding the impotence of our generation. I can say this to myself as well and thus can never quite take that final step. In the end, I'm only maintaining the status quo, even though I don't know enough to say that it ought to be maintained.

Cindy has related almost everything she knows. Resourceful as she is, she tells us that if anyone wants to drink coffee, he or she can go to a shop down the street to get it. All at once, everyone begins talking about coffee.

Now I regret that I didn't ask for a copy of the list of the 25 people. Even with only 25 examples it might be possible to come up with some sort of consistent information; even starting with only 25, four times 25 equals 100, ten times that amount is 1,000. All that is needed is the accumulation of simple efforts -- not an act of genius that transcends common sense. But I missed my chance. That's where I allowed myself to be vague, and that's exactly why and how I can't take that extra step forward.

When I went to the area for the second time at the end of October, this thing was weighing on my mind, but I was unable to make arrangements to meet Cindy again. Although this isn't confirmed, I have heard from others that Cindy and her friends perhaps felt that the results of their efforts were inadequate. I really wanted to go to that shop one more time... certainly, 25 is not a large number, but wouldn't that give me something to hold on to at first? I suppose that those symptoms they talked about were just dismissed as being nothing more than products of the imagination under stress. If there really had been nothing to worry about, it would be fine. But who is to decide whether the symptoms had nothing to do with the emission from the crippled plant? And wouldn't it be too bad if both their telephone survey and their methodology had aborted.

At a meeting held on October 23rd in New Cumberland, across the river from Harrisburg, Chauncey Kepford asked from the speaker's podium how many people there had suffered sore throats, nausea, diarrhea or headaches at the time of the accident. To my surprise, about one third of the two hundred or more in the audience raised their hands. At that moment, everyone looked deadly serious. At the same time, honestly speaking, I thought it was all very peculiar. Why then weren't they trying to determine the facts on their own?

To return to the conversations we had at *Sproutry*...

Fujita-san, the professional lawyer with us, asks me to inquire whether the dead pigeons were taken away somewhere; we have heard there were many people at the time who saw a lot of dead pigeons or who hit them while driving.

Cindy says, *"I don't believe they were. I mean it wasn't specific, it was quite generalized. I live about a mile, a little over a mile right out of town, straight south, and I noticed in the country -- I'm surrounded by trees on all four sides -- and I noticed, I walk my dogs, and noticed them here, there, and still... and I remember, I don't think it's my imagination, but I seem to remember lately, that is, since the occurrence, that more animals generally were being killed on the highways and I think that might be... 'cause they move slower. You know what I mean, they're just slowly being poisoned. Know what I mean? I notice that out in the country where I live, there's more dead animals laying around. They probably are hit by cars, but I think it's 'cause they're not moving as quickly, you know?"*

"I agree with you. There's an awful lot of road casualties," says Joyce Johnson.

Cindy continues, *"It's pure conjecture, but I noticed the difference 'cause I do travel that way all the time. I just noticed there are more animals. You expect some, but it seems to be a higher amount, a higher estimate."*

This is certainly not what would be called "hard information." Someone who has just received technical training would without doubt make quick judgment of it saying that it was useless, that it would not even be worth talking about. It's true that when Cindy says she wonders if it wasn't just her imagination, I too feel myself wavering. However, we must have some definite ground to stand on. Just as Cindy had wanted to have mountains between her and the power plant, in order to take action we need to rely on what our own bodies tell us, and our intuition which can provide a base for our suppositions, whether or not they are false or mistaken -- if we had been mistaken, it would be fine. Doesn't the situation clearly demand some action from the people?

The world reflected in our eyes might be warped with anxiety; but still, we see things that we ought to see. And to put it ironically, it is the work of the specialists to produce the crises

Most trees on the Goldsboro bank, across the river from Three Mile Island, have bare branches, regardless of age.

What Was That Metallic Taste?

" On Wenesday morning (March 28, 1979) we left the news conference that Met Ed had held there and I believe it was around 10:30, 11:30. As we crossed the street from the Observation Center to our car which was parked on the opposite side of 441, I noticed a sensation in my mouith, ah ... like a metallic taste, is the best I can describe it.

"One way I might describe it is at amusement parks they sometimes have bumping cars, and when I get off there I have this kind of bad taste in my mouth, and it would be similar to that. Anyway, I made a joke about something to the effect that that must be something to do with what they were releasing. At that point I felt because of my previous education and belief that you could not see, smell, or taste radiation.

"The following day I had started to hear area residents comment about such a similar metallic taste. And it made me start to wonder if something related to that... what they were doing there was causing this. It's possible that the air stored in that tank they were releasing had a taste to it. But nonetheless, if that was the case, I woiuldn't be anymore happy because it would indicate a fairly strong concentration of that air at that point.

"The following Friday, I had come back from my normal route of advertisers and it could have been somewhere between 2:00 and 5:00 that I was on a hill. I'll show you where, (spreads map out, points one mile east of the plant, near the junction of Locust Grove and Covered Bridge Road). I was taking pictures there and I saw a car, and I don't know whether I started to taste this first or whether the car was there first. I would imagine they were with the NRC, they had a federal government car. They had a driver and a passenger and the passenger was holding a probe out the window, monitoring. And I asked him what his reading was and he said, "Oh, we're not getting anything," or something to that effect and I pursued it further and it basically came down to he wouldn't tell me.,

" In fact, they had been driving back and forth, up and down the hill. Because generally if they get a reading and they don't, they'll usually go back and forth to try to pinpoint the reading; I know they were getting some type of reading. There's not much doubt in my mind.

" Also, while I was up there then, I started to get this very strong taste in my mouth. And after that, The whole way home, until I got home, which was a period of about five miniutes, I could tatse this in my mouth. And it was so strong at that point that I had to, I had to like ... I don't know if I ate something or washed my mouth out. I can remember doing this and telling my wife that, you know, about this, and the fact that I felt like something was happening there. The wind was coming directly from the plant toward the hillside. And that I was very upset that these gentlemen had not told me if the reading was low or what it was."

David Graybill
at the *Press & Journal*
Middletown, Aug. 27,1980

" I was on the way to a class with my neighbor (about 7:00 p.m. of March 28), and I said to her something about my lips felt really chapped and she said hers did, too. And we were even looking for herchapsticks at the time. And as we licked our lips, there was a strange metallic taste on our lips.

"But I really do not believe that it was something that I had made up because I didn't have any fears, you know. I had heard it (the accident) had happened, but it didn't mean a thing to me.

"We sent her (the oldest daughter, Meredith, who was then seven) to school that Friday morning, the 30th. And at that time, we still weren't that fearful because we still didn't understand it. An then, I listened to the radio and heard that they were going to evacuate some people or they were thinking about it. And so, one of my neighbors went to school and brought her home. And when she came home she said that a lot of

the children that were down on the corner (waiting for the school bus between 7:40 and 7:50) were very upset beacuse there was a terrible taste in the air. But she didn't describe it to me as mettalic. She didn't understand what it was. She just said there was a strong tatse in the air and that her mouth tatsted funny. That was all that happened then."

Terrel Stohler
Hummelstown, Aug. 17, 1980
(8 miles north of the plant)

and it is the general public who suffer the effects. What we need in order to deal with those crises is the point of view of the general public who are more than a little upset and bewildered. The accident itself was awful enough, but even more lamentable is the way our society has somehow led to such an accident. That action on the part of the common people should begin is only natural.

Wouldn't it be all right if the people themselves -- not the bloodless administrating agencies -- came to conclude how laughable it was that they had been mistakenly concerned about insignificant things? And there is absolutely no need for them to become part of the elite themselves. It is not those who seem to have their view of the world blocked by piles of papers who are neglected and to whom no attention is paid. Qualifications are something everyone has. There is no reason that they can't begin to do something themselves.

Even concerning the problem of just when and where the birds were hit by cars, if such information were simply gathered and plotted on a map of the area, some clues would probably emerge. Is there any way to make clear what those occurrences were, instead of just accepting the uncertainty that the word "probably" implies?

IV

Stephen Horcebein is another member of ANGRY; well, at least he came to the gathering on August 14 at *Sproutry* and was sitting there while we were talking. He had been

> **"Someone asks if I plan to publish the results of what I have gathered here;... What we have been able to obtain (working from Japan) ...is only a small part of what you could do if you would only start doing this yourselves."**

working at the State Department of Health until shortly before then. I never really asked him whether he had been a full-time employee there, just working temporarily as a part-timer, or what. In any case, while working for the Health Department, he interviewed about 150 families living within five miles of the Three Mile Island plant concerning the accident.

His accounts were not of his own personal experience, but rather of what information he gathered after hearing what the people said in those interviews. And as he didn't have the records there with him, no doubt there were things that he was unable to report correctly to us. Once again, however, we were surprised to hear someone describing what had happened in the area; once again we felt it somewhat peculiar that these things had not been spread in the news. ...

He says that those interviews were the informal part of the Health Department's survey -- but what the hell does 'informal' mean? And is the survey he is talking about the one that was conducted in June, jointly by the Health Department and the U.S. Public Health Service, the one intended to provide an accumulation of basic data for long-term studies on the state of health of all the residents there over a period of 25 years? Informed or not, if the Health Dept. has gotten hold of what he is telling us about, why don't they make it public? What? Because of the 'ethical' problems involved? He says that he doesn't have the records that he *"was not permitted to keep them because there are ethical problems involved."*

What really happened at the Pennsylvania Department of Health? In the little time that we have, we are unable to make much of anything clear; all that remain are the things he has said.

"I've had many people describe...like a sunburn feeling to me. You know, I didn't suggest anything to them; they mentioned it to me first. And it was like a sunburn feeling, a very dryness of the skin. It would be a drying of the skin, basically, a reddening of the skin, slightly. I don't...from what people have described that. Let's see. It lasted apparently between March 28 and I would tend to think about the 3rd or 4th of April. There were basically...all the people would have at least been outside their houses, or exposed, you know. You might, I don't know if you want to, call it exposure. But they would have been outside their houses, maybe just within the yard or doing something like that, umm, between March 28 and April 6. And really, I don't know if there's a whole lot else I could tell you about it, really.

"Apparently it just disappeared, uhh, after a few days. The same sort of thing happens when you get a sunburn. I guess that's why people describe it as sunburn.

"Some people that I spoke to had gone to see their family doctor. From what I understand, most people thought that it was something to be unconcerned about. But I don't know. I can't answer your question directly because I really don't know. Some people have gone to the doctor, I know that, but as far as the doctor's responses, all I can say is from what I understand, what they told me was the doctors didn't think it was that serious."

Kuman-san asks about the eggs that didn't hatch...

"From what I understand, the hens tended to the eggs, but they just did not hatch. I think the gentleman said it was something like four out of fifty-two eggs hatched, which is an incredibly large difference. Every person that I spoke to who mentioned anything regarding that said their animals wouldn't mate correctly or something.

"Well, every person I spoke to except one said that it was not happening prior to the accident at Three Mile Island. The only person that told me that the birds were dying off and things like this were happening the spring before Three Mile Island was an Amish woman, who umm—Amish is a very conservative form of Christianity in the United States, O.K., and other than that... I don't even know if her religion is relevant to any of this, but that's the only person who told me that it was even happening..."

Steve must have been talking about the same kinds of things I have written about in the first part of the article: that from years ago there had begun to occur abnormalities among animals on farms in the area. But it's odd that almost no one said that there had been such problems before the accident; rather, we can probably trust the story of the Amish woman which he emphasized was a unique exception. It seems consistent with accounts of the experiences of farms which Jane Lee has gathered.

"One other thing I don't know if I mentioned before -- I think someone else did -- was nausea. People were very nauseous during this. Let's see, people who left and also people who stayed here, who I interviewed, but In think both groups told me that. Also, diarrhea was a big thing. As far as I know... I don't know... but it seems right now, I'm tempted to say that it were small children who were most affected by that; school-aged children were more apt to get nauseous and diarrhea than adults. So, other than... I can't really think off hand of any other strange things..."

The clock already shows 6:25 and everyone has to eat something before the next meeting begins, so they bring us Sproutry's homemade pizza and we begin to bite into it. Steve looks a little left out and forlorn.

Probably he is adding his own interpretation to the things he is speaking of, to the material, as were the people who answered his questions when he inquired at each home as well as the farmers. Even so, these certainly are accounts with some substance. Somehow I should get hold of what this guy, who looks a little helpless is contrast to his strongly built body, has to say. And I just tell him I want to go with him and meet the people he interviewed, and he's saying he'll take me. At least I should get his address. But why should it be me who goes? It's true, we are eager to pull together all the things we can to take back home, but we will be able to get only as much as we can. Isn't that the way it is? How can you expect us to do more? Tomorrow we have to head off for Washington.

Yesterday, we also heard quite a lot from Jane Lee. These two days have kept us going at a high level of intensity, haven't they? It feels like all of a sudden we have gotten a handhold on the heart of the problem; yes, we have at last entered the territory of the Three Mile Island Plant.

But at the same time a great sense of dissatisfaction that is very hard to dispose of has come to plague, has risen up as far as the tip of my tongue. This dissatisfaction is neither with the government, nor with the power company. It is a discontent with the seeming fact that in this society ordinary people, the common people themselves, have almost no means of taking

August 8, 1979

Honorable Joseph M. Hendrie, Chairman
U.S. Nuclear Regulatory Commission
Washington, District of Columbia

Dear Chairman Hendrie,

I am entirely baffled by the apparent refusal of the U.S.
Nuclear Regulatory Commission to have extensively reviewed the
reports by hundreds of Three Mile Island area residents who,
during March 28-31, 1979 primarily, and at times subsequent,
experienced:

> (a) metallic taste in their mouth
> (b) metallic or Iodine-like odor in the air
> (c) irritated and watery eyes
> (d) moderate or severe respiratory inflammation
> (e) gastro-intestinal dysfunction and diarrhea
> (f) disruption of the menstral cycle in females
> (g) skin rashes (some appearing as radiation burns)
> (h) sharp, abnormal pains in joints.

The U.S. Public Health Service and Pennsylvania State Dept.
of Health are jointly conducting a survey of TMI area residents
to record medical histories so that the full health consequences
of TMI' radiation releases in the next 25 years will be documented.
That is all fine and should be done. But why is there a complete
dismissal by the NRC of any immediate indications of exposure to
levels of radiation higher than what were immediately thought the
first dates of the accident? Psychosomatically induced ailments
are possible with some, but not with hundreds or even more persons
and I suggest this matter has been conveniently laid aside.

The NRC is charged with ascertaining full details about the
TMI accident. You are further charged with knowing the full effects
of even low level radiation on populations near to nuclear reactors.
Failure to pursue the aforementioned reports from TMI area residents
is a dismal failure of your most important safety responsibilities
to the tens of millions of people living near reactors, not to mention the people around TMI.

I therefore recommend that all available expertise be applied
to ascertaining the cause of these physical ailments associated
with the TMI accident and a completely accurate public disclosure
made of its cause and the level of radiation or contamination that
people may have been exposed to. The inability of both Metropolitan
Edison and the NRC to know even to this day (or at least to have
disclosed if you actually do know) the levels of exposure is in
itself a major, most serious failing of pre-TMI accident obligations
by both parties. And if it is determined that the exact cause of
these physical ailments cannot be determined due to the lack of
adequate research on the subject pre-TMI, then the public should
know the extent to which we indeed are unprepared to deal with
nuclear plant emissions.

Yours sincerely,

STEPHEN R. REED
State Representative

State Representative Stephen Reed's Letter to the Nuclear Regulatory Commission (8 August 1979) ...

of the Three Mile Island Nuclear Power Plant. Technical staff members of the Nuclear Regulatory Commission (NRC), the Department of Health, Education and Welfare (HEW) and the Environmental Protection Agency (EPA), who constituted an Ad Hoc Dose Assessment Group, prepared the report. The report concludes that the offsite doses associated with the accident during the period March 28 to April 7, 1979, represents minimal risks of additional health effects to the offsite population. The projected number of additional fatal cancers due to the accident that could occur over the remaining lifetime of the population within 50 miles is less than one. This report, of course, did not address the immediate physiological reactions addressed in your letter. However, we have consulted with Dr. Marvin Goldman, a medical consultant for NRC and he has stated that at the radiation dose rates involved, as described in the report, none of the effects identified in items (a) thru (h) above can be expected to be caused by radiation.

If we can be of further assistance to you, please do not hesitate to write us.

Sincerely,

**Original Signed by
H. R. Denton**

Harold R. Denton, Director
Office of Nuclear Reactor Regulation

... and the NRC's Response (20 Sept. 1979) The above is a part of the NRC's reply to Stephen Reed which was written by Harold Denton on Behalf of Chairman Hendrie. It stated that "The only knowledge that we have of a large number of people exeriencing physiological reactions to the accident comes through Mr. Arnold of ParaScience International." Enclosed with the reply was a copy of Larry Arnold's letter to the NRC in which literally the same eight kinds of symptoms Mr. Reed had listed can be found. It is apparent that Mr. Reed's letter was based on Larry Arnold's report. It is as though the "information" went round that closed circuit: after sspending much time and energy, one throws one's results in front of those bureaucrats only to receive in return, if anything at all, the empty echo of one's own voice. No, it is worse than empty -- it echoes maliciously.

---

practical action other than lobbying activities; why do the people practice their own science and politics so little? With even this much of a grip on things, why couldn't you, of your own initiative, begin to make your appeal to the world by clearly identifying your unsettling experiences and the direct harm you received at the time of the accident?

No doubt there are people who, hearing me speak, become angry. What a self-centered and selfish person you are! What have you come here to do? Have you come to live off the sorrows of other people? Aren't you just making use of those events that occurred in a corner of Pennsylvania merely as a lesson? Aren't you only interested in your own country's welfare? At any rate, you'll soon leave here, so what's this about a 'people's science'?

Someone asks if I plan to publish the results of what I have gathered here. I am caught off guard, taken by surprise. What we have been able to obtain in this limited amount of time and with our limited language ability would only amount to a small part of what you could do if you would only start doing this yourselves. And in fact, aren't you listening right along with us? That's what I'm going to take back to Japan with me, as it is, and tell people about. Is it that you are thinking the knowledge about your own experiences, which has not been 'processed' by

specialists, is of no value? We are from a country far away from here and are only amateurs ourselves. Why is it that you are counting on our information? You are living here and are capable of many times more, many tens of times more than we are. Yes, I'll probably write down what I have just heard here; but you should write it down too, just as you have heard it.

"...Yes, I'll do something in some form or another, but why are you asking me this?" I ask them in return.

This time, they are surprised.

"Well! We live here. I mean, if you're going to put the information together..." says Beverly.

"... in Japanese."

"Yeah, in Japanese. We can even get an interpreter to translate it. But I mean it is of as much interest to us as it is to you, I believe."

She's right. I didn't really understand. More than anyone, as those most concerned, they have the right to know what information we have gathered. My saying that it would be in Japanese was senselessly cruel.

"Yes, we'll send it."

"Oh, well thank you. But why did you ask me such a dumb questions?" (Everyone laughs)

Nevertheless, I had already come to a decision: there

is no other way than for you to do it yourselves. Where are your journalists? Where are your intellectuals? There's no way we outsiders, though we do as much as we are able to do it ungrudgingly, can be expected to have any decisive influence.

Postscript (September 26, 1980):

At the time this article was written, I had not heard of any individuals who were conducting their own investigations except Mr. Larry Arnold, whom I mentioned in the comment on Mr. Stpehen Reed's letter to the NRC. Since then, I have learned of two independent researchers. The first, Norman Chupp, a commissioner of the Fish and Wildlife Service of the U.S. Dept. of the Interior, stationed then at Harrisburg, has been doing work on the contamination of animals. Mr. Chupp felt that the thyroids of animals in the area should be tested. Upon discovering that there wasn't a government agency undertaking to do such tests, he began them himself. During the period frm April 24 to May 1st, he trapped a number of specimens and sent them to the University of Missouri to be examined. The thyroids of the rabbits showed 161 picocuries per gram of Iodine 131. He presented this data to the NRC, but they dismissed it as not having any significance, pointing out that there was not a control group and other weaknesses in the study.

The figure 161 picocuries per gram is said to be considerably higher than the maximum of 5 picocuries previously reported possible. The figure 161 is from the time of detection in June, and the concentration in the earlier weeks should have been much higher. It is said that it might have been possible that some portion of this Iodine concentration in the rabbit thyroids came from the April 14-16 releases during the replacement of vent filters at Three Mile Island. But, besides the significance of those contaminations, it seems also significant that those studies on Iodine 131 in animal thyroids were not done immediately after the accident by those government agencies such as the Pennsylvania Bureau of Radiation Protection which had the most sophisticated equipment for this kind of test.

The other research I have learned of was initiated by a gradiate student of biology. I heard this summer that he found apparently high concentrartions of Iodine 131 in some wild animal thyroids. His study had been rejected by a couple of scientific journals because of methodological weaknesses. He and his colleagues, having refined their techniques, expected it to be published in another journal in a few months. The copyright laws and scientific procedures are such that if they talk about their own study ahead of its publication, they lose their rights to it; once the paper is accepted by a scientific journal, then the paper is the property of that journal, and only then can they make the announcement about their own study. As far as I know, the study has not been published anywhere. The reader might at least see the contrasting gap of pace between the never-hesitating technological threatening of man's environment and the scientific procedure of ascertaining facts.

(This article continues in Pulse of the Planet No.4, given the book title: On Wilhelm Reich and Orgonomy.)

Editor's Postscript: On 3 March 1983, Mitsuru Katagiri and Aileen Smith Katagiri testified before the Three Mile Island Commission, Philadelphia, Pennsylvania. This Commission is now spearheading a number of studies on the epidemiological effects of the TMI accident upon surrounding populations. The State of Pennsylvania also circulated a questionnaire to gather additional data on the subjective effects of the TMI accident. To the best of our knowledge, however, all research is centered upon searching for a chemical compound to explain the biological reactions to human beings. Wilhelm Reich's discovery of oranur is probably not being evaluated in any of these studies.

The following are extracts from a document prepared by Richarrd Martyr , Ph.D., for the TMI Public Heatlh Fund, 1622 Locust St., Philadelphia, PA 19103. (10 Jan. 1989)

Project 14
Taste and Smell Implications of the Hypothetical Emission of Nonradioactive Chemicals at TMI
by Richard Martyr, Ph.D.

"Now it is generally agreed that 'metallic' is not in itself a primary taste, but rather a combination of several factors including smell and taste... It has also been observed that the 'metallic' taste can be produced by olfactory stimuli alone..."

"In addition to oral and nasal stimulation, 'metallic' taste has been reported in cases that apparently have no association with sensory perception of the oral/nasal cavity such as: problems within the circulatory system; gastrointestinal disorders; iron and B-complex vitamin deficiencies; during menopause; during illness; and a variety of metabolic stresses. Thus, the experience of a metallic taste may originate from oral or nasal stimulation or from a systemic dysfunction."

"[It is suggested] that the experiences of a metallic taste following the TMI accident might have had an olfactory rather than a gustatory origin."

"Burning Mouth Syndrome: ...patients report a burning of the tongue or mouth and a distortion of taste which is frequently perceived as metallic in quality. "

"Following the accident at TMI on 28 March 1979, a number of residents in the surrounding area reported various symptoms... there is an apparent pattern among those who reported experiencing a 'metallic' taste. Some experienced the 'taste' before they were aware that there had been an accident at TMI. Some also reported that they experienced additional symptoms such as: a dryness of the mouth and throat; a 'sunburn-like' sensation; a sore throat; a choking feeling in the throat; a hot feeling in the throat and chest; headaches; dizziness; and nausea."

"There were 30 individuals included in the Katagiri testimony who mentioned experiencing a metallic taste, and Osborn and Molholt compiled separate lists totalling another 65 individuals who reported a metallic taste following the accident. [Ages] ranged from 28 to 53."

"...52% of those reporting a metallic taste on a specific date experienced it on 30 March. 26% reported the taste on 28 March and another 13% on 29 March. "

"In several instances the metallic taste...[was] accompanied by unusual atmospheric phenomena, such as heavy fog, mist or haze of varying colors."

Orgone Accumulator Therapy of Severely Diseased People
A Personal Report of Experiences

Heiko Lassek, Arzt. *

Introduction

The scientific work of Dr. Wilhelm Reich (1897-1957), who put the main emphasis of his research on the proof and scientific description of biophysical processes in the human organism, has not yet reached a wide public recognition, even by the late 1980s. In the 1920s, as a medical doctor and psychoanalyst, Reich was one of the closest collaborators of Sigmund Freud; he was the head of the Technical Department and Psychoanalytical Clinic of this social movement. His further development of psychoanalytic technique in the field of resistance and character analysis, presented at the beginning of the 1930s, was followed by his later suspension from psychoanalysis; yet only 20 years later his contributions were appreciated and integrated as fundamental to the theory of psychoanalysis. In the following period, his experimental psychosomatic research led Reich to the development of a somatically-oriented psychotherapy, which in time made him the target of discriminating attacks by former co-workers. The entire field of physically-oriented psychotherapy, which during the last three decades has been spreading through all Western countries, explicitly refers to Reich's findings; and the founders of the most influential branches of somatically-oriented psychotherapy, almost without exception, were collaborators or former patients of Wilhelm Reich.

Up to this point in his work, the formerly disputed doctor and natural scientist is today granted general acceptance. However, Reich's work was not restricted to the discoveries mentioned above, but also includes experimental research which provided a natural scientific foundation for the psychosomatic phenomena he had observed. In the 1930s, at the Universities of Copenhagen and Oslo, together with numerous collaborators, Reich carried out extensive biophysical experiments on the connection between basic functions of living substance and the existence of a specific biological, vegetative energy, as expressed in the emotions and drives.(1)

This experimental investigation extended from the recording of changes of the electrical potential of the skin and mucous membrane of human organisms, to the measurement of bioelectrical potentials of monocellular organisms. Reich was led to the discovery of a type of energy postulated by him and so far to a large extent not much investigated, which he called orgone (derived specifically from organism, as he first assumed it only existed in living systems). The scientific recording and description of this type of energy and its interaction with human organisms became the main focus of Reich's work over the following two decades.(2)

In the frame of his biophysical investigation, at the end of the 1930s, he repeatedly found phenomena of radiation in biological cultures, which only in single aspects could be explained by the types of energy known up to that time. In the 1940s, after moving to the United States, and while working as a professor of medical psychology in New York, Reich and his numerous colleagues and collaborators examined the basic physical laws and biological effects of orgone energy.

Reich succeeded in constructing a special chamber which acted as a screen against external electromagnetic radiation, but which at the same time selectively concentrated the orgone radiation: the orgone energy accumulator. Years of self-experiments by Reich and his co-workers proved distinctively positive effects of the orgone radiation and orgone energy accumulator on humans and other animals. According to Reich's research, the accumulator device charges the organisms situated inside with orgone energy. He and his co-workers began an experimental program of treatment of seriously diseased people, mainly free of charge to the patients.†

Dramatic changes in general condition, and positive influences were objectively documented in many patients, for such different disease processes as anemia, diabetes mellitus, various kinds of carcinoma and the acceleration of wound healing processes. Experiments with humans as well as laboratory mice proved the orgone accumulator's value for support of organic mechanisms of regulation and healing.

At the beginning of 1980, in West Berlin, a group of interested doctors and medical students began a private initiative to critically examine and experimentally evaluate the central

---

1. Reich, W.: *The Function of the Orgasm*, Farrar, Straus & Giroux, NY, 1973; Reich, W.: *Bioelectrical Investigation of Sexuality and Anxiety*, Farrar, Straus & Giroux, NY, 1982.

* Physician and Director, Wilhelm Reich Institute, ~~Delbruchstr. 4-C, 1000 Berlin 33~~, Germany

2. Reich, W.: *The Cancer Biopathy*, Farrar, Straus & Giroux, NY, 1973; Raphael, C. & MacDonald, H.: *Orgonomic Diagnosis of Cancer Biopathy*, Wilhelm Reich Foundation, Maine, 1952.

† *Editor's Note: From 1940 to 1957, Reich and his associates published dozens of experimental and clinical reports on the bioenergetics of cancer and other diseases, and on the therapeutic use of the orgone accumulator. These papers are cited in the online Bibliography on Orgonomy (orgonelab.org/bibliog.htm), and also are available from the Wilhem Reich Trust bookstore (wilhelmreichtrust.org)*

> **" I have been able to (achieve) great alleviation or even complete disappearance of pain.. with vegetotherapy supported by use of the orgone accumulator"**

experiments of Reich in the fields of cytoscopy and cancer diagnosis. After three years of research the results of this work were presented at numerous lectures in German and Scandinavian Universities. It was revealed that Reich had discovered and explained several phenomena that are still unknown today, and which can be demonstrated and documented by most modern methods such as post-contrasting video microscopy.

In 1982, the first self-experiments with equipment based upon the principle of orgone accumulation were set up; a mixed group consisting of doctors and university professors (of diverse subject areas such as mathematics, physics, and social science) began to investigate the effects of orgone radiation upon their own organism. It was made evident that subjective and objective influences of the accumulator were most clearly experienced in those individuals suffering from acute conditions of pain and disease. Many healthy test persons without disease symptoms needed longer and more frequent times of exposure in order to feel strong vegetative reactions in the body.

In the meantime, by letter and in lectures, hundreds of people have reported on their predominantly positive experiences with the orgone accumulator, which in most cases they constructed by themselves according to various instructions which became available since the mid 1980s.(3)

As an established doctor, who within the framework of my practice is including the application of the techniques developed by Wilhelm Reich for influencing the vegetative nervous system, I was consulted again and again by people suffering from cancer in the last stage of the disease, i.e. with an existing metastatic spread which by medical means could not be influenced any further, and mostly with a terrible condition of pain. My work consulting and caring for 17 so-called "terminal" cancer patients during the last 2-1/2 years has included both orgone energetic treatment and psychosomatically oriented vegetotherapy offered free of charge without exception. This work has demonstrated the wide range of possibilities, but also the clear limits of influencing cancer in such an advanced stage by use of the orgone energy accumulator.

With 2/3 of the patients, after instructions in the use of the accumulator and test sessions, a clear reduction of the consumption of analgesics, and in some cases even freedom from pain, was achieved on average after 20 exposures to orgone radiation. Almost without exception, their vitality was markedly increased, which was to be seen by the resumption of activities completely inconceivable before the beginning of the radiation therapy. Furthermore, the remaining expectation of life prognosticated by specialists was prolonged with most of the patients.

The following article is based on a lecture I gave on this part of my work at the end of 1989, on the occasion of the Wilhelm Reich Conference in Berlin. In its content, it gives a summary on the experience with orgone accumulator therapy on the most severely diseased patients. I have decided to publish this report because, after detailed consultation and personal care during the first two weeks of treatment, the

3. Freihold, J.F.: *Der Orgonakkumulator nach Wilhelm Reich*, Verlag Konstanze Freihold, Berlin; Gebauer, R. & Müschenich, S.: *Der Reichsche Orgonakkumulator*, Nexus Verlag, Frankfurt.

patients afterwards continued to use the orgone accumulator on their own responsibility. Therefore, the often discussed influence of the clinician was playing a very minor role.

Furthermore, since the first public presentation of these results I have received new reports previously unknown to me of similar experiences of cancer patients with orgone energy treatment. As a doctor, to me, it is the most important and most dignified aim of human medicine to decrease and, if possible, to prevent human suffering to the largest extent. If part of this aim can be achieved by the use of equipment that is constructed in such a simple way as the orgone accumulator developed by Wilhelm Reich, this possibility must not any longer be excluded from either broader public or medically specialized discussions.

Medical Experience with the Therapy Developed by Wilhelm Reich: Vegeto/Orgone Therapy

After several terminological changes, from the 1940s onward, Wilhelm Reich called his method of treatment ORGONE THERAPY. This kind of therapy can be divided into two subgroups which, however, in the practical work with the diseased person intersect or complement one another:

1) Psychiatric Orgone Therapy (character analytic vegeto therapy)

2) Biophysical Orgone Therapy (use of the orgone accumulator and medical DOR-buster)

The way I applied these types of therapy can be subdivided into three fields which also partly overlap:

I) Long-Term Therapy, i.e. character analytic vegeto therapy on patients suffering from diseases and somatic symptoms diagnosed by means of traditional medicine, and previously treated by conventional means without beneficial result. In this field, I work with the patient once a week, and with persons coming from outside Berlin, on average every three weeks for several hours on successive days. This therapy lasts 1-1/2 years on average.

II) Intervention Therapy, i.e. consultation and demonstration of self-aid techniques harmless to the patient, which they can continue independently after instruction and supervision. Later, if necessary, personal consultations and guidance is given.

III) Biophysical Intensive Therapy, i.e. consultation and guidance of most seriously ill patients mainly in the last stage of the cancer process; use of the orgone accumulator and of the diagnostic instrumentation developed by Wilhelm Reich for the follow-up observation.

I. Experience with Long-Term Therapy

Up to now, I have been able to treat the following diseases and symptom complexes with vegetotherapy, supported for a short time by use of the orgone accumulator, with very satisfactory results. By this I mean a great alleviation or even complete disappearance of pain, often followed by the complete discontinuation of all pain-reduction medications in patients with the following problems:

Trigeminal neuralgia
Chronic condition of pain of the locomotor system, especially of the spine
Chronic glaucoma
Relapsing gallbladder colic, also in cases of emergency
Bronchial asthma
Respiratory dysfunction with presence of a pulmonary emphysema
Patients with symptoms of relapsing angina pectoris
Meniere's syndrome (rotary vertigo)
Chronic lymphatic leukemia and chronic myeloid leukemia (The vegetotherapeutic treatment of these diseases will be described in another article. Use of the accumulator in this particular case is contraindicated.)
Schizophrenia, paranoid-hallucinative type
Chronic depression
Anxiety neuroses
Persons suffering from cancer (to the treatment of whom I give more details below).

II. Experience with Intervention Therapy

With this method I have been treating 41 patients during the last 2-1/2 years. With 17 patients, a complete disappearance of the main symptoms was achieved. With 8 patients, the present condition of pain could be reduced. From the remaining 16 patients I have not received any feedback for a significant time, or problems with practice of the recommended techniques were reported.

III. Experience with the Biophysical Intensive Therapy

Under this point, in the following, I want to concentrate on the treatment of most seriously ill people in the state of multiple metastatic spread, in order to demonstrate the wide range of possibilities, but also the clear limits of influencing the process of cancer in the last stage of the disease.

The patients coming to me in this terminal state had for years been trying all conventional methods of treatment, to include chemotherapy, surgical removal of the primary tumor, and radiation therapy. Also, in most of the cases, nature cures such as macrobiotics, homeopathy, fasting cures and treatments in private clinics had been tried without any evidence of even delaying the progress of the disease. From the end of 1987 until spring 1990, I was treating altogether 17 of such severely ill patients. As seen from a medical point of view, these patients had been completely "treated out" at the beginning of their orgone therapy; almost without exception, they had been given survival times of from one to three months by their treating medical specialists. To give an impression of the severity of the cases, I want to give an exemplary description of the condition of two patients before the beginning of the orgone accumulator therapy:

Patient A: My first patient at this time was a professor of economics and businessman, age 53. In 1984, he had been operated on for a malignant renal cell carcinoma (hypernephroma) without any signs of metastases, and he remained free of symptoms for 2-1/2 years. At the beginning of 1987, after six months of radiation treatments, seven pulmonary metastases were detected, bioptically identified as belonging to the primary tumor and diagnosed inoperable, since all lobes of the lung were affected. In September of the same year, two more cerebral

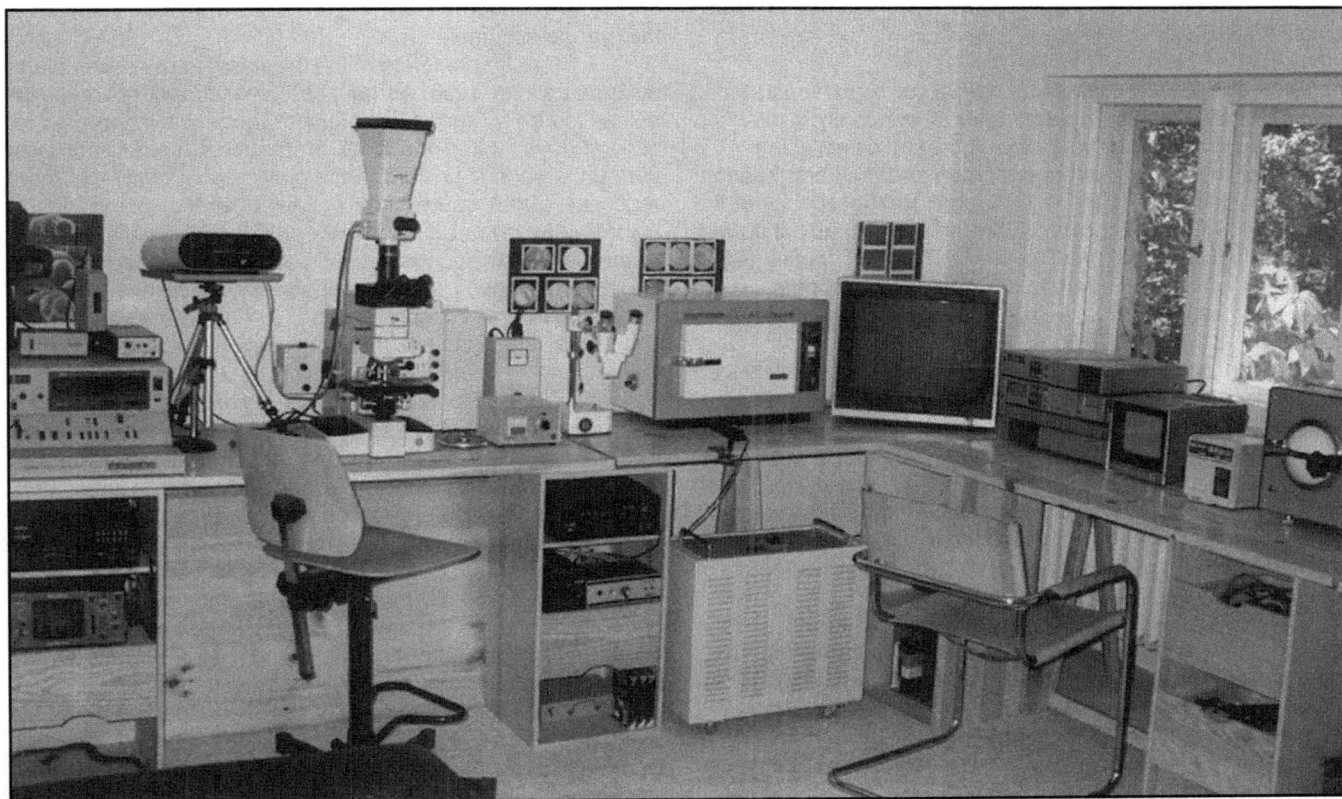

Laboratory facilities of Dr. Heiko Lassek at the Wilhelm Reich Institute in Berlin, Germany.

metastases were discovered, growing from 0 to 4 cm. on the right hand side within two months and from 0 to 3 cm. on the left hand side in the same time. Neither the pulmonary nor the cerebral metastases reacted on large-dose chemotherapy; further radiotherapy of the rapidly growing cerebral metastases could not be undertaken, given the involvement of neighboring motor centers. An implantation of radioactive cobalt into both temporal lobes was considered, but the patient refrained after consulting several specialists among others in the USA and USSR. He came to me in the middle of November 1987, rather being carried by two men, and showing a paralysis of the entire left half of his body. His paralysis, caused by the pressure exerted by the cerebral metastases on motor areas, had developed in the time of two weeks. Specialists had prognosticated him a maximum survival time of three to six weeks, and all medication had been discontinued at this time because of ineffectiveness, except for morphine sulfate and *Temgesic* (a precursor of opiate) for alleviation of pain.

Patient B: The second patient I want to introduce as an example was a 58 year old administrative employee who had an operation for a cancer of the gallbladder at the end of 1986. During 1987, he developed six constantly growing metastases of the liver. Another two vertebral metastases were detected via CT scan, after the patient had been suffering from increasing pain in that area for months. Altogether, five cycles of chemotherapy were undertaken without influence upon the continuous growth of the metastases. According to the patient's own report, he had been declining for several months, and his treating doctor had told him that very probably he would not live to see Christmas 1987. Our first meeting took place at the beginning of December 1987, and he was only able to come to me with physical support by his wife. I was shocked by his overall appearance, which made the opinion of his treating doctor appear to be a realistic prognosis.

The above two patients are held up as examples for all the other people I have treated: after all the conflicts they had gone through and the desperation on the first establishing of the diagnosis, the hope after the operation and their first freedom from metastases, then the terribly quick recrudescence of the spread of the cancer process These individuals had put up with their fate and were thankful to their doctors for frankly telling them about the short duration of life remaining to them. To us apparently healthy people, this attitude is probably hard to comprehend.

For their pain, both the above two patients received Temgesic and morphine sulfate, the last and strongest stage of analgesics that can be regularly prescribed to cancer patients in the so-called terminal stage. To them, as well as to all other patients in a similarly advanced stage of the disease, I had to tell them, in all clarity, that the process was far too advanced to be stopped. Nevertheless, the two patients wanted to try the orgone energy therapy, at least hoping for a slight reduction of their pain.

The Reich Blood Test of the two showed disastrous results: Immediately after the blood was applied to the slide, more than 90% of the erythrocytic membranes were in the process of disintegration; almost all red blood cells had become clumped aggregations; the preparations were full of t-bacilli. This kind of result was to be found repeatedly with almost every patient in the final stages of cancer.

Eleven of the patients and also their family members were present at the first blood test, which were observed simultaneously by physician, patients, and family members through a video system attached to the microscope. They had explained to them the easily understandable main criteria of the Reich Blood Test, and they were able to follow their own native blood picture on the video display during the entire period of examination. Thus, during the following weeks and months, most of the persons affected, apart from the subjective changes in their condition, could see the correlation with the objective picture of the blood diagnosis, and they could even partly evaluate it on their own. Without exception, they all appreciated this possibility very much.

The Course of Orgone Therapy

During the first two or three sessions in the orgone accumulator, most of the cancer patients have uncomfortable somatic sensations. In spite of taking morphine they often feel dragging or pulling pains at the locations of the metastases. However, this pain is described in their own words as "strange", "new", "being of a different, but somehow not alarming quality", compared to the well-known intense pains.

In several cases, one or two more areas of pain were felt and described, each time sharply localized by the patient, which later on were proved to be additional metastases not yet diagnosed at the time of the first sessions in the accumulator. Several patients also reported having a very detailed visual perception of their tumors or metastases while being in the accumulator.

Another less frequent observation was the reduction of pain during the first two accumulator sessions. Thirteen patients described a different sensation of pain, described by the patients as being "somehow beneficial" or "something moving in the body in the area concerned", which intensified while being treated in the orgone accumulator.

This first reaction of the organism to the accumulator disappears after 3 to 6 one-hour sessions, and will not reappear unless the daily orgone accumulator therapy is interrupted for several days. After one week of treatment, sometimes even during the second or third treatment session, more and more reactions of the entire body are experienced: sensations of warmth or even heat, dilation of the cutaneous vessels, increase of the peristaltic sounds clearly audible without stethoscope, and delicate tingling sensation especially in the limbs but also in the scapular and cervical region. Without exception, after a short initial astonishment, these perceptions were described by the patients as being very agreeable.

From this point on, the intensity of the continuous pain experienced by the patients decreased. This is the time when the patients began to use the orgone accumulator twice per day, for one hour duration at both noon and early evening. After three weeks of such large-dose accumulator therapy, the extreme pains of 9 of the 17 patients were reduced to such an extent that, to the astonishment of their treating medical advisors, they no longer desired the daily medication of analgesics. With 5 other patients, before the therapy the pain had been latently present even under strong long-term medication. However, under the influence of the orgone accumulator, the attacks of pain developed a certain rhythm: periods of complete freedom from pain alternated with periods of the old condition of pain. At their own request, these patients also carefully reduced the long-term medication.

In only 3 of the 17 cases did the accumulator therapy fail to achieve a reduction of pain: A 72 year old patient with a primary hepatocellular carcinoma with formation of pulmonary metastases could only use the accumulator for 30 minutes per day, due to the development of distinct hot flushes; A 47 year old woman with severely dedifferentiated mammary carcinoma was only feeling a slight reduction of pain during her stay in the accumulator, but she discontinued orgone accumulator therapy anyhow; A 61 years old patient with bladder carcinoma discontinued accumulator treatment after 6 sessions, even though it had given him a slight reduction of pain; he experienced anxiety attacks and complained of a strong feeling of restriction, as if being "locked up" while in the accumulator. From him, I did not get any further information.

An orgone energy accumulator at the Wilhelm Reich Institute in Berlin, Germany.

From 6 patients it was reported that they could manage with only aspirin in case of attacks of pain. The supposition expressed by the patients was that the accumulator and aspirin were mutually intensifying with regard to their pain-relieving affect, and this was later confirmed by many other patients.

It was especially impressive how the changes concerning the quality of life were described by the associated persons, the wives, children, and friends who could observe the patients: After a time of three to four weeks, normal appetite, joint walks, the resumption of old hobbies and independent car driving had become possible again. The general appearance of the patients had changed. Many of them reported a feeling of vitality which they had not felt for many years, as before the time when their disease symptoms first appeared.

Changes in Blood Diagnosis

The microscopical analysis of the disintegration of the erythrocytes, in sharp contrast to the changes felt by the patients during the first three weeks of treatment, scarcely showed any difference; only the aggregation, the tendency of red blood cell clumping, showed a clear decrease. Despite the still devastating cytolytic pictures, this could be interpreted as a reference to the fundamental change of the electrostatic fields of the human blood.(4)

During the second month of treatment, the blood picture began to improve constantly: The membrane cohesion, the internal pressure of the cell membrane and thus the resistance against the process of disintegration into bions and t-bacilli, drastically increased. Whereas at the beginning of the orgone accumulator therapy, only about 10% of the erythrocytes had their original shape immediately after withdrawal, now it was more than 50%. The way and speed of disintegration also changed towards an increased power of resistance. In Reich's terminology, an increase of the bioenergetic total condition of the organism.(5)

In several cases, to include the two patients A and B described above, central necroses with a decrease in tumor density were observed, even though metastases remained at constant size; in three cases, a shrinking of tumors was radiologically diagnosed. As described in Reich's work *The Cancer Biopathy*, edemas were formed around the tumors; Reich had explained this as inflammatory transformation of the tumorous tissue. He had confirmed this hypothesis by numerous experiments with laboratory animals, the metastases of which were bioptically examined.

However, with patient A above, who had cerebral metastases at the right and left paries, the process of the formation of edemas with central necroses around the secondary metastases caused a recurrence of just the paralysis that at first disappeared after only one week of orgone accumulator therapy! The patient, at this time apparently physically vital (instead of painfully dying in December), went, from February to

---

4. Lassek, H.: "Medizinische Aspekte der Orgonenergie", *Emotion*, 3, Nexus Verlag, Frankfurt, 1982.

5. Reich, *The Cancer Biopathy*, ibid., and Lassek & Gierlinger: "Blutdiagnostik und Bionforschung", *Emotion*, 6, Nexus Verlag, Frankfurt, 1987.

May, on business trips to the USA and South Africa equipped with only an orgone energy blanket (contrary to my advice, since the efficiency of orgone blankets compared to the larger accumulator is diminutively small). After May, he came back to me with completely different problems.

In spite of all his business activities, and without taking any further medication, all pulmonary metastases, which had been rapidly growing before the start of orgone therapy, instead remained constant in size. The left cerebral metastasis had shrunk by 2 cm., and the right one had remained constant, with a central necrosis but only insignificant edemas in the surrounding area. At first presentation, the pressure exerted by the tumor caused a paralysis on the left half of his body. The recurrence of paralysis after his business trips most probably had been *produced by* the orgone therapy, given the increased formation of edemas. On the other hand, in view of the cancer process, the orgone therapy could not be interrupted. So we decided together to venture an attempt to locally withdraw energy from the affected cerebral area.

As on our first meeting, he had to be brought to the clinic by two helpers. Based upon the Reich experiments with the medical DOR-buster, we aimed a water-grounded hollow metal draw pipe 2 cm. above the area where the deep metastasis causing the paralysis was located. Before setting the DOR-buster into operation, I placed the electrode of the Orgonometer by Marah SA in the patient's paralyzed left hand, and asked him to cover and press on his left hand with his right hand during the entire procedure. The Orgonometer showed an initial value of +114; immediately after putting the DOR-buster into operation, the value fell and oscillated between +65 and +72. The patient at that moment reported strong sensations of rotary vertigo, a phenomenon that is reported by most of the patients during use of the medical DOR-buster on the head region. During the following forty minutes, the drawing process continued at a minimum level, and the Orgonometer values rose to +190, with oscillations of +/- 10.

After one hour of this mild treatment, a reading of +210 was achieved, and I asked the patient to move his left hand. With an expression of unbelieving astonishment, which I will never forget, he raised his entire left arm, and all by himself, he sat up from the treatment table. For the second time, by means of orgone therapy it had been possible to effect a regression of the paralysis of the entire left half of his body. Like the first time more than four months previous, his very personal test was to try to use the remote control of my microscope video monitor by well coordinated movements of his fingers, and he managed to do this easily as well. He had come to the treatment being carried rather than being supported by two persons, and now he insisted on climbing the curved staircase leading up to the laboratory all by himself. The influence of the medical DOR-buster only lasted for eleven hours, and we had to repeat the treatment for four times altogether, until his condition stabilized without paralysis.

Further Developments

In the following months, the subjective freedom from pain and symptoms went along with dramatically changed pictures in the native blood diagnosis. The change was especially notable among those who at first had terrible pain despite the strongest medication, and had been prognosticated to have survival times of only 4 to 8 weeks. Contrary to Reich's publications, with two persons the blood picture even approached that of a completely healthy person.

With regard to the human encounter, this was the most difficult time for the patients, for their family members, and also for me. They had all come to me without any hope for a reversibility or even control of their disease, and now they were feeling as vital as if a mortal cancer process had never developed.

Some of them had taken up sports again, went for long walks and on short trips, and I had to assume the role of making them aware of the finality of the process of metastatic spread which was somatically too far advanced. With some of them, for the first time in the whole period of our cooperation, I noticed tendencies of evading the knowledge that their cancer long before the beginning of the orgone accumulator therapy had already been too far advanced as to be reversed or to be stopped for a long time. At such times, I was tormented by questions and doubts, which, being settled in the role of a white-coated doctor in a hospital, I would not have felt with such intensity: Was it right that despite the dramatic improvement of their condition concerning subjective as well as objective diagnostic criteria I constantly had to point out that some time they were going to die from their disease? The question was asked again and again "Don't you think that with the state I'm in now, a healing after all would at least be possible?" I had to deny again and again.

Most of my doubts were concerned with a question that seemed insoluble: What was to happen with the tumor masses that could possibly disintegrate, i.e. how could the tumors and their toxic disintegrating break-down products be removed from the body?

In the 1940s and 1950s, when Reich was investigating the possibility of influencing cancer in laboratory mice by orgone radiation, almost all laboratory animals experienced an inflammatory softening of tumors, followed by death from blockage of the renal transport system. It was the immense mass of cells of the disintegrating tumors and not the spreading of the disease process that had caused death by renal insufficiency. Later, some of Reich's most successful therapies on people also failed because of this problem. This possibility, that might put an end to the positive developments of the orgone treatment, I also discussed in full detail with the patients.

The Final Phase of the Therapy

After more than six months of experiencing physical well-being and freedom from pain, in all but 2 of the 17 patients, the following process began to appear. First, the picture of the native blood diagnosis began to deteriorate impressively within a few days. Apart from the erythrocytes showing well-shaped membrane coats and a normal disintegration process, single erythrocytes recurred which completely disintegrated into the corpuscles that Reich called t-bacilli within 20 minutes. The absolute number of these erythrocytes was small, about 15%. However, the correlated findings were alarming.

Immediately after the withdrawal of blood, and the beginning of the observation period, among the erythrocytes and leukocytes, more and more irregular cell fragments were to be seen, which most probably were to be ascribed to disintegration processes around the tumors and metastases. As I had feared, with the process of disintegration also developed an autointoxication of the body by tumor tissue.

I called several internists in order to discuss the possibility of an extracorporeal dialysis, to filter the blood plasma of this debris. After numerous discussions and demonstrations of videotapes on the constitution of the cell fragments, this possibility was excluded, because the fragments having a size of 1-12 micrometers were equally distributed. Even persons who claimed to have already treated cancer patients with orgone energy could not give any solution; I was especially struck by the fact that those people who did not even know the problem often reacted in an annoyed manner to my inquiry for help on the matter.

After the living blood picture had begun to deteriorate, the patients were still feeling subjectively very well, and the subjective condition and the objective findings were constantly diverging. The second alarming symptom was that at the same time, some of the patients' family members reported that each time after the patient had been in the orgone accumulator, the facial region of the patient turned grey, what sharply contrasted with their usually vital impression after use of the accumulator. This discoloration lasted for varying lengths of times, but disappeared in the course of the day. It was also observed by the patients themselves, but was not accompanied by any somatic symptoms such as circulatory distress or feelings of weakness. My advice to most of the patients was then to reduce their stay in the accumulator to only one 1-hour session per day, since the disintegration of the tumors continued and the blood pictures became more and more alarming.

Based on the patients blood pictures, and because of their acquired ability to assess for themselves the microscopical blood pictures on the video monitors, the patients completely understood the meaning of what was happening. None displayed desperation or emotional rebellion against the approaching end of their lives, but some of their family members did. I was deeply impressed by the thankfulness they showed with regard to the quality of the last period of their lives some of them still set themselves small goals: to paint the room in the basement, or to finish a certain business or personal goal which was of special importance to them. Neither did I observe any deep resignation in them, with perhaps two exceptions about which I am not sure if I observed, instead, something I call "gliding".

Two of my patients died of cardiac insufficiency during their holiday, suddenly and without pain, as reported by their wives. One patient still went on a business trip to Saudi Arabia where he carried out transactions, and after his return on the way from the airport he collapsed and also died without pain, of cardiac and circulatory failure in the hospital. During the last days he still managed his business from his bedside; we said goodbye to each other a few hours before his death.

Only two of my patients reverted back to use of morphine sulfate and Temgesic, and they died a few days afterward in the hospital. Others, after open discussion with their previous consulting physician, were given the opportunity to choose to take strong drugs which would, in the case of an overdose, bring about a painless death. This was extremely important to some of the patients who anticipated a return of the terrible cancer pains. Four of the patients made use of this possibility, after the time when even morphine sulfate would not relieve their pains. All other patients eventually died at home, having either no or only slight pain. With one exception, they all kept in contact with me until a few days before their death.

In summary, all 17 of these severely ill patients lived

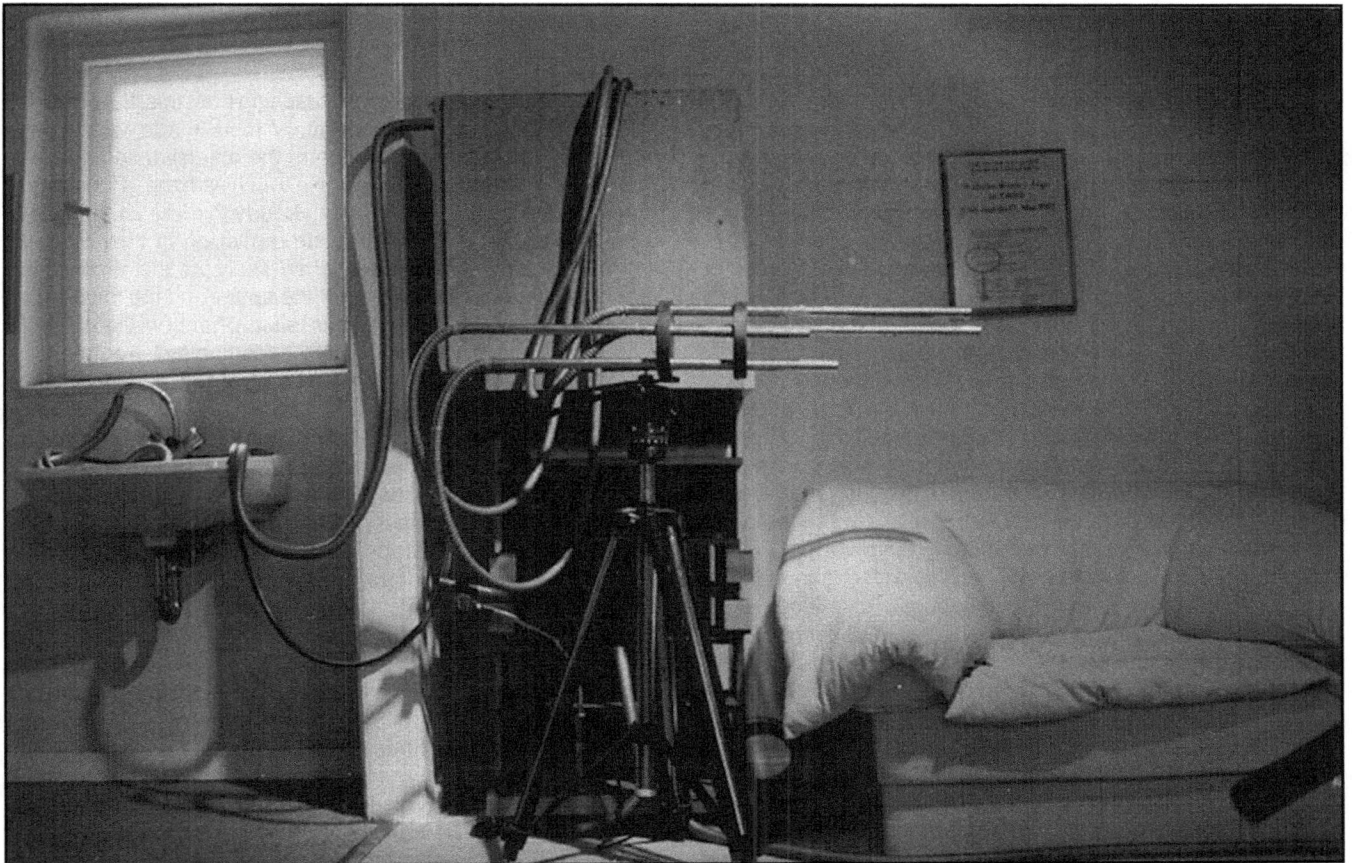

Medical DOR-buster at the Wilhelm Reich Institute in Berlin, Germany.

more than 5 months, with 50% living for more than one year, a significant positive extension of lifespan from the original "terminal" prognosis of only 1-3 months survival time. Regarding the two specific case histories given above, who had a very poor prognosis at the start of orgone treatment, patient A survived 7 months, while patient B survived 6 months.

From these and other experiences with both conventional and non-conventional methods of cancer treatment, the benefits of orgone accumulator therapy to severely ill patients, including so-called "terminal" cancer patients, are significant.

On the Problem of Auto-Intoxication (Self-Poisoning)

Months after the last so-called "terminal" patient had died, I was visited by Ms. Ursula Phillips, a former co-worker of Professor V. Brehmer (former head of the Berliner Biologische Reichsbundesanstalt, who fled Germany during the Third Reich) and close associate with the famous internist and cancer researcher Dr. Joseph Issels. From her former collaborations, Ms. Phillips was very knowledgeable about living blood pictures, including the essential features of the Reich Blood Test. When speaking with her about the possibilities and limits of the accumulator therapy on such seriously diseased people, she noted parallels to her experiences in the Ringberg Clinic of Dr. Issels: Each time when patients came to Issels for the vital radiation therapy, she found cytolytic products similar to those I had described in their living blood pictures. These were observed in connection with patients subjective complaints about radiation hangover, and also a grayish coloration of the skin.

Issels noted that radiation hangover often limited the course of the X-ray therapy he employed, which he interpreted as an expression of the autointoxication of the body by decomposition products of tumor cells. Based upon this working hypothesis, he instructed his medical and scientific associates as follows: From each patient who was given a course of X-ray therapy, living blood pictures were examined several times per day. If many cell fragments were found in the blood of the patient, the radiation was immediately reduced. If the number of irregular decomposition products in the native blood preparation decreased, the radiation therapy was continued.

According to reports by Dr. Issels' associates, this was one of the keys to the success of his form of radiological therapy, which in some cases healed very seriously diseased people. According to Ms. Phillips, to whom I owe a great debt of gratitude, after this method of blood analysis was introduced, there was not a single case of radiation hangover.

If we could ever develop a clinic with some stationary beds for the medical care of most seriously diseased people, and if we had experts for the permanent observation and analysis of the cancer patients' blood picture at our disposal, a constantly controlled large dosage of the orgone accumulator therapy could mean a great step beyond the limitations discussed above. Here, I want to point out again that all changes in the prolonged life span of these most seriously diseased people were achieved only by the technical use of the orgone energy accumulator. A report on the positive experience available up to now in the field of vegetotherapy in combination with the use of the orgone energy accumulator will be published shortly.

Epilogue, and a Note of Caution

The results documented in the prior article were gained by working with people belonging to an age group of 51 to 78. I place special emphasis on this fact because I know of several case histories where with young people different kinds of cancer showed only a short subjective improvement after use of the orgone accumulator, and the process of the disease was not affected at all. With several young patients (under 35 years) having acute myeloid leukemia with rapidly metastasizing processes partly combined with an unknown localization of the primary tumor I cannot close my mind to the terrible impression that the rate of spreading of the cancer was increased by the accumulator treatment.

Without exception, after diagnosis these patients had for personal reasons renounced all conventional medical treatments and based all their hope on alternative methods of treatment without prior consultation of any doctor experienced in the orgone accumulator therapy. Today, I am seeing 14 cases where people called me up on the telephone after, without any prior consultation, they had bought or built an orgone accumulator on their own, in order to "cure" their cancer disease. These cases reveal a complete misunderstanding of the facts revealed by Reich. In most of the talks, they describe a temporary remarkable improvement of the subjective condition after they had begun to expose themselves to the orgone accumulator radiation.

Animated by optimism resulting from this, they decided to have radiological and laboratory tests made which often showed a dramatic deterioration of the objective findings; and this was the reason why they got in touch with me. For many of these people, at this time, it was already too late as to achieve a far-reaching or even a complete remission of their disease, which might otherwise have been likely at the time of the diagnosis because of the kind of the tumor and the localization and manner of spreading. For this reason, here I wish to urgently warn not to employ orgone energy accumulator treatment for lymphatic or myeloid leukemia, or severely undifferentiated cancer in younger patients.

I wish to stress particularly that I have treated cancer patients not mentioned in this article, with a strong recommendation, often against the patients expressed opposition, to have the primary tumor surgically removed, or recommended treatment with radiation therapy or chemotherapy, in combination with daily large-dose orgone accumulator therapy. In such cases, there was not a single case where cancer symptoms recurred in an observation period of 3-1/2 years, nor was deterioration of laboratory values to be found. In these latter cases, the orgone accumulator therapy was started immediately after their operation. In cases of chemotherapy many straining side effects such as sickness, lack of drive and depressive mood can be remarkably mitigated by vegetotherapy and orgone accumulator therapy. On the other hand, the use of the orgone accumulator is strictly inadvisable during any cycles of radiation treatment, but can begin three days after the last exposure.

Unfortunately, in recent years I have seen dozens of

young and older people dying, after having completely refused, for ideological reasons, the temporary use of traditional modes of treatment, which are mostly very straining for the organism. They had called my position "not being in the sense of Reich" since I had urgently recommended them to a course of chemotherapy. For this reason, I recommend the following steps to every person suffering from cancer:†

1)   Don't let yourself be deprived of hope if the diagnosis is "cancer", but do try everything diagnostically possible to get a complete record of your disease (localization, spread, histology).

2)   Find a doctor who responsibly can give you information on the chances of a conventional treatment for your disease, or ask your treating doctor to get in contact with one of the main cancer therapy centers in order to be informed on the last state of the mode of treatment and the therapeutic results achieved by it.

3)   According to the above information you receive, contact the nearest center for consultation on holistic or alternative cancer therapy and make an appointment for consultation. Give the information you consider necessary in this context to your treating doctor, and discuss with him the decision you have made to possibly have another form of therapy. Contrary to the widespread opinion, many doctors are open to some, though certainly not all, unconventional methods of treatment; if your treating doctor is not open or sympathetic to your concerns, ask your friends and associates to recommend another doctor.

4)   Don't combine several natural or alternative methods of treatment with each other, but decide for at most two methods that do not interfere with each other but have a common starting point, such as the support of immunological defensive systems. Ask for detailed information on how these methods of treatment are meant to influence biological processes in the organism.

5)   Be cautious about interpreting initial improvements in your subjective condition, which can occur with many alternative methods. If after two months at the latest your condition does not improve objectively, change those aspects of the treatment which you feel are least helpful.

6)   There are many possibilities of animating and supporting the immunological resistance which I only do not quote here because I have no practical experience with them.

Based upon my personal experience I can recommend a pre- and post-operative orgone accumulator therapy when there is no contraindication; the combination of the use of the accumulator and large-dose intravenous mistletoe therapy has proved to be very efficient; however, I do not practice any additional naturopathic methods of treatment. At the beginning and during the first months of treatment, repeated personal consultations on the processes occurring in the organism, caused by the therapy, should be discussed with a doctor who is familiar with the orgone accumulator therapy. Later, this can be done by telephone.

A constant native blood diagnosis, with observations of the living blood picture, is indispensable in the case of multiple metastatic spread for reasons explained in the article. In all other cases medical observation and regular follow-up control by X-ray examination and laboratory diagnosis is sufficient.

If the reader decides upon the possibility of employing an orgone energy accumulator, they can find information on the construction and use of the apparatus in one of several books which are now available on the subject, and also may contact myself at the Wilhelm Reich Institute in Berlin, Germany.

Dr. Wilhelm Reich ran the Orgone Institute in the United States as a nonprofit organization. For more than four decades, he never aimed at making financial profit from his investigations. The Wilhelm Reich Institute, directed by myself and authorized by Reich's daughter and associate, Dr. Eva Reich, and the Wilhelm Reich Society, both in West Berlin, are working in this same tradition. The awareness of the possibilities and the limits of influencing the organism with the help of the discoveries of Dr. Reich must not be suppressed any longer, but must be further investigated and made freely accessible to every interested person.

† Editor's Note: Please see comments concerning the health care situation in the USA, and a list of clinics, on page 48.

Other papers by Dr. H. Lassek (not cited in text):

"Blutdiagnostik und Bionforschung nach Wilhelm Reich", Emotion, 6:101-141, 1984.
"Zur Bionforschung Wilhelm Reichs", Emotion, 9:128-153, 1989.
"Einführung in die Krebstheorie Dr. Wilhelm Reichs", Krebsforum, 17:5-10, Oktober 1990.
"Medizinische Erfahrungen mit der Therapie nach Wilhelm Reich", Krebsforum, 19:27-35, Marz 1991.
"Vegeto-Orgone Therapy", in Natural Medical Healing Methods for the European Community, Documentary Register Volume 5, Ministry of Technology of Niedersachen, University of Lüneburg, Lüneburg, Federal Republic of Germany, 1991. (Prepared on request from the European Economic Community, for existing and future administrative purposes.)

Medical Alternatives in the USA

Editor's Note:

Given the current war against unconventional, alternative health approaches by conventional allopathic medical practitioners and organizations, the likelihood of an open or public use of the orgone accumulator by trained medical doctors in the USA, as a therapy or adjunct for cancer treatment is, at present, very unlikely. Furthermore, epidemiological evidence demonstrates a generalized failure of conventional treatments (surgery, radiation therapy, chemotherapy), as practised in the USA, to significantly increase the survival times for most cancers. In fact, very few truly controlled studies exist to contrast outcomes of either conventional or unconventional treatments of cancer, and cancer incidence is today higher than ever before, with little or no progress having been made since the 1950s in the survival times of patients. It is even arguable that the retreat of the average individual from conventional treatments is the result of a quite valid and accurate perception that, as generally applied in the USA, conventional treatments for cancer do not help to bring the patient any closer to health than either no treatment, or unorthodox treatments alone. The recommendations given by Dr. Lassek must therefore be viewed in the context of the very different German health care system, which (unlike the USA) is rich with various non-conventional, holistic practitioners and health clinics, allows the experimental use of unconventional methods by health professionals, and does not use policemen, prosecutors and jails to harass and suppress health care practitioners who try unconventional approaches.

In the USA, most health practitioners who have quietly experimented with accumulator treatment of degenerative disease have found that therapeutic regimes which stress immune system enhancement and detoxification or cleansing of the colon and blood are excellent adjuncts to orgone accumulator therapy. The detoxification therapy developed by Dr. Max Gerson and offered by the Gerson Clinic is one such example.

North American Health Freedom Organizations, and Holistic Cancer Treatment Clinics and Institutes

Cancer Control Society and *Cancer Control Journal*
 2043 N. Berendo St., Los Angeles, CA 90027; (213) 663-7801
 - Maintains a complete listing of clinics and medical practitioners in the USA and abroad who employ non-toxic, holistic treatments for degenerative disease
 - Sells books on alternative health topics
 - Annual conventions on holistic health topics

Gerson Institute and *Healing Newsletter*
 PO Box 430, Bonita, CA 92002; (619) 472-7450

Bio-Medical Center, and Hoxsey Herbal Therapy
 PO Box 727, 615 General Ferreira (Colonia Juarez) Tijuana, B.C. Mexico; (706) 684-9011, 684-9132

Linus Pauling Institute
 440 Page Mill Rd., Palo Alto, CA 94306; (406) 327-4064

Livingstone Immunology Clinic
 3232 Duke St., San Diego, CA 92110; (619) 224-3515

National Health Federation and *Health Freedom News*
 PO Box 688, Monrovia, CA 91016; (818) 357-2181
 - Annual conferences on alternative health topics
 - Book sales on alternative health topics
 - Works to change repressive medical laws

People Against Cancer and *The Cancer Chronicles*
 PO Box 10, Otho, IA 50569; (800) 662-2623
 - Works to change repressive medical laws
 - Information on new methods of treatment

Project Cure: Center for Alternative Cancer Research
 1101 Connecticut Ave. NW, #403, Washington, DC 20036; (800) 552-2873
 - Works to change repressive medical laws
 - Publishes a newsletter
 - Information on new methods of treatment

Citizen's Alert!

The following quote from *The Cancer Chronicles* (Autumn 1990, p.1-2, New York) gives an idea of the continuing trend towards prosecution and jailing of biomedical pioneers, and erosion of Constitutional rights.

"California Victory -- For Now. Fast action in August blocked a California bill which would have allowed the state to seize the property of practitioners of alternative medicine. 'SB 2872' was proposed by Sen. Marian Bergeson of Newport Beach. Beefing up the Sherman Act, which already provides penalties of up to three years in jail for alleged health fraud, the new bill would have created a 'forfeiture scheme' to seize the property of anyone convicted under this law. It would have provided prosecutors with a financial motive for cracking down, since their offices could recover the victim's property.

According to the authors of the bill, 'this legislation is necessary so that California consumers can be protected from illness, injury and death resulting from reliance on the products of health fraud promoters and so that violators pay for the costs of the Department of Health Services and prosecuting agencies'. In other words, a witch hunt, in which the witch has to pay for the fuel.

Who would be subject to the bill? People who prescribe or knowingly administer an experimental drug; who sell, deliver, or even give away any new drug not fully approved by the FDA; who advertise any substance represented to have

an effect on AIDS, ARC or cancer; who use new or untested drugs, compounds, or devices invented by persons against whom injunctions or cease and desist orders have been issued. And so on.

Flying in the face of the 14th amendment to the Constitution, the bill specifically calls for 'seizure without [due] process'. According to 'SB 2872', seizure can occur if it is incident to an arrest due to a search warrant, if there is probable cause to believe that the property is directly or even indirectly dangerous to the public health or safety, or was used or intended to be used in violation of the law..."

Under the above law, a search warrant from a judge will allow policemen to force their way into a clinic or lab engaged in holistic medical approaches, or even the home of a holistic health practitioner or advocate, and seize almost any kind of property which they interpret as being "dangerous to public health" as defined by medical bureaucrats, including lab equipment, records, bank accounts, and personal effects. There would be no warning of such "midnight raids", as court action prior to seizure would become unnecessary, the burden of "proof of innocence" falling entirely upon the accused.

Orgonotic Devices in the Treatment of Infectious Conditions *

Myron D. Brenner, M.D. **

Introduction

The treatment of infectious diseases in classical medicine is mainly a matter of identifying micro-organisms, whose characteristics are generally known, and choosing chemical agents, whose modes of action are well defined. The control of microbial disease is truly a great accomplishment of modern medicine, and therefore it has been taken as proof of the validity of the mechanistic philosophy on which modern practice is based. As if confirming Pasteur's theses, the successes of modern microbiology and antimicrobial pharmacology have helped to bury the age-old intuitive perceptions that had been kept alive in the 19th century by the vitalists. Little attention is currently given to the inherent limitations of the mechanistic approach, such as the difficulty in accurately predicting therapeutic response to the use of an antibiotic.

However, Reich's discovery of biogenesis will eventually bring about a basic reorganization of thinking in microbiology, and his understanding of the somatic biopathies will some day lead to a reconsideration of microbial diseases. Of course, a positive therapeutic effect of orgonomic treatment on an infectious condition does not itself demonstrate the involvement of life-energy processes in the onset of the condition, but it raises the question of how an energetic process might have been an etiological factor.

It is generally acknowledged that the aspect of microbial diseases least well understood is that of the "host factors", those characteristics thought to make a person more or less vulnerable to the infectious agents to which he is exposed. As the field of immunology continues its rapid development, I expect it will uncover processes that we will be able to understand as demonstrating pre-existing disturbances in the pulsation of orgone energy in the affected person or population.

In the 15th Century, to help explain the viral pandemics that periodically swept Europe, Italian doctors coined the word *influenza*, referring to the influence of the heavenly bodies on these outbreaks. As we learn more about the interaction between atmospheric and bodily DOR (stagnant, immobilized, deadly orgone), we may yet discover that our distant predecessors had caught a glimpse of something quite real. Those few cases of infectious illness that workers in orgonomy have had opportunities to treat using orgonomic techniques suggest that a great amount of exploration lies ahead.

Case I: A Nonhealing Wound

In 1982, a healthy 33 year old woman fell and sustained a deep abrasion at the ankle, about 16 mm in diameter, and 2 to 3 mm deep. The wound was treated adequately, but the eschar (sloughed tissue) was slow to form, and the area of inflammation and tenderness slowly increased. When I observed the wound 4 weeks after the injury, the entire foot was a dark mottled red, and swollen. There was a slight limp, tenderness of the entire dorsal surface, and a moderately decreased range of motion of the ankle. The area around the wound was warm to the touch. Clearly, there was chronic inflammation and, probably, infection.

The regular application of moist heat, a conservative time-honored (and probably orgonotic) treatment having failed, I planned to refer her, the next morning, to an internist who would probably have prescribed an antibiotic. In the interim, I suggested a trial of orgonomic treatment.

First, in an effort to remove DOR from the whole body, she took a 20-minute warm bath with sea salt and baking soda.(1) This was deeply soothing to her. Two hours later, the DOR-buster (2) was used for 10 minutes, concentrating on the foot, and using lengthwise sweeps up and down the limb. During treatment, she experienced a cool breezy sensation in her limb. Five minutes later, there was markedly decreased strength, and a mildly decreased kinesthetic sense in the leg. She felt drowsy, and went into a deep sleep for half an hour.

---

* This paper was presented at the annual meeting of the American College of Orgonomy, Princeton, New Jersey on October 15, 1989. The author greatly appreciates the contributions of Mr. Stephen Dunlap and Dr. Peter Crist, who performed the Reich Blood Tests, and Dr. Samuel Schneider, who reviewed the paper and suggested changes.

** Medical orgonomist in Maryland, Diplomate, American Board of Psychiatry and Neurology. PO Box 588, Sparks, MD 21152.

1. This solution is a very dilute version of the body's extracellular fluid, though lacking that fluid's protenoids.

2. The DOR-buster used in this study incorporated 16 tubes of approximately 30 cm. length and 1 cm. diameter, arranged in 2 rows spaced 1.5 cm apart, center to center. The tubes were imbedded at a right angle in a larger tube of 4.1 cm. diameter and 15 cm. length, which was connected to a length of 2.5 cm. diameter flexible metal conduit. The conduit was placed into running water. No accumulator was attached, but the materials used make the entire device a one-fold accumulator, with metal inside, organic material outside: the drawing tubes are nylon and the base tube is fiberglass, and both are coated inside with silver metal-flake paint, while the galvanized steel conduit is covered with plastic.

Observing her foot six hours after treatment, I noted some decrease in the swelling. The range of motion had become normal, and there was no longer any limp, because the pain on weight-bearing had gone. By the next morning, I found that the swelling, heat, and dorsal tenderness were totally gone. Around the slightly concave eschar was a small halo of inflammation, and the entire foot looked slightly dark, but all the other abnormalities had disappeared.

After a slowly and steadily worsening condition for a month, such dramatic improvement only 12 hours after a brief treatment with the DOR-buster was nearly unbelievable to me, but I could not dispute what my eyes told me. No classically trained physician could have understood the rapid cure brought on simply by the induced movement of orgone energy.

A five minute follow-up treatment with the DOR-buster was then given. The next day, the color of the foot was normal. Ten days after the treatment, healing was complete, and nothing but a slight hyperpigmentation remained at the site of the lesion. Over the course of time, this too disappeared.

Case 2: Recurring Respiratory Diseases Complicating Orgone Therapy

In February 1983, a 59 year old divorced woman was referred for orgone therapy. She was anxious to the point of terror, and depressed to the point of suicidal thought, because of her illness with influenza and pneumonia earlier that winter. For as long as she could remember, she had feelings of physical vulnerability focused on the rhinitis, flu, sinusitis, and bronchitis that had made winters miserable for many years. Particularly severe sinusitis attacks responded to antibiotics, but only temporarily. She was not asthmatic, though she had been treated for respiratory allergies. She did not smoke.

The patient first had psychiatric treatment 32 years earlier. Over the course of years, a variety of approaches had been tried, including psychoanalysis. At presentation, she had been in group psychotherapy for 9 years.

I eventually concluded that significant characterologic ocular segment pathology was present. In addition to the early infantile trauma that was implied, there were repeated insults to the ocular segment throughout her life. For example, in late childhood, she fell on a nail and sustained a penetrating head injury. In her teens, she had mastoiditis. In earlier adulthood, she had encephalitis.

Treatment took into consideration the ocular block. Biophysical work included firm massage over the frontal and maxillary sinuses. The DOR-buster was used intermittently, starting in the fourth month of treatment. It always produced strong subjective reactions in the patient.

Her general functioning improved quite rapidly at first. Further progress, somewhat slower, was seen as therapy continued. There was some improvement in her general sense of health, but not enough. The respiratory illness, including sinusitis, recurred every winter. The origins of her deep dread of disintegration and death were explored, but the feelings persisted. By the summer of 1987, after 4-1/2 years of therapy, including 2 years with sessions of extended length, progress seemed to have reached a plateau.

Because her reactions to it had always been strong, I decided to begin using the DOR-buster more frequently. When I did, a pattern developed: recurrences of sinus symptoms coincided with one to three week lapses in the use of the DOR-buster. Therefore, by the winter I began using it every session,

usually for periods of 10 to 20 minutes. No other aspect of the characterological or biophysical treatment was changed.

The results were remarkable. First, the patient's usual minor to moderate physical problems, such as injuries and arthritic aches and pains, continued to occur, but no longer were accompanied by a feeling of dread, or a perception of physical frailty. Second, for the first time in years, the winter passed without sinusitis and flu, except for the most minor, transient symptoms.

These improvements were so encouraging that I had the patient obtain a DOR-buster of her own. She started using it on herself several times a week, in the spring of 1988. Her subjective sensations with self-treatment were not quite as intense as with treatment by me, but they were always present.(3)

After a while she felt ready to tentatively test her long dependence on therapy. We began a trial of increasingly long breaks from sessions. She was delighted to find herself remaining free of physical and emotional symptoms. By the winter of 1988-89, we had established a pattern of quarterly visits, for assessment and support. She was then using the DOR-buster one or two times per week. The flu season passed with only one period of mild intermittent flu symptoms, from which she easily recovered. The colds, sinus drainage, and headaches did not appear. The allergy season in the spring of 1989 was particularly bad in her area, but the patient's symptoms were minor. For the first time in 30 years, she did not need antihistamines.

Using the DOR-buster twice a week, the patient went through a third winter free of her old upper and lower respiratory illnesses. She has lost more than 30 pounds of excess weight, is very active physically, and is feeling better than she can ever remember. The best sign of the depth of her improvement is that she re-established her relationship with her childhood sweetheart and, after many years of living alone, she has married him.

Though psychiatric orgone therapy helped prepare the patient for this improvement, clearly it was the use of the DOR-buster that finally allowed it to happen. By altering the ocular segment stasis, it simultaneously relieved the emotional and somatic manifestations of that stasis, undoing a complex interaction of infectious and allergic illness, self-perceptions, and deep anxiety that had been maintained for decades.

Case 3: Recurring Cutaneous Infections

In January 1987, a 31 year old woman referred herself for orgone therapy. Her medical status and history were generally good. It is probably significant that at least once a year, for over 15 years, she had developed painful infections of the outer walls of the nose, and that her mother and sister had been troubled by lesions of the same type and location.

In early July 1987, she developed dozens of painful pustules on the buttocks. *Staphlococcus aureus* was cultured, and penicillin derivatives were prescribed. The outbreak took a month to bring under control. But the lesions kept recurring, in lesser numbers, throughout the rest of the summer and early fall, each outbreak following another course of an antibiotic. The patient was finally free of lesions, three months after the first

---

3. It could have been that her DOR-buster, which employed shorter tubes, was less potent for her, or that the state of receptivity involved in being treated by someone else was important. I wonder whether the most significant difference might have been the absence of my energy field, through which the conduit passed when I was holding and using the DOR-buster. There is a great deal to learn about how this remarkable device works.

> **"After a slowly and steadily worsening condition for a month, such dramatic improvement only 12 hours after a brief treatment with the DOR-buster was nearly unbelieveable to me, but I could not dispute what my eyes told me."**

outbreak. But by that time, two other types of lesions had appeared. These were considered by the treating doctors, including an infectious disease specialist, to be new loci of the same infection: a papular rash in the submandibular (throat) area, and pustules in the nostril walls. The facial rash was easy to tolerate. The nasal pustules were not. They were of the type that had been familiar to her for years, and were quite uncomfortable. Both types of infections would clear spontaneously, or with a course of antibiotic, but they kept recurring, about every two weeks. In addition, sties also began appearing in February 1988.

There was a particularly bad nostril infection in April 1988. In late June 1988, the patient developed otitis externa, an inflammation of the outer ear, which also proved to be staphylococcal. Then in September 1988, there was another quite serious lesion, a facial carbuncle, which required lancing, and yet another course of antibiotic. The three types of more minor lesions kept occurring.

It had become quite clear that competent conventional treatment was getting nowhere. Because her state of charge in general seemed low, I was not inclined to suggest frequent use of the DOR-buster. Because she lived in a big city, the use of an orgone accumulator device might have been problematic. Fortunately, a Greenray Lamp had become available, and I suggested a course of experimental treatment, beginning in mid-December 1988.(4)

Though the developer of this lamp offered guidelines, there was little to direct its usage for this condition, other than what felt correct to the patient. Experimenting with different exposure times, we gradually adjusted them downwards, from about 4-1/2 minutes a day, at the start, to 2 minutes a day, by early March 1989. By mid-June 1989, longer exposures felt better, and the patient began using the Lamp 4 to 5 minutes per day. The Lamp was used in a room with other light sources turned off.(5)

The first suggestion of a clinical response appeared after about 3 weeks of use. A sty appeared, in the same manner as those that had preceded it, but its progression was exceptionally rapid. From the first irritation to the complete resolution, only 24 hours passed, and there was no drainage. A second sign of response was that the nasal pustules which appeared were less tender, less frequent, and of shorter duration than previous ones.

---

4. The enclosure of this device is a one-fold orgone accumulator: a box of plywood, lined with galvanized steel. It contains a 175 watt mercury vapor lamp, a ballast, and a cooling fan. Directed by a reflector, the light passes through a dark green sheet of glass. Sensitive observers have seen wave-like pulsations emanating from the light, and changes in the appearance of the energy field of the subjects. For more information, see Blasband, R.A.: "The Orgone Energy Light: A Pilot Experiment", *Journal of Orgonomy*, 22(1):62-67, 1988.

5. By chance, the patient once failed to turn off a table lamp which had a peach colored shade. She felt a strange thickening of the atmosphere of the room, and that night she was unable to sleep. A month later, she purposely turned that table lamp on when using the Greenray Lamp, and felt the very same sensation. This suggests that color does play a role in the effect, and that the lamp does not work simply through an oranur reaction.

But in late February 1989, there was another particularly bad nasal pustule, and then another carbuncle which required lancing. I suggested that she move the lamp to a room that was larger and better ventilated, and that she add one accumulator fold to the outside of the Greenray Lamp enclosure. There is no way of knowing if these changes made a difference, or if we had been already gaining on the problem. As it turned out, the two lesions that appeared just before these technical changes were the last ones she was to endure, major or minor.

To summarize the course of the illness: in spite of aggressive medical treatment, the patient suffered a series of minor, moderate, and severe cutaneous lesions for about 1-1/2 years. Three weeks after starting the use of the Greenray Lamp, there was evidence of a clinical response. After 2-1/2 months of its use, and after the device was modified, the lesions ceased. The patient remained free of lesions for over a year, with minor exceptions that were consistent with her use of the lamp.(6)

Laboratory studies give an added dimensions to our understanding of this clinical course. The Reich Blood Test was performed 2 weeks before treatment started, 1-1/2 months after it began, and once again after 8-1/2 months of treatment. The correlation between the laboratory findings and the clinical course was quite strong. The overall assessment of the patient's fresh blood preparations before treatment with the Greenray Lamp was that the level of orgonotic charge was average or slightly below average. In the second test, the level of charge appeared to be higher. In the third test, she was found to have a state of quite vigorous charge. The specific findings are summarized in the Table on the following page.

An additional blood test was run 4 months later, after 12-1/2 months of treatment, and it generally confirmed the earlier findings. It should be noted that, compared with the findings in the fresh blood preparations, none of the autoclavation tests showed as dramatic an improvement. Perhaps this suggests something about a limit in the Greenray Lamp's therapeutic effect. The permanence of the Lamp's effect has yet to be determined.

Discussion

It is clear that orgonomic treatment can sometimes cure infectious conditions. This was known over forty years ago. In a 1945 article, Dr. Walter Hoppe mentioned his successful treatment of skin abscesses with an orgone energy accumulator.(7) Sometimes, cures have been dramatic, coming about after conventional treatment had failed. But many life-energy modalities, from acupuncture to homeopathy, have made simi-

---

6. In July 1989, the patient went without treatment for two weeks while traveling. About two weeks later, there was a brief outbreak of submandibular papules. In December 1989, she removed the additional accumulator fold, and then developed a very slight, but persistent, popular rash on the cheeks. It resolved after three months. An outer fold has been installed again, and no lesions have appeared since. Once the patient inadvertently sat in front of the lamp for ten minutes, and developed a tender nasal papule the next day.

7. Hoppe, W.: "My First Experiences with the Orgone Accumulator", *International Journal of Sex-Economy and Orgone Research*, IV:200-201, 1945.

Condition of Patient's Red Blood Cells (RBC) Relative to Treatment with a Greenray Lamp.

	Elapsed Time to 20% RBC Bionous Breakdown (8)	Percent of RBC Bionous Breakdown at 33 minutes	Width of RBC Frames (9) (As Fraction of RBC Diameter)	Width of RBC Energy Fields (10)
December 1988, Before Treatment	11 minutes	95% of RBCs	1/6	1/2
February 1989 1-1/2 months After Start of Treatment	26 minutes	55% of RBCs	1/5	1/2-3/4
September 1989 8-1/2 months After Start of Treatment	62 minutes	0% of RBCs	1/5 - 1/4	1-1/2

lar observations and claims. To progress beyond the status of a promising alternative therapy for infectious diseases, we will have to find answers to the questions presented by every successful treatment with an orgonomic device: Exactly how did a disturbance of life-energy contribute to the disease, and exactly how did treatment help?

In pursuing these questions, there are three domains to be explored. The first involves theoretical issues. Initially, infectious diseases were thought not to be somatic biopathies at all. In a 1942 article, Dr. William Thorburn contrasted the biopathies with infectious diseases, stating that the latter could cause only secondary disturbances of orgone energy movement.(11) Perhaps knowledge among the colleagues of Dr. Theodore Wolfe of the profound despondency that immediately preceded his recurrence of tuberculosis in 1952 prompted reconsideration of the matter. When Dr. Victor Sobey reported on his treatment of two cases of TB with the accumulator in 1955, he assumed that the disease was a somatic biopathy.(12)

One problem has been that the question has been too sharply cast: Is this disease a somatic biopathy, or is it not? A better question would be: How much of a biopathy is it? Rather

than separating biopathic from non-biopathic diseases, I propose we consider a continuum, extending from those conditions that are essentially the result of a derangement of orgonotic pulsation, at one extreme, and those conditions that essentially are not, at the other. The infectious diseases, in all their diversity, would cover a good portion of such a continuum.(13)

Another problem in considering how infectious conditions might be biopathic is that they are thought of as being caused by microbes, while one of the most consistent characteristics of somatic biopathies is that they are conditions of unclear etiology in classical medicine. But the matter of causation has not really been closed: Isn't it possible that, as with cancer, proximate cause has been allowed to obscure prior cause, and that the end-states or symptoms of the disease have been mistaken for the essence of the disease process?

Let us take what is usually seen as a clear example of simple contagion: One of those tragic episodes when a communicable disease carried by European settlers decimated a hardy but immunologically unprotected native population. It should be obvious that we cannot be sure how much their hardiness had been compromised prior to the epidemic. Can we measure how profoundly the natives had been shaken by their exposure to the alien culture, with its powerful technology, unnatural religion, distorted sexuality, and distilled alcohol? We cannot really understand the onset of a devastating infectious disease without fully assessing the impact of the prior stressors.

Most people living where the winters are cold have had the experience of an energetically contracted state preceding a microbial disease: a wet chill may be followed by a bad cold, or worse. Could it be that the sequence of contraction followed by infection happens in a hidden manner in other infectious conditions? If so, then the profuse microbial growth characteristic of the fully developed disease would be the end result, but not the essence of the disease.

---

8. The resistance of red blood cells to breaking down and starting to form bions within their membranes is a measure of their vitality. The longer the time to bion formation, the more energetic the cells.

9. The *RBC Frame* is the torus-shaped pigmented portion of the red blood cell. Expressed as a fraction of the cell diameter, it is a measure of the fullness of the cell, another indicator of vitality. It probably corresponds roughly to a standard laboratory measure, the mean corpuscular volume.

10. The *RBC Energy Field* is the portion of the energy envelope of the cell that is visible with the microscope. It is so obvious that the typical observer with classical training must dismiss it as an optical artifact.

11. Thorburn, W.: "Mechanistic Medicine and the Biopathies", *International Journal of Sex-Economy and Orgone Research*, I:257-258, 1942.

12. Sobey, V.M.: "Treatment of Pulmonary Tuberculosis with Orgone Energy", *Orgonomic Medicine*, 1(2):121-132, 1955.

13. A lattice, with diseases arranged by organ system along one axis, and a biopathic-to-nonbiopathic scale running along the other, could be the organizational scheme for a textbook on orgonomic medicine.

How might such a contraction-to-infection sequence actually operate? These are numerous possibilities that we could speculate about. Processes involving the energy fields of large numbers of people might be involved, such as the contraction of the atmospheric field preceding winter. One might "catch" a specific form of contraction, by field-to-field contact with an infected person, along with the microbe. The more virulent microbes might carry so concentrated a form of DOR that they would induce a biopathic state, immobilizing the host's field and impairing his immune response.

To reopen the issue of infection, the second domain for orgonomic science to explore is laboratory research. Reich's experiments with bionous breakdown and biogenesis would be one starting point. Another could be the work that has revealed the presence of characteristic bacterial forms in malignant tumors. Considered together, these two lines of research suggest the possibility of the biogenesis of pathogens, through bionous processes, within energetically weakened organisms.

Classical medicine assumes that the *in vivo* activity of micro-organisms is essentially the same as their *in vitro* activity, but this may not be very accurate. Orgonomic studies could be done on blood and tissue preparations from people with infectious conditions, and in infectious disease prodromes. Living preparations might disclose processes that have been missed in classical studies using killed, stained preparations.

The third domain for orgonomic exploration of infectious illnesses is that of clinical studies. The difficulties connected with doing such research are formidable. For one thing, a great variety of results is to be expected at first, because so many levels of orgone energy functions would be involved. For example the weather at the time of treatment, or the building where it is carried out, might prove to be significant factors. Two models of the same device, even if very similarly made, might not have exactly the same therapeutic characteristics. The orgonotic perceptions and nature of the energy field of the treating practitioner would probably affect results. And, obviously, different patients will tend to respond differently. To obtain statistically significant results, a very large number of cases of any one condition would be needed.

The only way to have access to large numbers of cases in the near future will be for orgonomic studies to be attached to conventional treatment or research. For example, a physician could randomly divide patients with the same microbial disease into 3 groups: standard treatment, standard treatment plus an orgonomic treatment, and standard plus sham orgonomic treatment.(14) In another type of study, orgonomic treatment could be evaluated for its effect on the incidence of a well known condition, such as post-operative wound infection.

Considering the present pressures on medical practice in the United States, research of this kind will probably be done elsewhere first. Until laboratory and clinical research on the orgonotic aspects of microbial diseases is undertaken, the entire field will be in the possession of the mechanistic practitioner.

---

13. A lattice, with diseases arranged by organ system along one axis, and a biopathic-to-nonbiopathic scale running along the other, could be the organizational scheme for a textbook on orgonomic medicine.

14. Though it would simplify the interpretation of results, failure to provide proven standard treatment would be unethical, unless the study were carried out in a place where it is not ordinarily available.

Personal Observations with the Orgone Accumulator *

James DeMeo, Ph.D.**

In the early 1970s, I met a young woman who had treated her ovarian cyst with an accumulator. Her doctor had urged surgery, but she did not have insurance or much money, and decided instead to try the accumulator. The woman had used the accumulator, a three-ply unit big enough to sit in, for about 45 minutes a day for two or three weeks. Around the middle of the third week, she had a vaginal discharge of blackish blood, which was the disintegrating tumor discharging into the uterine cavity. The woman felt completely healthy throughout the entire period, except for some discomfort during the time of the discharge. Some time after this, she went back to the doctor, who could not find a trace of the tumor. When told of the form of treatment, the doctor was derisive and uninterested.

Around this same time, I constructed a small but powerful accumulator, at a time when I was living only 8 miles away from the Turkey Point nuclear power stations, in South Florida. I had been advised not to build accumulators that close to a nuclear plant, and had read Reich's account about oranur. Still, I remember thinking to myself, "It's just a small accumulator, and can't do much harm". The accumulator was left in a garage, along with a number of large metal appliances and other objects, such as a clothes washer and dryer, refrigerator and filing cabinets. Within a week after doing so, the entire garage became so highly charged that it was impossible to stay in it for long. The sensible agitation and overcharge, which was provoked and amplified by the nuclear power plants, began to spread into the house, and the entire area often felt as if it were subtly resonating or vibrating. I still recall quite distinctly this phenomenon, which was most apparent at night, when winds ceased, and city noise was quiet. Meanwhile, plants inside the house began to die, and the white blood count of family members began to increase. A small Geiger counter began to yield erratic and racing counts for "background" radiation. In a bit of a panic, I dismantled the small accumulator, and removed other metal from the garage. A small draw-bucket was placed there, and the disturbance gradually quieted down. Still, the nuclear power plants were a constant worry, and we moved out of the area.

A few years later, I built another very powerful ten-ply accumulator, with shooter funnel... One day when I was working outside, barefoot, I accidentally stepped on a hot soldering iron that had been carelessly left on the ground. My flesh was badly seared, and I was in a great deal of pain. However, the new accumulator and shooter funnel were fortunately nearby, so I placed the burned foot into the shooter funnel. Within seconds, the pain receded, and in a few minutes there was no pain at all!

* Extracted from *The Orgone Accumulator Handbook,* Natural Energy Works, Ashland, Oregon. www.naturalenergyworks.net

** Director of Research, Orgone Biophysical Research Lab, Ashland, Oregon, USA demeo@orgonelab.org

Without further discomfort, I could clean the severe burn, which had taken away all the layers of skin. The wound healed very rapidly after this, and I subsequently learned that pain relief from burns, and the rapid healing of new skin, was one of the most powerful effects of the accumulator.

After constructing an accumulator that was large enough to sit in, I was able to confirm a number of subjective and objective measures that were first observed by Reich. It did indeed make one feel more invigorated and warmer, with a flushed skin. I no longer contracted colds or flu like before. I have never been sick in any major way, and so have no major "healing" of myself to report. Eventually I ceased sitting in the accumulator on a regular basis, as I just did not feel the need for it. More often, I use the orgone energy blanket. It is easier to store (on a hanger in an airy place) and can be retrieved for use very quickly. The most amazing effect of the blanket, I found, was its ability to stop a head cold, or at least to prevent it from developing into a chest cold. Prior to discovering the accumulator and blanket, all my colds or flu would spread from head to throat to chest. Since using the blanket, I rarely contract a head cold, and when I do, it can be prevented from spreading by simply resting with the blanket over my chest and throat. Over the years, I have had a variety of small cuts and bruises, or toes cracked from smashing them into table legs, all of which were treated with the shooter or blanket, with great pain relief and healing benefits.

Only on one occasion did the accumulator fail to help with a problem. I was bitten on the leg by a poisonous brown recluse "fiddleback" spider, the toxin from which killed a piece skin on my calf about 3 inches in diameter. I did not know about the dangers of that kind of spider, and only began treating the bite after the skin had turned purple and became numb. The wound was treated several times per day with the shooter, while sitting inside the large accumulator. These treatments did not restore feeling or normal color, and the entire depth of killed skin eventually turned black and hard, falling out of my leg, leaving me with a gaping open wound for several weeks. A secondary blood infection was treated with antibiotics, and I was on crutches for weeks. The wound healed over, however, and the leg functions today without a problem. Only a small scar exists to mark the bite. A survey of medical literature on this kind of spider bite indicates that, short of questionable cortisone shots into the bite shortly after it occurs, there is no known remedy.

On several occasions, friends of mine who knew about my accumulators would ask if they or their friends could use them. In one such case, a 19 year old female had a disc-shaped encapsulated benign tumor of the breast, measuring about 1" in diameter. The tumor first developed after she became pregnant out of wedlock several years before. Her parents had badly mistreated her for this, and called her all sorts of names. The pregnancy was terminated, but the emotional abuse she had gone through led to a powerful bioenergetic contraction, and to the development of the tumor. She understandably did not tell her parents about the tumor, and had avoided doctors, being afraid of losing her breast. She had been treating the tumor with a vegetarian diet for several years, and it had not grown, nor gotten any smaller. After we discussed the matter, she began the orgone accumulator treatment by sitting inside for around 45 minutes a day, with a large shooter funnel over the breast. After three treatments, the tumor began to break apart, and disintegrate into smaller pieces. She became anxious at this point, however, and was openly agitated and upset about the accumulator, refusing to sit in it any more. Upset feelings related to the treatment she had received during her past pregnancy began to surface. She was also a student of the biological sciences, and, while she had a feeling of desperation about her situation, she had maintained a jocular surface attitude, saying she would try the accumulator only to "humor" her concerned friends. The fact that the accumulator actually appeared to work, when nothing else had, was an intellectual confusion, and it simply became too much for her. She never sought additional treatment with the accumulator, but friends informed me shortly thereafter that the tumor had almost completely vanished. Here, it is important to point out Reich's observations that, in spite of the emotional components of the underlying cancer biopathy (which clearly emerged in the above case), certain kinds of superficial tumors, such as breast or skin cancer, could be effectively treated with orgone energy.

In another case, a 23 year old woman had been under conventional medical treatment for severe genital herpes for several years, but without any relief from the persisting genital lesions. She sat in the accumulator once, using a tube-type vaginal shooter wand. Within days of this, her lesions began to dry up and heal, leaving her symptom-free for the first time in years. She remained free of symptoms for at least several years thereafter.

I know of several cases where the orgone blanket was used for treatment, instead of a large accumulator. An elderly woman was given an orgone blanket to see if it would help her arthritis. She used it and found that it did provide relief from the discomfort and pain, and she regained a bit of movement in the affected areas. After this, she unfortunately used it with her electric blanket, after which all the arthritis symptoms intensified, back to their original condition. With great disappointment, she refused to have anything more to do with the orgone blanket.

In another case, a young woman treated her baby, which had a persisting slight fever and cold. She simply placed the child on top of the orgone blanket in the crib, and left it there for around 15 or 20 minutes. When she returned, the child had a temperature of around 102 degrees. She quickly removed the orgone blanket from the crib, and walked the child about for awhile. Its temperature soon dropped back to normal, but the cold symptoms also had vanished. Reich noted that orgone irradiation will increase a fever somewhat, even in adults, speeding the process of healing. Small children being treated for any kind of illness with a blanket or accumulator should obviously be watched closely. Also, no small child will feel comfortable being put inside a large accumulator all by themselves; but if mother will go with them, and make a game of it, they can sit on her lap, and this will be just as effective.

In another case, an elderly man with fibrosis of the lung, related to a life-time of smoking and emotional holding in the chest, was predicted to die within a few weeks. He was on oxygen, and could hardly speak a complete sentence or walk very far, given his inability to get a good breath. He began to use an orgone blanket and large box-type accumulator. Within a few weeks, he was up and about, rowing his small fishing boat. He reported that the only time he could get a good breath was when he was inside the accumulator, or when the blanket was on his chest. Many of his symptoms were relieved from the orgone therapy, and he remained active for many months thereafter. However, his condition worsened after he was put on an experimental medication by the doctors (prednazone). He died shortly thereafter. Again, no miracles were observed given his original terminal condition, but a good deal of comfort and relief, and an additional 6 months of life.

I once corresponded with a farmer who had a cow with a large gash in its side that had gotten badly infected and festered, refusing to heal. The veterinarians had tried all sorts of different treatments, but nothing seemed to help, and the poor beast was on the decline. Having tried everything else, the farmer made a four-ply orgone blanket, and secured it to the festering side of the cow with heavy tape. He left the blanket taped to the cow, not expecting to see any cure, and anticipated a sorry death for the animal. However, within a few days, the blanket had fallen off, revealing a large scab over the sore. He treated the cow a few more times with a new blanket, and says that today you can hardly find a scar on the lively beast.

Another farmer I met was diagnosed as having a fast-spreading form of liver cancer. The doctor told him to get his affairs in order, as he would be dead in 6 months. The farmer made an accumulator out of two steel oil drums, by removing both the tops and bottoms of the oil drums, sand-blasting the insides down to bare metal, and welding the two cylinders together, top to bottom. He then wrapped layers of steel wool and fiberglass around and around the steel tube he had constructed. With this tube accumulator laying on its side, he would go inside it, and take a nap from time to time. "Dr. DeMeo", he told me, "I object to your caution about not staying inside the accumulator for more than 30 or 45 minutes. I've stayed inside my accumulator for 7 hours at a stretch without problems, when I fell asleep inside it"! Well, I did not know what to make of this fellow, as when I met him, he was very weak and slow-moving. He seemed so low in energy that, in his case, the danger of overcharge did not exist. Still, he had lived for around a year beyond the terminal diagnosis of his doctor. About a year after my meeting with the man, I got a letter from him, saying he wanted to attend one of my workshops. When I finally met him again, I was absolutely amazed at his condition. He was about 40 pounds heavier, his face was ruddy and tanned, and he was literally bursting with energy. Sometimes, however, he would appear quite red in the face, as if he would explode, and once he started talking, you could not get him to shut up. Characterologically, he had gone from a situation of undercharge to possible overcharge. I pointed out this danger to him, and he did reduce his accumulator treatments. Anyhow, the story does not end here. It seems that he went back to his doctor, who saw his changed conditions, and could not find a trace of the liver cancer. The doctor got real mad at him, and accused him of going to some "big city hospital" for a "wonder drug". He told his doctor about the accumulator, but the doctor didn't believe him. Since this was in a small town in the Midwest, the fact that the farmer had survived the death sentence of the town's most reputable doctor, and had even thrived in spite of that death sentence, was the cause of considerable interest and discussion. Presently, I've been told, there's a shortage of steel oil drums, fiberglass, and steel wool in that town, as the man's friends and neighbors are very busy building their own orgone accumulators!

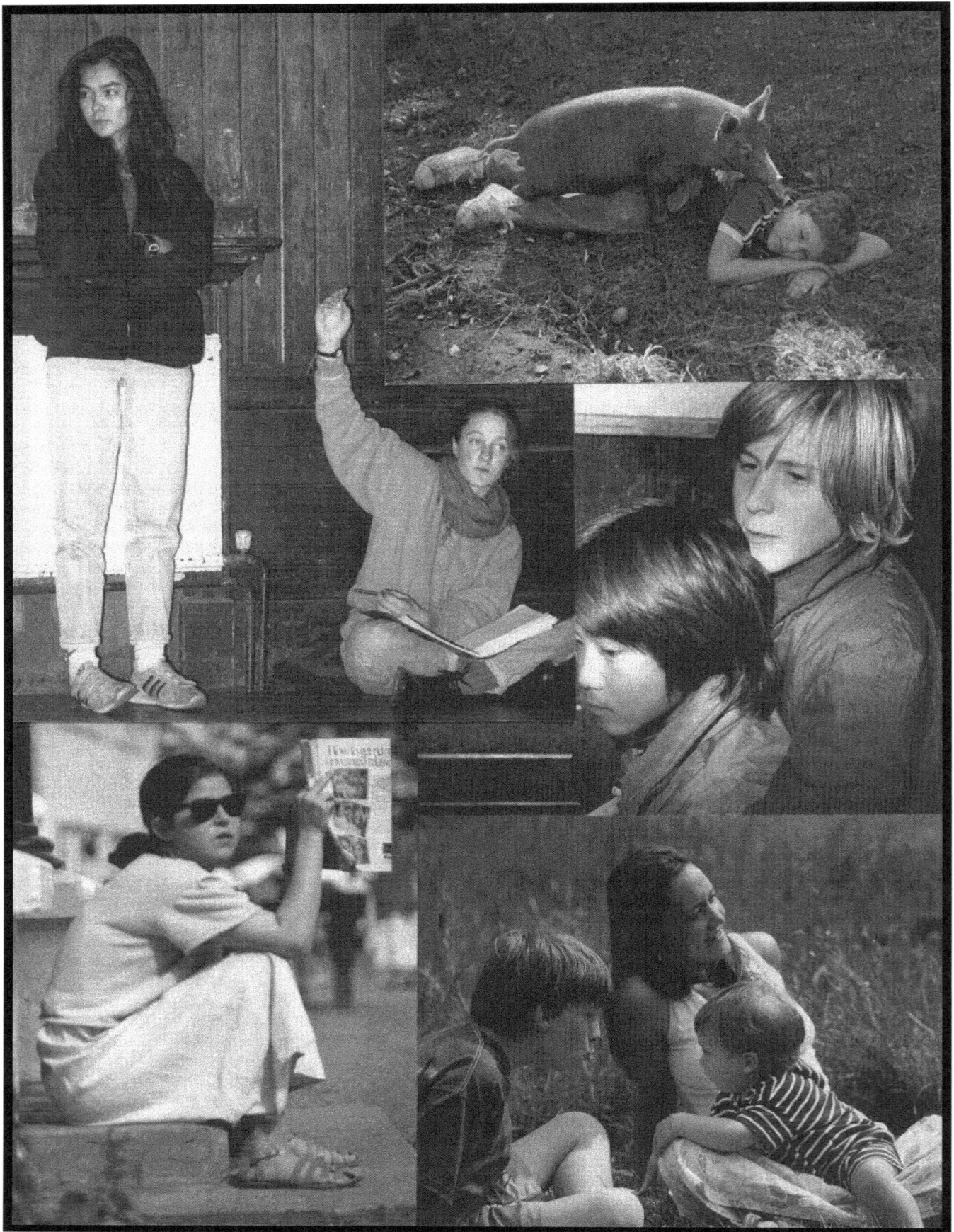

Children and staff at Summerhill School, England.   (All photos by Matthew Appleton)

The Ecology of Childhood: A View From Summerhill School

Matthew Appleton*

Over the past decade or so mechanized Western culture has begun to realize that it must work with nature, rather than against it, if humanity is to survive as a species. The stirring of a new consciousness is beginning to make itself felt, in small, somewhat stumbling ways, as we find ourselves facing a colossal crisis that we can no longer turn back from, but must begin to grope our way through as best we can. It is becoming more and more apparent that there is an inherent wisdom in nature which we have ignored for too long. Our ignorance has pushed us closer and closer to the brink of catastrophe. My concern here though is not with the environment or the atmosphere, which merely reflect the state of our own sickness. Our greatest stumbling block lies in our war with nature within ourselves. In particular I am concerned with our war with nature within the child, for it is in children that nature comes into our culture most immediately, spontaneously and alive. It is in our dealings with children that we wage this war with nature most arduously, and yet, most sublimely. In this arena we are probably more ignorant of nature's wisdom than in any other, and, in all good faith, we know what we do.

To look into a new born baby's eyes is like looking into the depths of the cosmos. All the wisdom of the great sages seems to gravitate there, unspoken, unchallenged, alive. Watch the rise and fall of the abdomen and chest; it rolls and ripples in a graceful, wavelike motion, unified and unlaboured. Look at children play, rooted in the newness and the nowness of the moment; lucid eyes and loose limbs, glowing with life and vitality. Look at what we make of them. Watch that glow begin to fade. See the sullenness and furtiveness that springs up in the eyes, the rigidity around the limbs that replaces natural, graceful movement with more angular expressions. See the breathing falter, become unsure of itself, grow shallow. Observe the absorption of the moment disintegrate into self consciousness, awkwardness and nervous listlessness. What are we doing? What has been done to us? Is this nature at work, or are we working against nature? If we are to begin to answer these questions we must move education beyond the contours of its present course; at one moment moving towards liberalism, the next towards authoritarianism, and floundering on both banks. We must consider, instead, a new course of questioning, concerned not with how we can best lead the child to conform to the equations of our culture, but, rather, with the ecology of childhood: How can we best nurture it and allow it to take its natural

course? At this point of departure we must give nature's wisdom the greatest scope possible, and in doing so admit our own ignorance. We must focus, not so intently on what we can teach children, but also on what we can learn from them.

One experiment which encompassed such an approach was Summerhill School, founded way back in 1921, by A. S. Neill, as a reaction to his own upbringing and his experiences as a young teacher in Scotland. Instead of being crammed into classrooms and having endless facts thrust at them, under the shadow of the strap and the stick, he wanted to create an environment where children could grow up free of fear, and enjoy their childhood in its own right. He saw childhood as more than an enslaved precondition to adult life. It was clear to him that children were emotionally crippled by enforced morality which created conflicts in them. By removing the taboos which caused them he sought to remove the conflicts. At Summerhill you were free to do what you liked as long as it did not interfere with anyone else. You did not have to go to lessons if you did not want to. That was no one's business but your own. The school was there to suit the child, rather than the child having to suit the school.

The school was also self-governing. There were weekly meetings whereby the laws were made by everyone, not just the staff. Everyone had one vote, from the youngest child to Neill himself. All had an equal voice. Rather than trying to guide the meeting with his own ideas, Neill would often sit back and see what the children came up with, or make silly proposals to see how the children would respond. Even his serious proposals were sometimes voted out. That was all part of self-government. People who broke laws or interfered with other people's freedom could be "brought up" at the meeting, and on the one-person, one-vote system, could be fined. These were usually small money fines, or a social fine, such as picking up litter. Neill noted that the children usually veered towards leniency, perhaps because there was no clear division between the miscreants and their judges; they were in a constant state of interchangeability. Neither were staff exempt from being brought up and fined at the meeting.

Given that the children has so much freedom and were able to guide the school in the manner they found most suitable, it was possible to observe much about the nature of children that is usually hidden or lost in an environment governed by adults. The Summerhill experience seemed to belie the generally accepted view that, without discipline and morality, an inherent, primeval savagery existed within children, as William Golding portrayed in his novel "Lord of the Flies". Indeed, Neill observed

* Houseparent at Summerhill School, Leiston, Suffolk, 1P16 4HY, England.

at Summerhill that so-called "lazy" children became actively involved in community life. "Insolent" children became tolerant and respectful of other people, and instead of reverting to "savagery", children reverted to what Neill believed to be an *inherent goodness*.

Although Neill died in 1973, Summerhill continues along the same lines that he established seventy years ago. It puts its faith in the goodness of the child, though I prefer to use the word *integrity*. By integrity I mean the integrated wholeness of the child, which, when allowed to take its own course, can regulate itself as is most fitting to its own needs, and function as a social being, responsive to the needs of others, and open to life and love. There is still much that can be learned from observing children in a Summerhill environment. As an ex-Summerhillian, now Summerhill parent, said at a recent conference, "You don't learn about the nature of chickens by studying battery hens". In conventional schooling, the ways in which we damage the integrity of the child is not always obvious, while at Summerhill, the hurt soon begins to surface as part of the healing process. This may manifest itself in many ways.

Left to their own devices children will follow their excitation wherever it takes them. They will express their excitation in sound and movement. The compulsory classroom, however liberal it may be, does not allow this natural excitation to flow as it should. To still their excitation, which is the wellspring of their well-being, children must stiffen and contract against it. They must tighten muscle, and breathe shallowly to quiet the thrill of life that pulses and streams throughout their bodies. In this way, children learn, literally, to cut off from nature in themselves, and live a facade instead. How, and to what degree this manifests itself is dependent on many factors. But manifest it does, when and wherever natural processes are sacrificed on the altar of culture.

When children arrive at Summerhill, for the first few weeks, they tend to maintain the superficial facade of the "nice", quiet, "good" boy or girl, which is the approved model of our society. When the realization breaks through that approval does not depend on maintaining the facade, then the natural excitation and integrity of the child begins to reinstate itself. At first this may take many bizarre forms, again depending on the form in and degree to which the child's integrity has been insulted. All the pent up excitation rushed to the surface, and with it all the emotions that accompanied and were held in check by the original blocking. The child may begin to express anger (especially against adults), or sorrow, or make various statements of independence, such as not washing. Some children become abusive, antisocial or obsessive in some way, for example, breaking into things and compulsive stealing. This period, in which all the pent up excitation and emotion begins to come out is the period in which Goldings' "Lord of the Flies" might justifiably be set. But the story does not end here...

No one at Summerhill tries to moralize or philosophize to the children, nor to politicize or spiritualize them. All problems are dealt with practically in the meetings. What we find is that in time, as the suppressed excitation is expressed, and the child's integrity is restored, s/he is able to regulate his or her life in a more harmonious and responsible way. Usually, this is with a maturity that children who have child's faith in his or her own

Summerhill School

> **"Allowed to define his or her own needs, the child often displays an instinctive intelligence that our cultural creeds do not recognize."**

natural core is reinstated, and s/he is able to act with a deeper self-understanding, giving a voice once more to the inherent wisdom with which nature conducts itself.

I am not proposing that conventional schooling is necessarily the source of the child's difficulties, but that it tends to compound them. More often than not the problem are rooted in the family. The attitudes, and inability of the parents to live their own lives fully, along with a myriad of cultural considerations, would the child's integrity. The most distinctive aspect of schooling at Summerhill is that it is based on choice. The school also has a wider function, though, as a sort of extended family, an international community of children. At present we have children from France, Spain, Germany, Morocco, Indonesia, Japan, England and America. Children come from many background, some well off, some not so well off. Some come with their parents blessings, and their integrity has been respected and nurtured throughout. Some have been SENT to the school as a last resort; their parents do not know what else to do with them. The integrity of these children is no longer intact and their confused attempts to reinstate it is problematic at home or at school.

The degree to which the family is able to embrace the integrity of it's children reveals itself in the ease with which the chil adjusts to freedom. Conversely, parents often state they have learned much about themselves by their children being at Summerhill. As the child's integrity grows more complete so does the family's. Where the family is not able to meet the emerging of the child's integrity a conflict situation arises. I think now of a teenage girl who has been with us for over three years now. Every time she tries to establish her integrity the parents pull the carpet out from beneath her feet. They tell her she is stupid. They belittle the school; the only place she has known any happiness, and threaten to withdraw her when she asserts herself at home during the holidays. In such a case the child is caught in a great gulf between the school, where she is always on the verge of opening up, and the home, where she is having to clamp shut again.

Usually though, there is a powerful healing period that takes place between the ages of seven, when we first accept children, and thirteen. Experience has taught us not to take on new children over the age of twelve. The release of pent up excitation, combined with the powerful biological drives of puberty, is a recipe for disaster. By the time children have reached adolescence at Summerhill they have already lived out their "delinquent" period, and begun to take on the responsibilities of running the community. It is the big kids who sort out the disputes amongst the smaller kids, put them to bed, organize social events, chair the meetings, and generally have the strongest voice in the community.†

There is something to be gleaned from this. It tells us something about the ecology of childhood that could be of great significance in a society where adolescent discontent and delinquency is such a problem. Just as we have begun to realize the ramifications of the abuse of nature around us, so we can begin to tackle the problems of the abuse of nature within us. In

all areas of life our injured nature is showing us the same signs. The booming industry in therapy is another facet of the same picture. People are beginning to voice their sense of loss, to articulate the emptiness they feel inside them, and rediscover the hurt child frozen within. Moreover, this new therapy industry is not so much aimed at the people society would deem as "victims", but at those, who by societies measure, are successes: The up and coming, the well to do, the professional.

The question of health is a far reaching one. We have learned enough to know that wherever nature's wisdom is ignored then nature's discontent finds a voice. My definition of health here is not simply a lack of the symptoms of sickness, but the ability to fully establish ones own personal integrity. When Neill founded Summerhill he stated he wanted a school that would "follow the child". Allowed to define his or her own needs, the child often displays an instinctive intelligence that our cultural creeds do not recognize. A young boy came to the school suffering from chronic asthma. Back in Japan his mother did not acknowledge his problem and would leave him alone in the house without medication, so that he had a great deal of anxiety about returning home for the holidays. At Summerhill he continued to have very severe attacks, but was able to enjoy a fairly full life, playing and socializing with other children. He also came to develop very trusting relationships with the adults at the school. As I got to know him, and he became more relaxed with me, he would take my fingers in his mouth and suck on them. As he did this he began to spontaneously kick his legs and move his arms like a baby would, and make a deep gurgling sounds in his chest and throat. It becomes clear to me that he was living out an earlier phase in his life that he had missed out on, and in doing so was re-establishing his integrity, and with it his potential for health. He asthma has not disappeared, but since that time he has not suffered from the severe attacks he had before.

On another occasion a child seeing a young goat being fed with a baby bottle asked if he might have one himself. Within a week half the community had baby bottles, and visitors to the Saturday night meeting were surprised to see even big sixteen year old lads sucking away. Apart from the odd bout of leg pulling no-one was really derisive about this temporary reversion to infantile desire, and a definite air of contentment permeated the community at this time. As the main purveyor of bottles though, I had attracted a somewhat notorious reputation at the local chemists.

Parental anxiety is a great source of injury to the integrity of the child. Children with over protective parents inevitably spend their first few weeks at Summerhill bumping into things, falling over and generally injuring themselves. They are uncoordinated, clumsy and have lost their trust in their own ability to function coherently. In time their integrity heals itself, and they are off in the woods, climbing trees and running around, as nimble and agile as nature intended. The children are not supervised in their activities as they are in most schools, and yet the accident rate is very low, despite the potential dangers that exist in such exhilarating adventures. A teacher, visiting the school recently, became so anxious that she nearly had to leave a room in which children of all shapes and sizes were milling around with hot mugs of tea in their hands. This is an everyday events at Summerhill, and as yet, not one has been scalded.

---

† Editor's Note: This is precisely the same as seen within the *Bukumatula, Ghotul,* or other *children's democracies* within peaceful, unarmored, matristic cultures. (See J. DeMeo's forthcoming *Saharasia*.)

> **"... It is time that education began to function around the desire to learn, rather than the fear of not knowing."**

Equally, parental anxieties about learning poses problems for children who, if not interfered with, would learn much quicker. Many children do not attend lessons with any regularity for years, and yet, when they are motivated from within, they learn quickly and efficiently. When children are allowed to follow their excitation things happen naturally and spontaneously. One of the greatest sources of anxiety of our times is the fear of what we do not know, and yet this is one of the foundation stones of our education system. If you do not know it is because you are stupid, or lazy. If you do not know, you will not pass your exams. Motivated by the fear of not knowing children are continually being stuck in frames of reference that are quickly becoming outdated. At a time when humanity is making vast leaps into the unknown (especially in the sciences), surely it is time that education began to function around the desire to learn, rather than the fear of not knowing. Again, child nature is not understood, and both educationalists and parents are stuck in the belief that children need to be pushed to learn. This inevitably damages the child's natural desire to learn. Freedom to not attend classes implies a certain faith in the child's inherent will to learn, and with the confidence which arises from this faith, children at Summerhill tend to learn in a relatively short time what it takes conventionally raised children years to learn.

Summerhill has been accused of neglecting the academic, and concentrating on the emotional. This is true only in the sense that, as Neill stated, "If you look after the emotions, the intellect will take care of itself." What is important is not how much a child can learn in a given time, but that the child's desire to learn, when s/he is ready, is not damaged. The emotionally whole child learns at a ferocious speed what is relevant to his or her own needs at the time. Even if the bulk of academic learning occurs after the conventional period allocated for education, it will always be achieved more fully if entered into wholeheartedly. Children who have been pushed into the academic labyrinth before they are ready spend more time and energy trying to thrash through dead ends than making the progress they would if nature was allowed to take it's own course.

An eleven year old boy, whose parents were very anxious about his inability to read, came to me for private lessons. A series of bad experiences with teachers at previous schools had left his self-esteem very low, and he believed himself to be stupid. After a few lessons I realized he was insincere in his desire to read, but was more concerned with trying to please his anxious parents. I suggested to him, that instead of laboring the point, I would read him a book by an American Indian Medicine Man, which he had shown interest in before. Soon, he was looking over my shoulder and picking out words that he would never have deciphered where the motivating factor was the pressure of HAVING to learn to read. Another boy, of about the same age, whose parents were both ex-Summerhillians and supported him for who he was, decided, completely of his accord, that he wanted to learn Japanese. He asked one of his Japanese friends to teach him. Within a couple of months, he was able to read, write, speak and understand large chunks of the Japanese language. It is my own belief that if children were allowed to follow their excitation unhindered by unnecessary adult interference, they would each find the natural genius within them. Genius has it's root in genuineness, and if children were allowed to do what they were genuinely interested in their genius would emerge. If doing and being were not so severely segregated, nature would be more fully able to express it's inherent wisdom.

Given that we live in a consumer society, and that children at Summerhill are able to dictate their own laws by which to live, it is also interesting that expensive toys and television play a very small part in the children's lives. There are various self-imposed regulations about the viewing of television, and it would seen that most children would prefer the real contact of living human beings to the pseudo intimacies of the T.V. tube. Television is watched, and enjoyed, as are computer games, but they do not take on the all pervasive distraction to real life that they do in so many homes.

Although Summerhill plays a therapeutic role for many children, it's primary function is prophylactic. The principle of the school is to protect and nourish the integrity of the child before the damage is done, although in practice this is rarely the case. Experience has shown that even though children whose integrity has been severely undermined can benefit from the school, it may distract the community from meeting the needs of the other children, to whom freedom comes more readily. The school's commitment to children with "problems" must always be relative to the make up of the community as a whole. Essentially though, Summerhill is Utopian in it's approach. It demonstrates a natural wisdom, an inherent integrity in children, that goes way beyond the vision of contemporary society. It exposes our ignorance in such matters, and raises questions that many would prefer not to ask. Often it reminds us of the forgotten pains of our own childhoods, which we have learned to sublimate into the social fabrications of our culture.

What Summerhill advocates, it has consistently lived throughout it's seventy years of existence. It is no mere theory. Even in this, though, Summerhill has had to accept many compromises, which inevitably limit it's scope. If possible, the school would only accept the children of parents who fully believe in and support the integrity of their children. As yet though, such parents are rare. To meet the demands of the society we live in, the school provides exam courses for it's pupils. The children are well aware that they must pass exams to make their way in the world. Mostly they do well in their exams, but this emphasis on education through fear of not knowing, cannot but have some effect on their natural desire to learn.

As well as endowing children with inherent wisdom, nature has also endowed children with sexuality, which, in adolescence, is at it's most intense. Unfortunately, the law of the land has decreed that this sexuality should not express itself in the fullness it deserves. Anthropological evidence suggests that in cultures which have been affirmative towards childhood and adolescent sexuality, there is a distinct lack of promiscuity, sexual perversion, venereal disease, rape, or the subjugation of women and children. Our culture is riddled with all these things, along with a generally immature, unhealthy attitude towards sexual feeling. Nature expresses itself sexuality in it's young. We are the only species to deny that sexuality, and the only species to suffer from chronic and widespread sexual anxiety. Perhaps, again, we should start to respect nature, rather than work against it.

Matthew Appleton surrounded by Summerhill children at a school meeting.

We live in a time when the educational world is moving more and more towards rigid, academic standards, metered out by consistent testing at ever younger ages. The ethic of the fear of not knowing is becoming more deeply entrenched than ever. It is ironic that at a time when the British education system is moving ever closer to the Japanese system, nearly half the pupils at Summerhill are Japanese. If it wanted to, the school could fill itself twice over with Japanese children, and is frequently visited by Japanese educationalists, who are looking to Summerhill to solve some of the grave problems they are now beginning to admit exists in their schools. Socially, children in our culture have more the status of commodities than living, feeling beings in their own right. They must be "presentable". They must be "sweet" and "lovable", like E.T., or Bambi. Children's clothes become more expensive, as they become more geared towards adult aesthetics and less to the needs of the children. The demands to stay clean, and "be good" are more palpable than ever. Even if this is not the everyday reality, it is the model by which success is measured.

The way of life that Summerhill demonstrates cannot be simply reduced to yet another form of "alternative education". It is an attitude towards children, and ultimately, an attitude towards life. Personally, I would no more desire to impose my values on a child than I would lock up a homosexual, deprive women of the vote, or subjugate another race because it's beliefs or skin color were different than mine. Better a child is totally absorbed in reading Dandy or Beano (comic books) than

forced to read a Shakespeare play s/he is not interested in. The Bash Street Kids have as much a place in the scheme of things as Hamlet. Let the child follow his or her own excitation and an interest in and love of life will always be there. When we consistently interrupt the flow of their excitation we fragment our children's integrity, we cut them off from the nature they are rooted in. When we begin to study the ecology of childhood we find a deep wellspring of wisdom that the over-cultivation of conventional education largely ignores and obscures. Throughout the planet, nature is protesting our ill treatment, not only in the atmosphere and environment, but in our schools and homes too. When we have learne to acknowledge the wisdom of nature in our children, our understanding of nature's wisdom in the world, and in ourselves will deepen of it's own accord. We have already made the small step of advocating free range chickens, when will we make the great step of allowing *free range children!*

Publications of Interest:

A. S. Neill: *Summerhill: A Radical Approach to Child Rearing*, Hart Press, NY 1960.
A. S. Neill: *Freedom, Not License!*, Hart Press, NY, 1966.
J. Croall: *Neill of Summerhill*, Pantheon, NY 1983.
Friends of the Summerhill Trust Journal, c/o Hylda Sims, 280 Lordship Lane, London SE22 8LY England ($25/ year).

Wildfire Interview with James DeMeo*

Conducted by Matthew Ryan **

At Wildfire, we first became aware of Jim DeMeo in the fall of 1986. The summer of that year had seen the Southeast section of the U.S. devastated by drought. Following its sudden and unexpected ending we discovered that DeMeo, with others, had been involved in turning it around.

In 1987 DeMeo again came to our attention. This time in a series of articles linking the development of deserts with the origin of the kind of behavior many of us associate with — and blame on — "Western Civilization." DeMeo's focus was much broader and provided startling insights into the origin of destructive cultures, the relationship of Earth and climate to human behavior, and the reasons why almost all the peaceful cultures worldwide have been destroyed.

We were impressed, intrigued, and wanted to know more. So in September of 1988 we invited Dr. DeMeo to present a two–day workshop here in Spokane. Following the workshop we knew that this man had something very important to say; more, that he said it in such a way that scientist and layperson alike might hear. A rare combination.

We hope you will join us for this wide–ranging discussion with a unique and caring scientist as he talks about weather, trees, people, and the force in nature that moves them all. M.R.

MATTHEW RYAN: *In the last ten years you've been involved in a variety of projects and I'd like to at least cover some of them here. So why don't we just start with what you're doing today and see where that takes us.*

JAMES DEMEO: Right now I'm trying to publish my writings from two major research areas. The first concerns atmospheric energetic questions — the weather changes we're seeing, the droughts, and primarily the experimental field work I've done on Wilhelm Reich's cloudbusting techniques. In this connection, I'm also trying to make myself useful in droughty situations by actually going into these areas and doing cloudbusting operations to relieve the droughts.

And then I'm also trying to publish my work on what I call *Saharasia*, which has to do with the origin and diffusion of patrism — patriarchal, authoritarian culture — from the vast desert area that extends from North Africa through the Middle East and up through Central Asia.

* Reprinted with permission from the December 1988 issue of *Wildfire* magazine, PO Box 9167, Spokane, WA 99209.

** Editor of *Wildfire*. Matt Ryan has a long time familarity with orgonomy, and has worked with both Jerome Eden and James DeMeo on various research projects, including CORE operations in the Northwestern USA. Under Ryan's editorship, *Wildfire* has carried many articles on natural childbirth, sex-economic issues, earth changes, renewable energy, Native American and environmental issues, and about the life energy.

MR: *You say "trying to publish..."*

JD: I'm having great difficulty in the sense that the traditional outlets for this kind of work — scholarly journals and book publishers — seem to be absolutely scared of all this. I mean, I get the most incredible knee–jerk reactions from people. With academic journals, for instance, I'll send the same article to four different journals and every one of them will reject it, but for entirely different reasons — not one of which has any common theme except that the name "Wilhelm Reich" bothers them; or the idea that my work on Saharasia demonstrates our awful treatment of babies and children.

MR: *You mean the name Wilhelm Reich is still a kiss–of–death in academic circles?*

JD: Almost. There is only a bit more openmindedness today. Back in 1975 I finished an environmental studies degree at Florida International University, and by then I was so sick and tired of hiding my research interests that I decided I would search for a university that would let me do legitimate research...testing of Reich's work on orgone energy. I mean this is what science is *supposed* to be about.

So I started writing letters to every person in academia who sounded like they were openminded: I wrote to all the parapsychologists; I wrote to all the physicists and biologists who were doing unorthodox research. I was almost a straight–A student so I came highly recommended, and I figured I would adapt myself to whatever program allowed me to do this work. Well, after two years of searching and getting back the most incredible array of letters, all negative, I finally made contact with a group of professors at the University of Kansas who were interested in Reich's work and they allowed me to proceed with my research. Only then was I able to do two years of field work with a large cloudbuster and produce the first master's thesis...and as far as I know, the only one...in the U.S. that addressed *any* aspect of Reich's biophysical work.

MR: *Aside from your review board, were there any other reactions?*

JD: Some of the scholars at the University thought it was going to prove a negative thesis: They thought the experiments would demonstrate that the device didn't work. So some people were *very* surprised when the data demonstrated a clear set of affects... others that were less clear...but by and large it vindicated Reich's position, that this instrument — the cloudbuster — had an influence on weather.

MR: *Enough so that if this research was in another field, or not associated with Reich, it would have prompted further research?*

JD: Absolutely, no question. In fact I was *supposed* to do a Ph.D. dissertation on the cloudbuster that would have had a much more in–depth and rigorous program of testing. But then the critics got a hold of what was going on, and some people at the University, who hated Reich's guts, started prodding and sticking their noses into what my professors were doing…I won't go into a lot of detail about this but by then I had put in another year–and–a–half worth of field work preparing for the dissertation, and I put through a research proposal to the National Science Foundation for a few thousand dollars for equipment. Well, somebody at the National Science Foundation called somebody at the University and said, "What the hell are you guys funding down there? This guy Reich is a quack," or whatever. The next thing I know the program I was doing was in total jeopardy, research support was pulled out, and some of the people on my committee got scared and withdrew. Now I want to make clear that some of the people, such as Professor Robert Nunley, fought like hell for my program, but there was just too much against us.

MR: *So after 2 years of finding a university, 2 years on a masters, a year–and–a–half preliminary work on the Ph.D., nothing?*

JD: As it turned out, Professor Nunley…who helped me out in many, many ways…also helped me to develop an alternative research program — something we thought would be noncontroversial. It would have been a disaster if I was forced to start from scratch at some other university, but more, I didn't *know* another university that would allow me to continue with this work.

So I thought, well, I'll make some maps of the status of women and the treatment of children and babies, because I'm interested in human behavior… so I'll just make some world maps showing where people live who treat their babies one way, as opposed to another.

MR: *Something noncontroversial.*

JD: Well….

WF: *In contemporary terms or historical?*

JD: I wanted to do a global study that reflected conditions before the influence of European migrations in colonial periods. So originally I used data from anthropologists who had gone out in the last 100 years and lived with, or reconstructed the culture of, what are known as primitive subsistence level peoples — I focused on these cultures to minimize the effect of European diffusion.

Then I used my microcomputer to print out maps showing where certain kinds of behavior occurred. Initially, I was looking for places where people circumcised their baby boys or their female children, where they inflicted trauma on their babies as opposed to where they didn't, where they suppressed the sexuality of their children as opposed to where they allowed their children free expression of their sexual drives, and where marriages were arranged by the parents. And the patterns that started coming out on the maps were just absolutely *astounding!*

Now in the back of my mind at this time was something Reich had said, that it was likely that the armoring of the human animal had something to do with deserts and desertification. He had pointed to things like lizards with thick skins and cactus plants with spiny skins — things that are very thick and bristly — and he made a correlation between that kind of *biological* characteristic of life in the desert, to the bristly, thick–skinned nature of the armored human *character* structure. So I thought if I could do a geographical study on behavior it might show a link between harsher types of human behavior and deserts; *and,* where people lived in rainforests, I expected that they would be relatively softer, characterlogically.

MR: *Was there such a link?*

JD: Yes…although eventually it wasn't just simply that. But what this mapping technique showed was the major desert of this Earth, this big, huge desert area that goes across North Africa through the Middle East and up through Central Asia — I named it *Saharasia* — which is not just only the largest desert but the *harshest,* turns out to be the place where human behavior is the most harsh, warlike, anti–child, anti–sexual, anti–pleasure…extremely patrist.

MR: *By patrist you mean…*

JD: Patriarchal authoritarian cultures that inflict pain on their babies, demand obedience and punish the children who disobey, are more manipulating in terms of the sexual lives of the children, punish transgressions of a sexual nature very severely — such as killing your daughter if she's not a virgin before she's married, because if so then she's not worth anything, or because it's a shame on your honor…which is *still* the cause of death of many young girls in some societies today. Patrist cultures often have special mutilations of the genitals, they manipulate the marriages, the women are subordinated to the men, and the men are wrapped up in a highly structured caste system with the strongest, most ruthless male at the top. In terms of religion, they tend to have male anthropomorphic gods. The people are very warlike and they have a lot of sadistic energy — and sadism, according to the arguments of Reich, is developed by virtue of the very real traumas that the young people experience in growing up, which essentially thwarts their capacity to function: the orgastic discharge is severely disturbed; they can't experience deep genital gratification and discharge, and this drives the behavior towards masochism and sadism.

So this is what I mean by *patrist,* and the peoples within the great Saharasian desert belt expressed the most extreme patrist behaviors.

MR: *But certainly it isn't unique to Saharasia. Degrees of what you're describing exist everywhere in the world today. And in the past I'm sure many Amerindian peoples, for instance, considered the European migration an extreme patrist invasion.*

JD: Yes, but again, my original work focused on subsistence level cultures, globally, using data that minimized the influence of European diffusion. So the question became: How could it have gotten that way and what was the relationship between history and the archaeology we see?

To understand that, I had to first create my own systematically–derived geographical database on the history of human behavior in different areas around the world. Additionally, I started examining the natural history of this Saharasian environment, as well as the migrations of peoples, looking for changes, because there was obviously a connection between behavior and climate being expressed on the maps. And what emerged from this analysis was even *more astounding!*

First of all, paleoclimatologists... who study and interpret what ancient climates were like...demonstrate very clearly that the Saharasian desert belt, for the most part, was once a wet, lush semi–forested grassland which around 4,000 to 3,500 B.C. began to dry up — first in Arabia and parts of Central Asia and then eventually across the whole area we now call the Sahara Desert. This big, huge dry belt used to be quite wet and lush.

When I looked at what the behavior was like at those times when it was wet and lush, there's *no indication* of warfare, subordination of women, or the chaotic social conditions leading to fortifications; there were no weapons of war and no indication of male gods nor any other male–dominated, warrior–cult oppressive type of situation.

But, as Saharasia dries up you begin to see the development of the social complex that builds temples devoted to male gods — or subordinates female goddesses to a male god. And they start sacrificing people; they begin to ritually murder women so that when a man dies, they murder a couple of his wives, stick them in the grave with him...generally younger women too, indicating arranged marriages as well as polygamy.

MR: *What is it about the desert that made these people change, and become so violent?*

JD: This change, from peaceful to violent social conditions, took many years or even generations, but appears to have been initiated by one of two processes. First, the severe drought, famine and starvation conditions of the post–4,000 B.C. climate change got the process under way. Drought is the number one cause of starvation and famine, particularly among subsistence–level peoples, and can utterly destroy all social and family bonds. We have contemporary and historical examples of this happening. The starved and famished babies of these peoples, whom we have seen on TV in Africa in recent years, in their helpless, miserable state, live through a hell of severe emotional and physical trauma. These children literally shrink or wither away, both physically and emotionally, and if the starvation is severe enough, they never recover to full potential later on. They exhibit life–long symptoms similar to abused or neglected children within our own culture, and their behavior later on as adults is changed away from that of their peaceful ancestors. They are made more anxious about basic biological functions, particularly relating to strong pleasure, and become impulsively violent when confronted with pleasure–charged situations, even if it is someone else's pleasure. Having been denied it in infancy and childhood, they simply can't stand it anymore, not even in someone else. And so they feel great anxiety, and strike out,

or run away from pleasurable experiences, and eventually begin, after generations of such parched, withering conditions, to change their social institutions into mechanisms for the avoidance or crushing of pleasure. Secondly, I found many examples of peaceful peoples living on the borders of the new desert regions, being overrun, conquered, and devastated by violent invaders who had themselves come from the desert regions. I found many examples of this, especially regarding the invasions of areas with secure food and water. But all of it started with the formation of the great Saharasian desert belt, around 4,000 B.C. As desert conditions got worse, over the centuries, these violent behaviors became more and more apparent in the archaeological strata, and also spread more widely around the world.

MR: *For example?*

JD: Europe. The first traces of warfare appear in Europe with the arrival of the Battle–Axe people...archaeologists call them "Kurgans" because of a special kind of grave they built, but historians refer to them as the Battle–Axe people because everywhere they went they left behind these battle–axes. They conquered what were, essentially, very peaceful, unfortified agricultural and trading settlements that were widespread across Europe. The Battle–Axe people introduced the first armored, warlike culture into Europe.

MR: *And this was around what time?*

JD: This started around 4,000 B.C.

MR: *And the Kurgans, or Battle–Axe people, came from...?*

JD: Central Asia. They were displaced out of the Caspian–Black Sea region when the weather changed and desertification began. The archaeology shows that these peoples were in chaos; you can see the gradual abandonment of peaceful agricultural settlements across Central Asia, and their subsequent replacement by "pastoral nomads," or more appropriately, *warrior* nomads — who began to take up weapons, form huge armies, and conquer the surrounding cultures.

MR: And they brought the first male–dominated warrior culture into Europe?

JD: I'll say even more than that: There's no clear and unambiguous evidence for the existence of warfare, sadism, traumatization of babies, subordination of women, nor any of the trappings of patrism *anywhere in the world* prior to around 4,000 B.C. *None!* And where it begins to appear first is precisely in those areas of Saharasia that were drying up. Patrism later spreads outward from Saharasia with the migration of peoples who, abandoning their dried–up lands, invaded the surrounding wetter territories.

It spread into European culture out of the deserts; it spread into Chinese culture out of the deserts; it spread into the cultures of India, and tropical Africa, out of the deserts.

And then over a period of hundreds to thousands of years, warrior/nomad kingly states developed to the point of ship-building, to eventually spread patrism across the sea into parts of Oceania, to the coasts of South America, Meso-America, and the Pacific Northwest. My maps suggest that patrism arrived in the New World through pre-Colombian contacts, after around 3,000 B.C.

Patrism started in the Old World Saharasian desert but spread outward with the migrations of peoples. But it began in the Saharasian deserts and Saharasia remains the region with the clearest, most widespread expression of the problem.

MR: *So after having to stop cloudbusting research, you decide to 'just make some maps' of human behavior, and they in turn bring you right back to the weather, or climate. Obviously, a connection is begging here.*

JD: Well, the two tie together because the deserts are spreading. They're spreading at a rate of 70,000 square kilometers of new desert land every year *globally*. It's *the* major atmospheric pathology, overshadowing ozone, global warming and everything else, and yet it's being almost totally ignored by both the media and the academic climatologists.

And the deserts are expanding because populations are expanding; because people are deforesting the planet like mad, like there's no tomorrow. They're going to cut down every last tree on the planet, burn up every bit of grassland, plow it up, tear it apart. We're devegetating the planet and that's what's causing the deserts to spread in the modern era.

What we're also seeing is that in areas surrounding the deserts, droughts are breaking out. This is an observation I've made, that droughts are intrinsically related to the deserts. Dry, hot, stagnant air masses are periodically breaking out of the deserts. They move over wetter areas, triggering droughts. Then the wet areas stagnate and become just like the desert.

MR: *So as the Southwest desert in the U.S. grows, it will affect weather patterns in the Midwest?*

JD: Exactly. And the most telling proof of that was last summer's drought in the Midwest (1988). I've been saying for years that droughts in the Midwest and the Great Plains were caused by outbreaks of desert air from the southwest deserts, and I predicted that if you had a drought in those areas you could get rid of it by breaking up the stagnation in the desert to the southwest. So in August of last year I did a cloudbusting operation in the Phoenix area and within a week the major aspects of the Midwest drought were broken up. The responses were quite significant.

MR: *What is the function of cloudbusting in a drought situation?*

JD: Well, it has to do with this problem of stagnation, and also with the question of the life energy. Dr. Reich clearly discovered a pulsatile energetic medium in the atmosphere which he called orgone energy. It's a free form of the same energy that charges up our organism — charges up all living creatures, including plants. This energy is spontaneously pulsatile...in a healthy organism it moves, flows and pulses; it has a certain pulsatory rate.

Now when the trees are cut down and the vegetation is wiped out in an area, the life energy in the atmosphere over that region tends to suffer, it goes stagnant, dies, becomes dead...becomes what Reich called *deadly orgone*, or *DOR*. Deadly orgone is characterized by an absence of the normal rich bluish color in the atmosphere. It tends to make the atmosphere very hazy, even darkish, and under these kinds of conditions rainclouds stop forming — you get a decline in rainfall. So if you wipe out the vegetation in an area you kill the life energy over that area and you don't get natural rainfall.

This decline in rainfall is being measured in the deforested areas of Brazil, and in the areas just south of the Sahara Desert, where in some areas the desert is spreading south at the rate of *5 miles every year!* And what we're seeing there is an increase in social violence between the various tribal cultures, with guerrilla warfare focusing on water rights and ownership of fertile land.

Some environmentalists are talking about the whole continent of Africa becoming a desert in 50 to 100 years. And part of this is because they're overpopulating like crazy. Black African culture, for the most part, is largely patrist culture that doesn't believe women have equal rights or should be allowed access to birth control. So population rates are skyrocketing: 4.5% population growth in Kenya, which means a doubling of the population every 15 years...there's no way that can continue.

MR: *To go back to the cloudbuster for a moment, what is the influence on the atmosphere from cloudbusting?*

JD: When you go into an area with a cloudbuster, you have a means to eliminate the atmospheric stagnation... restore some of the normal pulsation that would exist in that area if it were more covered with trees and vegetation. By doing so you can, at least temporarily, restore rainfall to that area.

MR: *And this is what you're attempting to do with the Drought–Abatement Program?*

JD: That's right. I've been doing it informally with some enormous help from different people over the years, and now I'm working with a group of responsible individuals and organizations to try and develop an organized and funded program. Then when a drought *starts* to occur, we can go in with a team of trained people and put an end to it with this cloudbusting technique.

You know, it's a tragic situation. In the 1986 Southeastern drought, and the droughts this last summer in the Midwest and in the Pacific Northwest, we had the equipment, we had the people, we had everything ready to go to end those droughts in the very early stages, but we didn't have the money. We were getting calls from people, from desperate farmers who said, "can't you guys get your equipment out here and help us?" And we wanted to, but we couldn't afford the gasoline, or to defray other basic expenses. It takes money to launch these cloudbusting expeditions…anywhere from two to fifteen thousand dollars…so we really need to be funded, and we're looking for help and ideas on this as well as contributions, no matter how small.

MR: *I want to go on record as saying, in my opinion, the cloudbusting operation you did up here in Washington last September was responsible for breaking the devastating drought the whole Northwest was in. People may remember this drought because of the huge fires at Yellowstone, but in fact there were forest fires burning right on through to the Pacific coast — northern Wyoming, Montana, Idaho, Washington, parts of Oregon, were all excessively droughty. At Yellowstone, for instance, it was the driest summer in 112 years of recordkeeping.*

Here in Washington — east of the Cascades — places hadn't received rainfall in almost three months. Yet five days after you started working, the rains started falling and fires across the Northwest went out or were contained. In Spokane, rainfall for September ended up being twice the average and the effects persisted to where we recovered by the end of '88 to our yearly average level — despite being 2 inches below when you began and in an area where we usually get less than 17 inches of precipitation for the entire year.

So this was dramatic as well as being very needed and I just want to acknowledge it, and thank you publicly.

JD: Up in the Pacific Northwest they're cutting down trees like there's no tomorrow! People have got to realize that if they continue on that path they're going to have more and more drought in the Pacific Northwest, which means more forest fires, and, in the long run, even fewer trees. They've got to *stop* cutting down trees…it's as simple as that.

Last September's operation was another situation where we just didn't have the money to go in when the drought started, in the early stages before the major damage.

It was the same during the drought in the Southeast in 1986. We wrote letters to the Governors of all six affected states in early June and said we have this technique, we're ready to go, please help us with the money. We didn't get a single return letter or phone call. Nothing. So we called all kinds of people to try and get money and it wasn't until early August — after the drought had done most of its damage — that a couple of people…they didn't even live in the Southeast…were so touched by the situation that they funded us. On August 6th we started our operation and the month of August turned into one of the wettest months ever in that area, with 200-250% of the normal rainfall for that month. And that was the end of the drought.

MR: *So it wasn't just a shower here and there?*

JD: No. It just completely turned around, though the rains were not heavy, or flooding; it was a slow, incessant drizzle that lasted for two weeks and very slowly saturated the soil. It couldn't have been better.

You see, when a drought ends naturally it's more common that it's quite violent. You get big, huge thunderstorms that push into a large stagnated area, and finally break it up. A gentle drizzle terminating a drought is somewhat abnormal, but it tends to be the kind of thing that happens from a correctly performed cloudbusting operation — the drought ends with a nice, gentle, persisting rain.

MR: *There's a joke among firefighters that goes: "How does the Forest Service stop a wildfire?" — "It pours money on the flames until it rains." On the Yellowstone fires alone last year, they poured some 120 million dollars on the flames until the rains came, and here you're asking for only thousands…do you anticipate any funding from government sources?*

JD: No. I don't see any way for government or other public funding for many years into the future, because the hatred for this work and for Reich's findings is so great among the academics. Many of them would rather see drought and widespread economic devastation, than to admit their theories might be wrong, and Reich was right.

I mean, this is an expression of patrism in our culture; it's a sadistic energy that all this intellectual facade covers up. One of the groups that knows this in a real way is the Animal Rights people. In a large measure they see that a lot of these biological experiments are really nothing more than an excuse for some brainy scientist to dump sadistic energy on a helpless animal. People have to realize that a lot of what the scientists do is pure sadistic hatred, and has *nothing* to do with science! It's similar to the way the inquisitors of the Middle Ages would talk all flowery words about how they were saving someone's soul, at the same time as they were pouring molten lead in their ears and preparing to burn them at the stake.

We have to look beyond people's words; we have to look at what they're doing. The classical scientists who oppose all this stuff I've been talking about are the same ones who have been on watch…they had the responsibility to sound the alarm when something was going wrong with our environment and they didn't. For the most part they sat back, they kept quiet, or they actively interfered with people who were trying to really do something helpful …now we've got this enormous devastating situation on our hands.

MR: *Orgone energy — life energy — a primordial energy in the atmosphere that Reich claimed to have demonstrated visually, thermically, electroscopically, as well as in biological functions…would you say, as an independent scientist, that you've verified his findings?*

JD: Yes. I'll go on record. I've verified some of his experiments, and am convinced that the orgone energy exists. I've also put together a bibliography citing evidence from 1934 to 1986, with 400 different citations by over 100 different Ph.D. or M.D. research scientists, and they've also confirmed this energy principle time and time again. So in my view it's not a question of the absence of evidence — it's the presence of blatant narrow–mindedness and hostility on the part of the people who have power in the classical institutions, and who adamantly refuse to seriously examine this evidence.

I've also come across, in my readings and research,

dozens of physicists, biologists, chemists, and people in many natural scientific branches who have discovered a principle very similar to orgone energy...many of these people don't know about, or even like Reich, but nonetheless, their experiments show the existence of a dynamic, pulsating cosmic energy.

MR: *Can you give a few examples?*

JD: Halton Arp, the dean of American astrophysics, keeps making these photographs of energy/matter bridges between galaxies out in deep space. But the galaxies, according to classical ideas, are supposed to be millions and billions of light–years apart. So how can there be these energy/matter bridges intertwining between them? Dr. Arp says it's because they're not really that far apart and that the distance indicator being used is inaccurate. Now if that's the case, virtually every single major theory in astrophysics, from the Big Bang to Relativity, has to be wrong.

And you know how they finally dealt with Arp's disturbing pictures? They told him *you can't make anymore pictures! You're forbidden from using the big telescopes!* Now he's over at the Max Planck Institute in Germany... couldn't get an honest break here, so he went abroad.

Then there's Dr. Robert Becker, an extraordinary scientist who found that by using very subtle pulses of DC electrical currents you could get mammals to regenerate amputated limbs! But just as he's ready to apply this technique to human beings — which would be an incredible breakthrough — he's attacked by the biomedical community. His funding is cut off, he's denied access to his own laboratory, and his research team of exceptional young people is scattered to the winds.

Now that's average, that's typical, that's going on in universities and research centers across the country. And of course there's the example of Dr. Reich...the FDA invaded his laboratory, put the instruments to the axe, burned his books and threw him in jail. This is 1957 I'm talking about, and the FDA burned Reich's books in New York City because they "decided" that orgone energy didn't exist... Reich died in jail.

Today, the biomedical community remains almost as viciously hostile to Reich's findings as in 1950. Any doctor who attempts to reproduce Reich's experiments on people in the U.S. is taking a very great risk. Now in Europe, it's quite a different situation. Over there, doctors use Reich's approach; they use the orgone energy accumulators openly in hospitals to treat people, and they get pretty good results.(1)

MR: *What about Reich's orgone energy principle compared to, let's say, the "chi" in Chinese medicine?*

JD: In my view it's the same energy, people are just defining it according to what they know about it. In acupuncture you have this energy that flows through the body according to this very delicate circuit pattern. When it flows freely and moves, the organism is in a state of health; when it gets blocked or dammed up, it results in formation of symptoms. Now this is very much in keeping with Reich's formulation.

MR: *What about the images of an energy produced by Kirlian photography?*

JD: It's the same thing. Dr. Thelma Moss has made Kirlian photos...actually the better term is electrophotos...of a leaf. What she does is take a leaf and cut part of it off, and if you do it fast enough, in the right kind of environmental conditions, you will get a photo image of the entire leaf. Now Thelma Moss was also viciously attacked and harassed, and lost her job for a time, as I understand it, because of her findings. It was a horrible mess, which she suffered through right at the time she figured out a way to make *nonelectrical* energy field photos...the criticism of electrophotos is that the photos are just that, electricity — not life energy. And you know how she made them? She took a leaf, put it on the film, and put it inside of an *orgone accumulator*. The accumulator enhances charges enough to make them register on film. So very clearly, what she's showing is the image of the human orgone energy, or life energy.

You see there are many, many people in entirely different fields who have verified the existence of this energy, time and again. But these points of view are not allowed expression in the major journals; there's a rigid censorship involved and a suppression of most of the research. Most of the researchers I'm talking about were fired from their jobs, or had to go overseas. Part of it is narrow-mindedness, hatred and fear, but a major part of it is economics. If you consider that if these points of view are correct, then literally billions of dollars of research, into things like these particle accelerators or nuclear energy, and also the black hole of modern cancer research, are in error in a fundamental way and they ought not to be allowed...what they're doing is likely to be far more damaging to people and the environment than they realize. For instance...and I know *Wildfire* has published some reports on these geophysical disturbances from underground nuclear tests...in the first issue of *Pulse of the Planet* we reprinted an article by Dr. Yoshio Kato of Tokai University in Japan, who found that following underground nuclear tests the upper atmosphere would soar by about 100 degrees, and the Earth would wobble on its axis a little bit...not every time, but there's a correlation between nuclear testing and those phenomena.(2) Now this is a *very serious* problem we're facing.

MR: *I want to put the other oar in the water for a moment and get back to this problem of patrism, particularly its diffusion. From what you're saying it's obvious that it's had a major global effect, and I can't think of one major area in the world today where some kind of widespread destructive behavior isn't rampant, and chronic. But historically, if patrism originated with the desert in Central Asia, then spread out, wouldn't it follow that the geographic locations farthest from Saharasia would have shown the least evidence of it?*

JD: That's absolutely correct. And one of the telling things is that the behavior, for the most part, of peoples in Oceania...in the South Seas...and in North and South America was very gentle and what I call *matrist*; they were reasonably nonviolent people.

---

1. See the articles by Lassek and Brenner in this issue of the *Pulse*.

2. See *Pulse of the Planet*, Vol 1, No.1, p.4.

Now there are exceptions to this and they are the exceptions that prove the rule of the Saharasian thesis. For instance, let's look at the Inca culture. This was a male–dominated warrior caste society in Peru. They committed massacres at various times; they constantly attempted to dominate and subordinate all local peoples in the area surrounding the Inca central state empire. The progenitors of the Incas appeared on the coast of Peru, around 2,500 B.C., and quickly developed many hundreds of settlements... settlements which were ocean–looking. They also built pyramid mounds and temples and big tombs, and they interred sacrificed women in the graves of their dead kings, and they deformed the crania of their children — all in the manner of the Central Asians. So what does that suggest in terms of their possible origins? They arrived by *ship,* from a distant patrist empire.

If you look at the archaeology of the peoples who immediately preceeded them, they were simple hunter/gatherer people without any trappings of a central state, no warrior class, no traumatization of women and children. And this is an argument I've developed out of my Saharasian studies: Sadistic behavior doesn't just *spontaneously* appear in people; the urge to suppress the sexuality of children or to crush the life out of babies doesn't just spontaneously appear. It's *alien* to a society that doesn't already do it.

There's an anecdote underscoring this point that concerns a Nez Perce Indian chief who, while on a peace mission in the 1800s, was riding through a white settler's camp and witnessed a soldier beating a child. Reining in his horse, he's reported to have said: "There is no point in talking peace with barbarians. What could you say to a man who would strike a child?"

MR: *Any other exceptions to the rule in the Americas?*

JD: The Meso–American peoples of the Maya and Aztec did a lot of sacrificing of their enemies. They built pyramids and their rulers were divine kings, absolute rulers; they sacrificed women, deformed the crania of their babies, and performed genital mutilations. Their behavior was essentially Saharasian — they were like those bloody kings of Egypt and Mesopotamia and Central Asia.

Then you can see intrusions of Meso–American peoples all the way up into the Mississippi River valley... these peoples were not as bad as the Meso–Americans, but you see some patrist tendencies in them.

You know, Barry Fell and scholars like him have argued for years that there were large movements of peoples from the Old World to the New World in times before Columbus, but long after the Bering Strait migrations. And in the Pacific Northwest we can see that some of the groups were clearly affected by more recent influences coming over from East Asia. For instance, the Athabascan peoples, which include some of the Navajo and the Apache, were warrior–nomads who preyed upon the other Amerindian peoples. When the Athapascans arrived in the New World in the Pacific Northwest, they headed southward through the Great Basin, eventually arriving in the areas of the Anasazi.

Now the Anasazi were very peaceful people, no weapons of war, beautiful architecture with no fortification walls in the earliest period...Chaco Canyon, for instance, speaks of great peacefulness. But then something happened that is very revealing in terms of what I'm talking about: At one of the major pueblos at Chaco Canyon archeologists found a grave with several dead women who had been murdered. And right after that event, they determined that the entire place had been abandoned: The Anasazi abandoned their unfortified dwellings on the open plains to build fortified settlements in cliffs, and there was a shift in living conditions away from the fertile irrigated valleys, into the cliff faces, and up onto the mesa tops as well. This happened around the same time as the arrival of these Athabascans who kidnapped women, stole crops, murdered, deformed the crania of their babies...they didn't build pyramids and that sort of thing but they were warrior nomads nonetheless.(3)

MR: *So geographical barriers could provide a resistance to patrism?*

JD: In the sense that in North America, the more peaceful Indian cultures persisted by living in fortified cliffs or up on the mesas, and by doing so they could survive against these nomadic bands of relative newcomers to the New World. And what you see today in the areas of mesa–top living are the Hopi and the Zuni, who are among the relatively more peaceful, female-oriented Amerindian peoples.

It seems to me that whereas in Central Asia, the Huns and the Sythians and the Mongols would ride down on an enemy, take whatever prisoners they wanted and kill everybody else, causing a widespread and uniform change in the culture, here in North America you had little roving bands of warrior–nomad people and there weren't enough of them to really wipe out the pre–existing cultures. So there's a greater variance in culture here than in Central Asia, which is uniformly very patrist.

MR: *Were there areas of resistance in Europe?*

JD: Yes, in fact, the thick forests of Europe tended to take away the military advantage the huge Central Asian horse–mounted armies had on the grasslands...and we see in other areas that the rainforest acted as a barrier. But Europe was essentially transformed by these invasions, beginning with the Battle–Axe people, whom historians also call the first wave of Indo–Aryans...they were a little bit taller peoples, heavier skeletal structure. Now some people say, "Oh, this is the first civilization of Europe." But I say absolutely not! These were, instead, the first *sadistic brutes* to show up in Europe.

We have this tendency to define civilization on the basis of technology or the central state. That's a mistake, because obviously our technology is a double–edged sword, and if we're not careful — and we haven't been — we could kill everything and everybody. And think of what Hitler did, what

---

3. Another migrating patrist Amerindian group, the *Numi* peoples, may also figure into these particular cultural changes

> **"Democracy was invented by primitive peoples all around the world... if you go deep enough into the archaeology... people lived a rather democratic, weapons-free, peaceful existence."**

Stalin did; that's what a "central state" can do — and it has *nothing* intrinsically to do with being "civil."

We have to recognize that the European heritage carries negative baggage, but we also need to recognize that European peoples still had a trace of their early, matrist orientation. The Common Law of England formed a basis for the later Magna Carta, and from there to the Declaration of Independence, the Bill of Rights — all this stems from an ancient tradition in European peoples that predates the Battle–Axe people...a freer more democratic tradition. It's a *myth* that the Greeks invented democracy. Democracy was invented by primitive peoples all around the world. Everywhere you go, if you go deep enough into the archaeology you get down into this layer of peaceful traditions where people lived a rather democratic, weapons-free, peaceful existence.

MR: *And these would be more matriarchal societies?*

JD: Absolutely. But we need to emphasize that they were not societies *dominated* by women in the sense that males dominate societies today; rather, they were societies where "domination" per se, did not exist. If you look at the Iroquois before the white man...and we now know that Benjamin Franklin and Thomas Jefferson got at least some of their ideas for the structuring of the American state from the Iroquois society...they had a House of Men and a House of Women... The women and even the children had a certain say in society that was generally equal to men, with the exception of conditions having to do with battle, defensive war, and so on. Actually, the more you get into conditions of warfare, the less women have a say in running things.

MR: *Which brings us back to how warfare and military castes developed in societies in the first place.*

JD: Exactly, and again it's very much wrapped up with the development of this Saharasian desert belt and the effect it had on people.

MR: *I guess the question that's looming here is why did patrism persist outside of the desert? Why didn't the behavior, if it was tied up with the desert, stop when they left...went into Europe, North America?*

JD: That's what makes this work so controversial because it overthrows many different social science theories, and resurrects Dr. Reich's sex–economic theory. The sex–economic theory, basically, says that neurotic, psychotic, and socially destructive tendencies in human beings arise in the individual by virtue of the emotional contraction stimulated in the organism through painful traumas, purposely inflicted on babies and young people, and by the suppression of the sexual feelings and impulse of both adolescents and adults. The accumulated, undischarged emotional and sexual tension -- a bioenergetic force in the body which can only be held in by armoring up -- drives the individual towards sadistic, compulsive, self-destructive, and antisocial modes of behavior. Social

institutions such as hospitals, schools, Church and State also blindly participate in the process of repression and armoring, oftentimes creating the very social problems they seek to resolve.

MR: *Well I can see why it's so controversial. I don't know of anything that's so defended as this rough way we routinely treat children or suppress their sexual needs...though in another context, there's unanimous agreement that similar behavior is destructive, if not insane. I mean, if I bend a young seedling tree, it will never grow straight; if I keep sunshine and water from it, it will suffer. But why did this kind of behavior persist, generation after generation, out of the desert?*

JD: Chronic pain and trauma, as Reich pointed out very clearly, have specific somatic effects...they have an effect in the musculature and nervous system, and can lead to a chronic *neuromuscular armoring*. If people are punished long enough, severely enough, the biophysical contraction, or armor, will become permanent. Then the neurotic components associated with this kind of upbringing...you know, "my father raised me with a strict hand, so..." are passed along, generation to generation. And the disturbance of the sexual drive of young people through the instigation of various social institutions — arranged marriage systems, compulsive marriage, certain virginity taboos, and genital mutilations — add to the trauma and armor, and instill various anti-pleasure attitudes and "traditions" which are unquestioningly passed on from one generation to the next. While these patterns started in the desert, once underway, they persist on their own, irrespective of climate factors.

This is what my global study demonstrates, that you can trace patriarchal authoritarian aspects of culture, all around the globe, by virtue of the diffusion patterns coming from the Saharasian region.

MR: *The thought that keeps occurring to me is that our "socialization" process might be totally backwards. That is, the very behaviors that threaten society — warfare, destructiveness, child abuse, sexual violence — and which it is supposedly mitigating through certain practices, are actually being created by those practices, and perpetuated.*

JD: Absolutely, and it's been going on for 6,000 years. This is very disturbing...no one wants to look too closely at it.

MR: *Well it's certainly consistent with our position that many traditional societies often were healthier, saner than any we see elsewhere...though I can't say I've ever been able to present as clearly why there's a difference, or how it came to be. What's a clear example of a culture where patrism didn't intrude and where that made a difference in people's behavior toward each other, toward children, toward nature...?*

JD: In the modern era there were perhaps several dozen very peaceful societies left in the world, although most are under severe attack...like some Pygmy groups in the rainforest regions of Africa. But these areas are being deforested, and the Pygmies are being displaced. So I don't know the exact conditions today. But as recently as the 1920s and '30s many of these societies were intact, such as the Muria of the rainforests in India, who are a prototype of what people were like all around the world before 4,000 B.C.; and the Trobriand Islanders in the South Seas, near New Guinea.

Among the Muria, the babies had an exceptionally close bond with their mothers; there was no pain inflicted upon the babies in any purposeful way, no social custom designed to separate the baby from the mother, and no ritual mutilations of any sort. Nothing was done to disturb the maternal/infant bond. The babies were also looked out for by a number of adults, male and female, who developed a warm and loving bond with the child. When they got to be about five or six, the children would begin to explore their village environment, and make more of a bond with other young children and adolescents.

They had what we call a *children's democracy*...an age group of children who would determine their daily activities as a group...and they were not subject to the same kind of adult pressures that our children are. For instance, if the Murian kids wanted to go into the forest and look for bird's eggs, or climb on the hills, that's what they did. If the adults said, "No, you've got to come help in the garden," and the kids didn't want to, that was the end of it: The kids didn't have to do it.

Now as it turns out, the kids were very curious and they *liked* helping out with things, particularly the exciting kinds of activities...like when the Trobriand Islanders were netting fish, the kids loved to help. The children in the children's democracies would very much participate in what the adults were doing, but it was never in a coercive way.

MR: *What was their sexual behavior like?*

JD: Well, most importantly there was no suppression of the sexuality of these kids. When they would begin to express sexual feeling towards each other there would be no inhibition or restriction of it. The kids would begin sleeping with each other...not always with intercourse, sometimes they just wanted to hold each other at night. And they had a special house for the young people which provided privacy for them, as well as to get them out of the hair of the parents. In many cases, the adults were forbidden from going into the this house.

The younger children were initiated into sexual activity by the slightly older ones, and after a certain, rather promiscuous, period of sexual exploration, the older children would begin to develop a bond with a special favorite person of the opposite sex. These young people would later spontaneously develop love–match marriages, and these marriages were quite stable and enduring.

Simultaneously, there was none of the kind of sexual pathology that we normally think of here in the West — no wife–swapping, no polygamy, no homosexuality, no adult–child sex, or rape. Among the Trobriand Islanders for instance, there was no word for "rape" or "stealing" because nobody did it... people had no desire to do it. Nor was there any evidence of sadism.... except where Christian missionaries had worked a sex-negative influence.

So what we see is that the maternal/infant bond, undisturbed, led to a strong male/female bond, which was also undisturbed by the adults as it expressed itself in the very

early stages. The result was a very healthy society that did not make war, had very low levels of violence, was basically non–neurotic...very happy, sincere, honest people.

Regarding the Muria, a missionary named Elwin was sent to convert them by one of the Christian churches, but when he got there, he said: "This is the most loving, happy, and peaceful society that I've ever seen in my life, and I'm not going to do a thing to change them. I'm going to devote my life to protecting them from outside influences." And that's what he did.

WF: *What about pregnancy? Didn't a lot of these young girls get pregnant?*

JD: No, and this has been called the "riddle of adolescent infertile promiscuity" by the first anthropologists to observe it. The unmarried girls simply didn't get pregnant in any significant number. As it turns out, the natives, when asked about this, would tell the anthropologists about special contraceptive herbs the girls ingested, often known only to the women. But at the turn of the century, there were so few female anthropologists who could gain the trust of the native women...and this was before the birth control pill...so the natives were simply not believed. After all, if we "civilized" folk could not control conception, how could a lot of "ignorant primitives" do it? I have since documented a large number of cases where these contraceptive herbs were in use, and some biochemists have even studied the plants, confirming their contraceptive effects.†

MR: *And what if a marriage broke up for one reason or another, what would be the routine then?*

JD: The young man would go live in something on the order of a bachelor's hut; the young woman would maintain control over the small hut that had been built...the woman's role in childbearing gave her certain privileges. But both man and woman would be free to seek out somebody else after this separation.

MR: *And if children were involved?*

JD: Well, as I said, the caretaking of the older child was more of a collective responsibility. They had extended families and village groups, so at any given time the children would have several adult caretakers, and the full weight of childrearing did not fall on any one woman. The men also participated very much, they helped out a great deal.

But mostly the women were not so disenfranchised from the economic resources of the society that they needed a man to look out for them, you see? In Western culture women are so economically disenfranchised... they can't get equal jobs, pay, or education and they're *not* the predominant holders of wealth. So when a woman separates from her husband, particularly if she has children, it's generally a severe economic blow and she has greater difficulties going out into the workplace in the same way a man can.

We've seen some changes here in the U.S., but not

---

† A separate Chapter by Dr. DeMeo on the subject of contraceptive plant materials is available in his new book *Saharasia*, from Natural Energy Works, Ashland, Oregon. www.naturalenergyworks.net

compared to these societies, where there was more of a collective ownership of the land and means of production, and a much greater role for women. In land tenure, for example, the land passed on through family clans that traced the lineage through female ancestry. A man would say, "I belong to the clan of my sister or mother," and by virtue of that, when he went to work in the fields of his clan he was helping out his mother and sister.

And in these cultures the fathers very rarely participated in a lot of the education and upbringing. That was left to the mother's brothers. The child would look to the father as someone who cared for him, as a good friend, a buddy, and an intimate bond often formed that would last a lifetime; but the uncle was seen as the primary educator and caregiver. Neither men nor women could ever be "divorced" from their clan, nor from the land, food, or wealth held in common by the clan. This is quite a different situation than we are accustomed to, where compassion–less economic competition against our own countrymen is encouraged, and poverty and homelessness of even women and children is now widespread.

MR: *So the success or continuation of a marriage would be based only on personal, emotional and sexual satisfaction?*

JD: Right.

MR: *You mentioned the Iroquois and their contributions to the U.S. Constitution, and interestingly, among them the women owned the crops and the women's clans were charged with the responsibility of appointing the chiefs, or sachems. But Paula Underwood Spencer points out that our present constitution is still lacking in two important rights that the Iroquois had: The Rights of Women, and The Rights of Children. And she points out that the children's rights included pointing out to the adults when they were not acting in accordance with tribal beliefs. Apparently it wasn't a gratuitous thing either, but was acknowledged and respected.*

JD: That fits very much with the functioning of a children's democracy. In Murian society the children had a special house, called the *Ghotul,* and they revered this as a religious sanctuary. They believed that anything the kids could think of, or do, in terms of festivities or social functions or rituals, was somehow given a certain sanctity and they allowed the kids the freedom to do this. So the kids were always coming up with dances, or a new this or that, and the adults would say, "OK, that's great."

MR: *And these things were incorporated into the religious practices?*

JD: Well yes, but their religious practices were not as encoded as what we're used to thinking of as religion. Their's was more one of song and dance and festivity — a celebration — as opposed to something solemn and ritualized.

MR: *Do you find this to be true in matrist cultures generally?*

JD: In general, yes. The religions tended to acknowledge a Great Spirit at work in the here and now — present☐ — as opposed to some anthropomorphic deity off in a heaven and hell, divorced from the real world.

The spirit force of the matrist people tended to be female, or at least have a female aspect, and was something quite tangible in the sense that it was perceivable, and "charged up" the trees and forests and the air and the water. It was something that they could appeal to directly without the need for a religious specialist. So they didn't have a full–time priesthood; they didn't have temples. They just had selected spots, you know, like an area of the forest that was exceptionally charged, a beautiful spot...a spring where the water was exceptionally charged, and sweet.

MR: *This sounds very similar to orgone energy.*

JD: Precisely. And what Reich argued was that people in these cultures, because of the absence of trauma and the lack of armoring, had much more direct contact with it: They could *see* the life energy, and *feel* it, and their religion oriented itself around it.

MR: *There's a criticism I've heard about Reich, from some anyway, that he wasn't very "spiritual."*

JD: Well, he wasn't "spiritual" in the sense of the modern mystic who wants to get out of their body, or who needs strenuous spiritual exercises to regain contact with the life–force. His whole argument was that *that's* a perceptual problem. The orgone, life energy, or Great Spirit is in the here and now, very tangibly, but people, because of their armoring, generally can't feel it; and can't see it.

But the point is, the unarmored peoples of the world did feel and see it, and when they talked about it, it was in the here and now. They didn't have to get a religious specialist or go through some ritual — it was just there.

Part II:
Wildfire Interview
with James DeMeo*

Conducted by Matthew Ryan**

MR: *I'd like to talk about some current events on the national and international scene from the perspective of this problem of patrism that you've worked out in the Saharasian Principle. But before that, I'd like to review some aspects of the work, and address some comments that readers have had. One of them is that your model for the diffusion of warlike cultures from Saharasia after 4000 B.C. doesn't jibe with the accepted notion that this hemisphere was settled exclusively by peoples migrating over the Bering Strait thousands of years before that, so that any patrist cultures here would have been, let's say, "home grown."*

JD: Well, if we just look at native people around the world, before the period of European colonial expansion into the New World, the farther away you get from Saharasia on the planet, the more likely cultures are to maintain their original, peaceful characteristics. This is what I was able to demonstrate with global maps of human behavior. In terms of large geographical areas, Oceania – the South Seas –vi and North and South America stand out in the anthropological record as areas of *relatively* peaceful conditions up until the 1700's and 1800's. However, there are some glaring exceptions to this view, like the Incas and the natives of Meso–America, as well as a few North American groups that were quite violent. But I would say they are the exceptions that prove the rule.

Now, the most "popular" theory says all of the peoples of the New World arrived by coming over the Bering Strait tens of thousands of years ago. But the behavior data refutes that. The behavior data implies that the harsh, patristic, violent cultures must have come to the New World only *after* 4000 B.C., because before that date there simply is very little unambiguous evidence for violent cultures in the archaeological record, *anywhere* in the world. That's true in both the Old and New World. Since that's the case, and since we see in the archaeology that the violence starts after 4000 B.C. in Central Asia and the Middle East and then spreads, the riddle is: How did that violent characteristic in culture transport itself from those areas of the Old World to the New World? This is where the Saharasian findings tend to support the advocates of pre–Colombian contact, that these people, who were already quite violent, came from Mediterranean or Persian Gulf regions, or perhaps China, somewhere around 2500 to 1000 B.C. And they came in ships.

MR: *Are you familiar with reports that the Spaniards allegedly found the members of the Incan ruling class to be people of fair skin and blond and red hair?*

JD: I believe I've heard of that. What also comes to mind is Amerigo Vespucci, the man for whom America is named.

* Reprinted from *Wildfire* magazine, 5(2):18-31, Fall 1990.

** Editor of *Wildfire*, ~~PO Box 9167, Spokane, WA 99209.~~

While he was in the Americas he wrote letters home saying he saw people with red and blond hair, he saw blue–eyed people, he saw black–skinned people. He saw groups of these people, tribes and so forth, that supposedly only existed in the Old World. Apparently there were a lot of observations made about people who did not fit the stereotypical North or South American aboriginal appearance. For instance, black people were not supposed to have been here until the coming of European slavery. But there are these huge stone heads carved from the early Olmec period in Central America that look very much like the peoples of Africa; there's also some evidence from African tradition about these kinds of contacts. We could also talk about the large stone anchors from early dynastic Chinese ships that have been found off the coast of California, or the fact that certain Native American tribes speak languages quite similar to Central Asian dialects . . . I mean, we could spend an hour just citing these types of examples.

But these kinds of things are not discussed in most textbooks because everyone wants to say Columbus discovered America, and the peoples living here arrived only via the Bering Strait. It's a catechism that you have to recite and recite and not question if you want to get through the university system. Personally, I don't believe it anymore.

Certainly, my work doesn't explain everything. But in terms of being a *global* theory, my work on Saharasia is quite robust and does explain an enormous number of things . . . one of which is that sadistic, violent behavior just doesn't spontaneously appear among peoples. It doesn't now and it didn't in the ancient past. So to the extent that we find it in the Meso–Americans, for instance, it didn't just spontaneously appear; it wasn't "home grown." Rather, it can be very clearly traced back in time to this Saharasian area and to the 4000–3500 B.C. development of the desert there. Both the violent culture and the extremely harsh desert environment of Saharasia developed synchronous with each other.

MR: *Developed among peoples whose societies were peaceful, in a land that originally was fairly lush?*

JD: That's right. As the Saharasian desert begins to spread in 4000 B.C., we see more and more evidence for mass migrations, conflicts, weapons of war, and fortifications to protect what they had from someone else invading. That's what the archaeology tells us. It all seems to start with the desert, first in Central Asia and Arabia, and later spreading to other regions.

Prior to that, there were large, peaceful, agricultural settlements throughout Central Asia and the Middle East. The land was somewhat like the African veldt today, which has rivers and streams and big trees and grasslands.

Peaceful, non–violent matrist cultures were found globally in those earliest times, and matrism was ubiquitous among humankind.

MR: *You didn't just stumble onto this discovery of a specific origin for violent, destructive cultures?*

JD: No, no. It was based squarely upon Wilhelm Reich's prior research and findings, his development of the sex–economic model to explain the source of human neurosis and destructive aggression. I became interested in Wilhelm

Reich's work and I was simply one of the first scholars to take his findings at face value, and test them systematically.

MR: *Would you say the sex–economic model predicted that patrism would have a developmental as opposed to an instinctual or innate origin?*

JD: *Absolutely.* Reich was very clear about that. He had speculated that human armoring, as he called it, had an origin at some ancient point in time, but he also said that it was so lost and obscured in ancient history that one could never sort it out. However, I used some approaches to the problem that Reich either didn't think of, or didn't have access to. Also, there was another worker before me, by the name of James Prescott, who subjected Reich's sex–economic model to a cross–cultural test. So I had Reich's clinical findings and original suppositions, and I had Prescott's work as a foundation from which to start my geographical investigations.

MR: *How would you explain sex–economy in lay terms?*

JD: Reich's sex–economic theory indicates that basic instinctual impulses have a biological origin, and the goals of all pleasure–directed impulses are rational, with an inherent capacity for self–regulation. The emotional expressiveness of the individual can be encouraged and protected by society, from infancy through adulthood, in which case one will not observe signs of neurosis, or psychotic, violent or destructive behavior. Societies that, for example, breast–feed all infants to the child's satisfaction, and allow them to toilet–train spontaneously, and which allow and even look favorably upon the sexual play and expression of children, and which also allow their unmarried adolescents the full rights to a private sexual life — these are the truly non–violent, unarmored, matrist cultures which once characterized the whole of humankind.

So even though they clearly constituted a minority of all cultures, worldwide, over the last 100 years or so, Reich did not develop his views on these societies as a fantasy: They are true and real cultures with decent, self–regulated social conditions, free of the sexual misery, social chaos and sadistic violence of our "civilized" world; their very existence challenges the popular notions about a supposedly *unchangeable* violent nature in *Homo sapiens.*

In an armored, patristic culture, the basically good and decent feelings of the child are bowed and crushed by the adult world, and sadistic impulses develop in the child by defeating its natural and healthy proclivities towards honesty and decency and love. Parents in patristic cultures essentially raise their kids less by instinct than by the demands and rules of their culture, which are anti–life, anti–sex, and suppressive of everything spontaneous and instinctual. The child very quickly and painfully contracts, like a whipped dog, and then you end up with another generation of fearful, corrupted and hateful adults. And it's been going on for 6,000 years now.

MR: *Another question that comes up is whether all deserts give rise to patrism?*

JD: What we're talking about is *extreme* desert conditions that cover a large territory, and which change from a wet to a dry condition with such rapidity that people are pushed to, or over, the brink of starvation. They're starving. The family

bonds begin to break apart — husbands abandon wives and children; mothers' milk dries up and babies starve. Eventually, even mothers will abandon their children. A single–minded search for food takes precedence over everything else, and every family and social bond completely degenerates and is destroyed. This is what happens under conditions of severe famine. We know this because we have plenty of evidence from more recent drought conditions.

MR: *Such as in Africa?*

JD: Such as in Africa, although nowadays we can get some relief to these people. But in 4000 B.C. there was no relief and we see the results in the archaeological evidence.

For example, there is a ruin — one of the largest ancient ruins in the entire world — located at a site in Central Asia known as Altyn Depe. It's a huge city–state built on the edge of a big river which used to provide an enormous irrigation system for agriculture. It sprawls over an area much larger than the ruins of Teotihuacan in Mexico. Just *enormous.* Now this civilization was completely abandoned after approximately 3500 B.C. when the climate of Central Asia changed and this big river dried up.

Altyn Depe was abandoned after a short period of chaotic conditions. Then we see these enormous migrations — one wave of people after another — invading into Eastern Europe. They're on their way out of the desert into the wetter lands of Eastern Europe, and by the time they appear in the East European archaeology they are extremely violent people.

Then we begin to see the destruction layers in the pre-existing settlements in Europe. We see the people who occupy the land change from a shorter stature to a taller, heavier-boned people — the Central Asians. We begin to see the ritual murdering of women in perhaps 80 to 90 percent of the European graves after this period. This is literally where the old man dies and his favorite young girl has her throat slit and is put into the grave with him. Now that kind of evidence shows not only this grotesque brutality against women, but it also shows arranged marriage systems. The fact that you have older men and young girls is a clear indication the young boys and girls are not being allowed to live out their own attractions. We also know these were mostly military men because they had weapons which previously didn't exist. Shortly after this, fortifications begin cropping up all over Europe as violence becomes more epidemic.

MR: *Which is the point at which historians say "civilization" begins in Europe?*

JD: Yes, but in reality it's the beginning of the *craziness* in Europe. The places where these early violent invader people settled can be traced all the way into the modern era as places where military glory and warfare is a dominant emphasis of the people.

Bavaria, for instance, is one of the noted areas where these early peoples settled. They were known to linguistic scholars as Indo–Aryans . . . and this is where there's a certain kernel of truth to the white supremacist view that the Aryan peoples were the original race that settled Germany. The original Germanic tribes were from Central Asia — regions where Indo–Aryan or Indo–European dialects prevailed. But you know, nothing *else* of that mythology has any truth to it. They weren't superior in anything except that they were superior butchers, superior sadists, and the native European people whom they killed off or subordinated tended to be softer, much gentler people.

MR: *The question begs as to why, once the droughts and chaos are left behind, does the behavior persist?*

JD: What we see is that once the process is underway, it becomes uncoupled from climate concerns. After a while, it doesn't really matter what the climate is anymore; instead of people becoming aggressive and violent by virtue of extremely powerful environmental affect, they recreate the violence and aggression by virtue of *new social institutions* that they developed under those early oppressive conditions of desert and starvation.

MR: *What are some examples of practices that developed in this period that were institutionalized, and then acted as mechanisms for perpetuating the harshness?*

JD: Well, we can talk about two major classes of abusive social institutions: genital mutilations, and infant cranial deformations and swaddling. It's clear that genital mutilations first originated in the deserts of northeast Africa or Arabia and the Middle East. The earliest evidence shows the Egyptians circumcising boys shortly after the age of the great pyramid, and my global maps of this practice imply an origin of the genital mutilations in that desert region, with a diffusion of the practice to very distant lands.

Swaddling and infant cranial deformation — practices which cause infants much pain and inflict mild to severe emotional stress and resignation — these got started in Central Asia, probably as a by–product of the back–pack cradleboards employed by Asian nomads during their long marches. This practice also was transplanted to distant lands by migration and invasion.

There are numerous other examples that could be made, but a central fact is that something happened to these early peoples to cut them off from deep emotional connections with their kids, and with each other. Because after the harsh desert was formed, after they lived through a period of famine, starvation, and social chaos, then all of these new, very painful and frightening social institutions and practices appear. At that point, they are violent, not only to strangers and other cultures, but also within their own families.

Generally, it's axiomatic that wherever you have a culture that's punitive toward its children, that manipulates and dominates them, and where the sexual feelings of those children are suppressed so that kids from the ages of 10 to 18 are forbidden from having any sexual interest — those kinds of cultures are invariably very violent. In fact, you can't find an example of a culture like that that is peaceful. They are always, *always* violent.

On the other hand, in a culture where the children are raised with a great deal of love, where the maternal–infant bond is preserved, where adults allow the children the freedom to express themselves, their emotions, explore their sexuality — where budding romance between young girls and boys is not suppressed — in those societies always, *always* the people are peaceful and don't have any interest in warfare.

The sad thing is that in the modern era, the migrations of people with this warlike character — and the accompanying social institutions — have spread nearly all around the world, so that the truly peaceful cultures which persisted into the modern era are small in number. They tend to be located in areas off the beaten track, away from trade routes, deep in the forests and so on — such as the Trobriand Islanders in the South Seas or the Muria in the rainforests of India.

MR: *You've talked elsewhere of a gradual lessening of the patrist influence as it diffuses great distances from Saharasia. But how do the social and political structures of the peoples who remained in the Saharasian deserts compare?*

JD: Well, the people who stayed behind developed a nomadic, pastoral lifestyle; or they engaged in trading and raiding and set up predatory systems of having slaves work the land around any available water. This is where you have very hierarchical systems with the big guys — the warrior kings and their armies — riding around on their horses terrifying people with swords.

Remember the empires that were located in Central Asia, not only Babylon and Hammurabi and those of Old Testament history, which are borne out in the archaeology of the area, but all the way up to recent times — the Mongols and the Turks. These empires were centered upon secure water supplies in Saharasia, or at the very edge of the Saharasian deserts.

The social and political structures in these areas were much harsher than in the distant ones. For example, the region stretching from Turkey eastward through Iraq, Iran, Afghanistan and Pakistan is an area that has a nonstop history of incessant warfare, bloodletting, massacre, building up of gargantuan kingdoms and the crushing of those same kingdoms by a neighboring one. It starts around 4000 B.C. and it just keeps going and going all the way up to the modern era — without a break.

MR: *Which brings us to the current situation there. How do you see the role of the United States in the Persian Gulf?*

JD: Well, projecting American military into that area with a foreign policy that imagines, somehow, that we are going to change 6,000 years of bloodletting is a real danger. Our foreign policy isn't to try and help educate these people, or help them raise their kids differently so that the next generation might be more friendly and peaceful, it's that we're going to use superior military technology to threaten them or twist their arms into submission. You know, we're talking about a culture that prides itself on its compulsive, obstinate nature.

MR: *President Bush says we're fighting tyranny and defending our way of life . . .*

JD: Well, Saddam Hussein is indeed the Butcher of Baghdad, who's murdered thousands of his own people, not only Kurds... But when Hussein was gassing civilians — men, women and children — our top leaders didn't say anything and didn't do anything. Why? Because American, German and French companies were making money hand over fist. Iraq's poison gas plants and nuclear facilities were constructed with Western help. Our companies were also selling things to the Iraqis while our government was secretly selling sophisticated weapons to the Iranians *and* secretly feeding military information to the Iraqis so that they could kill more Iranians. We were feeding both sides of the vicious, bloody Iran-Iraq war. It was disgusting, absolutely disgusting, and nobody in our government really objected.

MR: *What about the way of life in the countries we're defending, Saudi Arabia, Kuwait?*

JD: If you go back to the turn of the century, all these lands — Saudi Arabia, Kuwait, Iraq, and Jordan and Israel — had been under the boot heel of the Turks for hundreds of years. And the Turks were ruthless, as bad as the Nazis. After the Turks were defeated, new nations were created, under French and British mandates, by the League of Nations.

Now there is truth to the assertion that the Saudis and Kuwaitis today are not as violent as they were at the turn of the century. Their situation, their societies, have *softened* a bit with the long–term influence of Westerners and oil money. But it's a *relative* softening, you understand, and many of the things they continue to do, we would consider unacceptable in any Western country.

For example, freedom of the press or of speech do not exist in either Kuwait or Saudi Arabia, and political dissent is strictly regulated or repressed in both nations. They do not have democratic institutions of any significance, and are ruled by hereditary monarchs of an authoritarian nature. The wealth and natural resources of those nations are treated as the personal property of the ruling families. Both Saudi Arabia and Kuwait have jailed women who were calling for greater freedoms. Women's freedom is a big issue in Arab countries because they have the custom of *purda* — the seclusion of women. The men have absolute legal rights and dominance over society to the extent that a woman has to legally ask her husband or father for permission to work outside the home, to open a bank account, or travel and so forth. Little things that we take for granted in the West for women, over there they either don't have these rights, or the women must constantly ask their husbands or fathers for permissions. So if a woman steps out of line, she can be legally beaten up, thrown in jail, or worse . . .

MR: *What's an example of "worse?"*

JD: Sexual transgression is still a cause for execution by knife, stoning, or other means. Israeli feminists have an underground railroad where they take young Arab girls who have a sentence of death on their heads, for having violated "family honor" taboos, and try to get them out of the Middle East. A young girl might go off to college, for instance, and

fall in love with a young man, and then her father finds out. He sends the brothers or the uncles to kill the girl. That still happens on occasion.

But let's contrast this with what existed in Saudi Arabia at the turn of the century with the Wahhabis. The Wahhabis are a dominant Islamic sect who wish to rid the faith of what they consider "corruptions." Their view was that, if you were not a Wahhabi Muslim and they encountered you, they were obligated to kill you! I mean, if you somehow stumbled onto them as you were crossing the desert it was, "Oh, you're not a Wahhabi Muslim? You're dead." That was it. They simply killed all outsiders.

Now the Wahhabis were totally isolated from the rest of the world in the Central Arabian deserts until WW I. Then, all of a sudden, they were confronted with British rifles and tanks. At that point, the Wahhabis had only swords and flintlock muskets. So shortly afterward they became, let's say, reluctant to act upon their religious beliefs quite so enthusiastically.

But the point is, it was from that particular culture that the Saudi kingdom and its Royal Family had their point of origin. So if you think of the change from *that* to the modern era where Saudi women, even though they're still segregated from Saudi men, are now being educated and have their own universities, well, that's an *incredible* amount of change for such a short time. And, as Reich pointed out, you can't legislate out of existence centuries–old habits and traditions regarding sexuality and freedom. It takes time, and a great deal of social work and education.

MR: *Do you see any fundamental cultural difference between Iraq and Saudi Arabia?*

JD: The major difference is more political than cultural: for the last several decades Iraq was aligned with the Soviet Union, while the Saudis were aligned with the West. Has that made the Saudis more democratic than the Iraqis? A little bit. Yes.* But the secular state of Iraq in fact outpaced many other Muslim nations in reforms of family law, the veiling of women, and so forth, and they also limited the power of the Ayatollahs.

But Saddam Hussein has visions of the great Babylon of old. He's the most dominant warrior king of the region, and a sadistic brute as well. But you know, that's not saying *much* in that part of the world. Look at Saddam Hussein and the Ayatollah Khomeni during the bloody war between Iraq and Iran. Two vicious madmen followed by millions of *crazy people*. They both gleefully sacrificed their children, sending their sons out to be "martyrs." I'm talking about giving a little eight–year–old boy a plastic key to heaven and a picture of the Ayatollah and telling him to go out on the battlefield and jump on a land mine and blow himself straight up to Islamic heaven. The child becomes a "martyr" and his mother gets a medal. It's insane. Total mass psychotic insanity!

Fueling these social outbursts are enormous amounts of pent–up emotional and sexual tension that's expressed sadistically. In these cultures, the young boys and girls are not allowed to have anything to do with each other — not even a simple conversation. The girls are locked away from the

---

* In later years, with better information, I realized my ideas on Saudi "reforms" were fully wrong and way too optimistic.

boys when they are still toddlers, and they are constantly denigrated and suppressed. The boys, when they're between five and eight, start spending most of their time with the men, and eventually are completely cut off from the women, and brainwashed with anti–sexual, Islamic male superiority. So there's no dating, no sexual contact, none of the simple and pleasurable things we take for granted between boys and girls. That's completely out of the question. In fact, a young girl caught having a mere *conversation* with a young boy would *literally* be at risk for her life in many places. The taboo is that strong.

MR: *But it's a taboo that's enforced by the women also. I mean, where sex suppression exists to any degree, the women are also directly involved, aren't they?*

JD: Certainly the women are directly involved. They play a powerful role. But I don't want to just confirm that and lose sight of the fact that men, being the larger of the species, oftentimes play the fundamental role. The men's role is providing the police force, the enforcement, the beating up of women, and the execution of women in the extreme cases. In Iran, Saudi Arabia and elsewhere in Saharasia, you have special "religious police" who do nothing all day except go around and poke and snoop into the everyday affairs of people, making sure everyone is "properly" dressed and sufficiently "moral," and so forth.

The woman's role is more in the context of child rearing. The women are the enforcers of traditional morality in the child rearing practices. In all sex–repressive, male–dominated cultures, the young males are handled and raised almost exclusively by women for many years — and the feelings they have about women afterward don't come out of nowhere. They come partly out of the physical repression of the child by both parents, and partly out of the cultural brainwash that men are superior to women.

MR: *So you wouldn't support the view that patrism is simply an action of men against women?*

JD: No, it isn't. In the most extremely brutal cultural rites I can cite that are directed at women, such as the female genital mutilations, it's the older women doing it to the younger ones. The men play a role in that they talk about how this is good for the women . . . although that is changing. In her new book, *Prisoners of Ritual*, Henny Lightfoot–Klein interviews a lot of men and women in the Sudan, where they do these cruel female genital mutilations. A lot of the men complained that they wished the practice would stop. But the women are the primary enforcers of it, mostly grandmothers cutting the genitals of their granddaughters with razors.

The problem of patrism, of patriarchal, authoritarian societies, is partly men and partly women. It's not any one sex. But the men tend to be very active in hunting down and punishing dissenters, and so, particularly when outside a culture looking in, it's much easier to *see* what the men are doing as public enforcers of the status quo. One doesn't see so often the mothers' role in demeaning and emotionally castrating their own children, separating the young girls from the young boys. You have to be inside the woman's world to see how they also clearly enforce the cultural status quo on the young.

MR: *In terms of men and women in our culture finding a way to decrease the patristic influence without becoming polarized . . . well, I'm trying to remember something the anthropologist Margaret Mead said about feminist politics having its place, but science having the answers.*

JD: Her point was that political action was necessary for making change, but that women must look to science for deeper answers to the questions they are confronting. I would say that people who want to change things for the better today, in dealing with social issues, can also be acting out of cultural brainwashing *they* have received. Therefore, they may not have anything more functional or productive to offer than what they are attacking, what they want to replace. It's very important, always, to work through whatever kind of replacement ideas you want to have.

For instance, much of the emphasis on rape prevention, on trying to eliminate rape, is focused on talking and education. Now I want to be very clear that there is a pressing need for shelters for abused women and children, and dealing in a helpful way with those who have been raped and abused; that's not the question. But to say that just by talking to and educating people and appealing to their intellect alone, you're going to prevent rape — like it was the result of a lack of education or a lack of perception — that's irrational. It completely ignores the anti–sexual repression that goes on throughout society, as if that had no bearing on rape. Yet, the scientific evaluation of sexual violence from a cross–cultural perspective indicates that rape is *rooted in the anti–sexual moralism of the church*. The suppression of adolescent sexuality leads to frustration that is laden with a lot of anger: rape is, indeed, a product of sex–repressive child rearing.

I think the feminists are partly right, but they're partly wrong in the sense that they aren't focusing on the physical sex–repression the child experiences. If it were only a matter of "education," most social ills would have been eliminated a long time ago because we have a *lot* of education. In fact, the school systems *themselves* tend to reinforce some of the same so–called "traditional values" of patriarchal sex repression that lead to the problems of rape, spouse abuse, and child abuse in the first place.

MR: *You just mentioned the Sudanese people changing their attitudes. It was very hopeful this past year to see the president of Kenya enact a law requiring anyone performing a genital mutilation on a woman that resulted in her death be prosecuted for murder.*

JD: Yes, it was, and it was also quite a switch. When Jomo Kenyatta was first trying to raise opposition against the British back in the 1950's, the only issue he could get people to rally on was that the British were trying to ban circumcision of women. When Kenyatta objected to British interference with this native "custom," the people rallied around him and it was one of the central issues on which the British were driven out.

You see, sub–Saharan Africa is one of the wetter borderland areas where these Saharasian peoples spread into. It has elements of patrism similar to the Saharan regions to the North.

"...(the drought) out West may be related to the cutting down of all the trees in the Pacific Northwest."

MR: *You've described the Pygmy peoples of the rainforests in the Congo as one of the peaceful, matrist cultures that made it into this century. What's their situation today?*

JD: The Pygmies are being badly affected by the increasing deforestation. The rate of tree cutting in the Congo is so great that most of the big forest areas in which they lived are today wiped out or in the process of being wiped out. It's very sad because the Pygmies have made attempts to preserve their forest; they lived in symbiosis with it. But the taller Bantu people who run the government often look at the Pygmy as an "animal," in the most pejorative sense. And they look at trees as something to cut for money, something to cut down to make way for cattle and grassland. The Pygmies are being very badly affected by this and it's so, so sad.

MR: *I'd like to talk about some other aspects of deforestation for a moment, such as its connection to droughts and deserting. I know a major part of your work involves use of Reich's cloudbusting technique for breaking droughts, and that you are attempting to develop a desert regreening project in the American Southwest.*

First of all, what's your opinion of studies released this year that challenged the notion that desertification is increasing worldwide?

JD: The estimate that 27,000 square miles of desert are being created every year comes from the United Nations. However, there's some recent disputes about whether that figure is accurate. It may be that there's a pulsing quality to the deserting so that in one series of years the deserts spread, but in the next series of years they don't spread so much. But taken over the longer term period, there's no question that the deserts are spreading and spreading. There's nothing really to dispute that.

MR: *You just recently returned from doing cloudbusting work in Europe. What is the drought situation in Greece and Southern Europe . . . or for that matter even in England, which experienced very dry conditions this past summer? Is there an encroachment of the Sahara northward?*

JD: What we're seeing in the last five years or so is that the droughts of sub–Saharan Africa have diminished; and the droughts north of Africa in the southern Mediterranean and Europe have increased.

When I was in Greece in the early part of this year, there was no question that the atmosphere over Greece looked like the atmosphere of the Sahara desert. Other researchers who understand Reich's principles of atmospheric health and who have long–term observations over there, say they are shocked when they visit the islands of Greece today, compared with even just ten years ago. They remember then a crisp, blue, sparkling atmosphere, with cool conditions and green vegetation. Now, the atmosphere is a dark gray or brownish haze — what we call dor — and the vegetation is like Tucson, Arizona. It's very hot, dusty and dry. So I think we can say,

clearly there is a short–term climate change of northward spreading deserts into Europe.

We did one series of cloudbusting operations in Greece this April which resulted in solid rain for two to three days over the whole of Greece. A very, very gratifying operation. But we went back again later in July, which is the driest time of the year, and we didn't get any results. So on the one hand, we have the knowledge and the tools to be relatively effective against droughts of even this extent, but on the other, there are limits to what can be done.

MR: *If the desert pushes into Europe and food production drops, if famines begin, could we be looking at a reenactment of the chaotic conditions of 4000 B.C.?*

JD: Well, if that happens you would start to see the same mechanisms at work. You know, it's not often talked about, but part of the pressure on the Soviets the last several decades has been desert–related. The closer you get into Central Asia — the heart of the Soviet agricultural system — the more you're looking at an area that's constantly being subjected to episodes of drought. But I think we're still a ways from that in Europe; I don't see it happening quite so fast there.

But let's take a look closer to home — the western United States. We're now in our fourth year of drought and water supplies are being diminished. A fifth year of drought will have even more severe consequences.

Also, we've seen massive forest fires these past few years. In 1988, over a million acres of U.S. forest burned to the ground. Now it's struck me that under severe drought conditions, you don't need 100 years to see a major change — you can lose a tremendous amount of forest within a few years from fire alone.

MR: *And when the trees fall the rainfall decreases.*

JD: Exactly. You literally begin to change the characteristics of the atmosphere. In part, what is going on out West may be related to the cutting down of all the trees in the Pacific Northwest. These huge, ancient trees are being wiped out, and the forests are converting to pasture type landscape. People need to realize there's a real risk of climate change toward drier conditions if this continues.

MR: *What about your cloudbusting operations out West? Have they stopped? The last time we talked you had had a number of successful drought breaking operations — the Southeast drought in 1986, the Northwest drought of 1988. Hasn't there been any continuing support for this?*

JD: Some, but not enough. For example, a group of farmers in Montana brought us up for cloudbusting work earlier this year, and we were able to bring several weeks of cool, rainy weather into an area which had been quite dry for several years. However, there were only enough funds to keep us working in that area for a short time, and nothing could be

"Big Political Power people are scared witless by renewable energy, because look what changes..."

done to help when drought conditions later reappeared.

We're trying to find a way to do cloudbusting on the West Coast but we haven't found a sponsor yet. So we continue to do what we can, and we've been able to be helpful in certain situations. But it's very clear we need to do more, and that is going to take some serious financial support.

MR: *It's my experience that most of us just don't realize how many alternative approaches there are to many of our most pressing problems. Unless you're directly involved with one of these alternatives, you don't know about them. The suppression of them by "authorities" is so great, it's as if they didn't exist or were just child's play. Makes it very difficult to get serious support.*

This includes not just revolutionary discoveries such as orgone energy, or Zero Point energy, but solar energy — something even the "experts" can't deny exists. You've also done research in solar energy. What's your estimation of its utility presently? Could it supply a significant amount of our energy needs?

JD: Absolutely. But we're doing damn little about it. There's no political leadership in Washington that's really addressing this, and there hasn't been since Jimmy Carter. Before I got involved in studying Reich's work, in studying life energy, I was very involved in solar energy research. I had a small company in Florida. So I *know* it's workable, but . . . well, it's very interesting if you look at the first act of a political leader when they get into office. If we look at when the Ayatollah took over Iran, his first act was to undo the Family Reform Law that the Shah had previously enacted. That Law granted women legal standing in the courts, the right to divorce, the right to children in divorce, and rights to inheritances. This law came about because the Shah's wife was Western and she pushed him for legal rights for women. And the Ayatollah *hated* this and so undid it as his first official act.

Now when Reagan took office, *his* first official act was to fire Dennis Hayes as director of the Solar Energy Research Institute in Golden, Colorado. The reason Reagan fired Hayes was because Hayes had recently published a study on *The Renewable Energy Potential for the United States.* This was an official document of the Research Institute, and Reagan asked Jimmy Carter to tell Dennis Hayes not to publish . . . incoming presidents will ask the incumbent to do these kinds of things, and Reagan said, I don't want that published. But Hayes published it anyway, so Reagan fired him as his first official act because he didn't want the American people to know that they could get *significant* percentages of their energy from renewable sources.

Nothing was done under Reagan, and now Bush, to support the alternatives *we know* work. Where is the national solar water heating program? We should have a national program to install solar water heaters in every American home by the year 2000. That's workable. We could do that and it would give a lot of people employment in jobs that many are already trained for — like plumbing, carpentry, and electrical.

Plus, every solar water heater would help reduce America's energy burden — both consumption as well as pollution — not only for the decade but, essentially, forever.

And where is the national photovoltaic electrical energy program, where the government spends a little money to help these photovoltaic companies tool up for mass production? I mean, the sunlight that falls on the roof of the average home is enough to provide that home with all of its energy needs. All. *100 percent.* That's what the potential is.

But where is the political leadership on this energy issue? Where and how are we currently focused on the energy issue?

MR: *On oil in the Saudi desert.*

JD: Exactly. Now personally, I don't want anybody to go to war and die so that I only have to pay a buck and a quarter for gas. I would be willing to pay $3 for gas or $5 for gas. I'll ride a bike and I'll put a solar water heater on my house. Yet, you go into any business in America today and in the middle of the day they have all the lights on, the doors wide open while the air conditioner is running. Outside of any city at night you see lights *blazing* across the whole night sky. We have homes in America — wealthy people's homes — that consume more energy than some Third World *villages.* But somehow we're supposed to sit back and say, well, in order to preserve this "lifestyle," we have to sacrifice some of our children?

You see, we talked earlier about the sadistic sacrifice of Iranian children sent out to stomp on a land mine. Well, this is *our* sadistic pathology, to take our young people and say — Go sacrifice yourself so I can continue to drive my Cadillac. Go sacrifice yourself so I don't have to worry about running my air conditioner while the door is open. See what I'm getting at? This is *our* pathology.

Actually, if Iraq stays in Kuwait and even provokes an end to the Royal Family in Saudi Arabia, yes, they would have the West at their mercy; but, maybe that's what we need in order to provoke us into a real change in our energy use, our energy supply, and all the attendant problems of pollution.

Here in California, we get a very large percentage of our electrical power from wind generators and solar energy sources. We have geothermal power. We have conservation of energy. And the potentials for development of these kinds of approaches on a national level is *enormous.*

MR: *But the current reality is very local. And certainly big business is not getting involved in it.*

JD: No, and one of the reasons is the Big Money people, the Big Political Power people, are scared witless by renewable energy. Because look what *happens,* what changes, when you go to renewable energy systems. Instead of putting all the money into huge power and oil companies — money which tends to concentrate in fewer and fewer hands and then transmutes into political power — your money goes to support local businesses who employ local people in jobs that are down to earth, that are not exotic, that don't have to import strange technology. It's fairly simple, everyday stuff. And, you subsequently pay *nothing* for "fuel," as it comes from the sun or wind.

It also works to decentralize economic and political power. That's why Big Money continues to try and push nuclear power or coal or oil down the throats of the public. But if you go to renewable energy, you not only solve with one blow the problem of fuel supply and pollution, you also with one blow begin to restore America back to its decentralized, egalitarian roots. And that scares a lot of people in the political establishment.

MR: *So we're looking at the problem of patrism again in highly centralized government or social structures?*

JD: Very much so. Big, hierarchical, centralized political states are always patristic. That's a danger in America. We want to think we can maintain a strong democratic tradition and still have this *huge* government bureaucracy. They're totally incompatible.

We have a real need to get back to a decentralized, local decision making process for the things that affect everyday life. Now it may be that something like the United Nations is necessary, and that the U.S. would play a lead role in that — I'm not saying that wouldn't be the case. I am saying that the more you have different kinds and numbers of institutions that become highly centralized, the more they begin to overlap and interact with each other in a domineering, authoritarian manner. That's dangerous for any democracy . . . and I think it's already taken a significant toll in America, in reducing our freedoms, and eroding democracy.

MR: *An example that comes to mind is the AMA and the FDA making it illegal to treat cancer except by surgery, chemotherapy or radiation, which causes people seeking alternative treatments to either go underground or out of the country.*

JD: Well, we could go down the list, couldn't we? The point is when these different functions become centralized, the tendency is to ignore the more fundamental needs of people. So we see the needs of mothers and babies neglected more and more each year. We see 25 percent of the children born below the poverty line, significant numbers of mothers on welfare, the fathers skipped out. Yet, the huge government bureaucracy is essentially not interested in these social problems, or it helped to create them.

The more centralized the school system, the more unresponsive it is to the wishes of the people and the more authoritarian it is to the children. You're not going to get an A.S. Neill Summerhill type school — a free school — coming out of a system dictated by Washington. Now, you may not get such a school on a local level either, but at least you have a *chance* where the methods are determined by local school boards and average people.

Wasn't it William F. Buckley who said he would rather trust the affairs of American government to 400 people selected at random from the telephone directory, than the current crop? There's a fundamental truth there. I mean, there are times when the people can be very rational — like if you're charged with a crime, the dilemma is whether to put your fate in the hands of single judge, or 12 jurors.

MR: *What's your evaluation of America today on a patrist/ matrist scale?*

JD: Well, America has gotten a bit tighter over the years. After WW II, our military people tended to take over the government. Today, anybody who has a strong dissent from the "official position" on anything will catch flak, and possibly lose their job. You can see this clearly in politics, but also in the news media, corporations, and in academia. Nobody dares to question the guy at the top. It's like he's a god or pharaoh — you don't question. Doesn't matter whether he's right or wrong, whether his opinions are founded in fact or fiction — it just matters that he's at the top and if you disagree, you're at risk.

But on the other hand, we have some laws now that are much more matristic in orientation, even though the conditions of everyday living are more difficult. We've had the Civil Rights movement in all of its expressions, the women's movement, legalization of contraception and abortion, and other positive things which came out of this period. And yet, the power of our patristic Church institutions have not weakened very much. Socially, we are also an emotionally resigned society, where murder, suicide, alcohol and drug abuse, child abuse, wife-beating and degenerative illness are epidemic... these factors are also related to deteriorating economic and social conditions. *Obedience* is still a major emphasis in our schools, families, and workplaces, and yet, there's a strongly energetic, rebellious quality that's present .. .mostly in the subculture of American youth, which because of its strongly sexual nature is widely emulated and copied by young people all around the world. So for America, we exist somewhere in the middle of the patrist/matrist scale. Though it's difficult to see where we will be in 50 or 100 years, it does seem like we're sliding towards increased armoring and patrism in many ways.

MR: *Are there any examples of a strongly patrist society that's softened recently?*

JD: I can't give you an example of a *strongly* patrist society that's softened . . . but let's look at Europe. It seems to be more relaxed now than in the 1930's. Of course they lived through a bloodbath over there and it seems that now there's a great deal more potential for some kinds of freedoms. I think part of that is due to the number of fascists who were killed off or removed from power during WW II, which allowed the more democratic elements to come back. So what we're seeing in the scientific area, for instance — which is, again, less centralized than here — is more freedom to explore, investigate and use new approaches. For example, there is a lot of interest in Reich's work over there, and they use the orgone accumulator openly in clinics. Whereas here, if you were to use the accumulator in a medical context, you could be thrown in jail because the government has such power to regulate scientific activity and medical practice.

It's still too soon to tell, but I think the Soviet Union has softened a bit since Stalin. Certainly under Gorbachev, the Soviet Union has evidenced a softening. The Western European parts of the Soviet Empire evidence more of this softening, of course, than its Central Asian — that is, Saharasian — parts, which are badly afflicted with ethnic violence and other sadistic acts.

> **"America couldn't _stand_ the expansion and pleasure of all the hopeful possibilities. If Saddam Hussein wasn't there in the Gulf to be the bastard, we would have to find someone else."**

You see, patrism in a political system comes from people who have been raised with a great deal of suppression, and they duplicate politically what they experience in the family during childhood. Basically, freedom frightens them. As Reich pointed out, the emotional turmoil that's created by the constant trauma, and the sex repression — the preoccupation with the unsatisfied longing — creates such anxiety on a daily level in people's lives that they just can't think or behave rationally. They are emotionally tied up and bound. They can't really work through their needs or their daily social difficulties with any degree of effectiveness. So they get lost in some mystical philosophy or are constantly falling prey to one or another plaguey politician who comes along and promises to take away their responsibility and just _give_ them the solution. You know, _"Trust me."_

MR: _Yes, it was very alarming to watch national attention to our major social problems with the environment, the S&L bailout, the budget deficit, disappear when our leaders decided we just had to focus on the real problem over in the Arabian desert._

JD: In 1970 we had the first Earth Day. There was this enormous outpouring of positive feeling, of warm feeling, that was a very hopeful expression of people who were going to band together and solve their problems. Then what happened? A short time later we were invading and bombing Cambodia. And everybody forgot about Earth Day.

The _same_ thing happened this year. Earth Day 1990 comes along and there's all this legislation for positive social change: people banding together to deal with social problems and environmental problems and very creative, lively people talking and working together. Then a few months later our military moved into Arabia.

Now there's no national discussion about the environment or social conditions _except_ in the context of "Sacrifice for the National Emergency." We're told we must now drill offshore oil in California because it's a national emergency. Forget about an effective Clean Air Act and emission control laws, because we need to rally around the flag.

MR: _Which gives rise to a point of view, which many hold, that this kind of response is a manipulation, a conspiracy. Trouble at home? Go to war somewhere._

JD: That's an appealing explanation and I think it arises because people can't see any other alternative to explain why at almost every crossroads the human race has moved in a life–negative direction. This war situation does seem to be totally created and planned, certainly by stupidity and blunder, if not by design. . . But I think there's another explanation that is much more functional, having to do with the way people live.

People around the world, Americans included, cannot tolerate pleasure; they can't tolerate responsibility; and they can't tolerate too much freedom. So here we have Earth Day

1990 and it was the first mass expression of all the hopeful possibilities that people could have . . . I mean, I've been doing environmental work for decades and I've never seen anything so expansive and publicly supported. Every night on the news there was coverage of the problems and solutions and ideas that had been gathering dust on the shelves for a decade or more, ideas that had been suppressed or laughed at by the "experts," but were finally getting some open, rational discussion. It was this huge outpouring of hope, of very positive concern for life. And now it's all gone. Just swept away by the new war.

Because America couldn't _stand_ the expansion and pleasure of all the hopeful possibilities. If Saddam Hussein wasn't there in the Gulf to be the bastard, we would have to find someone else. Part of what leads me to believe this is what in fact happened is that so _few_ people who were on the Earth Day bandwagon two months ago, today seem the _slightest_ bit concerned about it. It was like they could just touch it briefly, but then it had to be put aside like a child's toy — the more "serious" matters of the adult world are, after all, war.

I can give you an example that kind of sums it up. I was a speaker at the Seattle Medicine Wheel Gathering in September, and during the gathering I set up a little table with some books on it. These books were authored by some of the most innovative scientists and life–affirming people on the planet, books with practical solutions to pressing problems. Toward the end of the gathering, two guys came up to the table — local fellows who wandered in to see what was going on — and they're wearing their feed caps with advertisements for popular herbicides on them, and they look rather worn by decades of hard work with very little pleasure in their lives. So they stop at the table and look down at the books for about five minutes. They never touch them, they never say a word — just look in total silence for five minutes. Then one of them, out of the blue, says to the other, "I don't believe a word of it." The other one says, "Nope. Me neither." Then they both turned and walked away!

That summarizes, in my view, the whole Persian Gulf situation. Let's get back to things we can believe in, like M–1 tanks and fighting Saddam Hussein and defending our "lifestyle." All this orgone energy and solar energy stuff, which demands active social participation, hard work and personal change, well, this is just a little too much!

MR: _I have to say it's been almost too much to talk about this problem of patrism. It's so ingrained and widespread; it's not something that is going to change tomorrow. How do you stay focused on this without being overwhelmed? How do you present solutions to it in our society, or elsewhere?_

JD: Well, I think you have to keep an eye on what's going on around the world, but you can only start locally. It's getting to be a cliche, but we must _think globally but act locally._ Start with yourself, start with your family, workplace, local school, hospital, and community. Find like–minded people and work individually or collectively on those problems which most immediately confront you, within your own sphere of influence.

There are a lot of routine but fundamental things that support patrism, like the abusive neglect of infants and children, the misery of children in schools that are compulsive and demanding of absolute obedience to angry teachers. There's the constant denigration and putting down of natural, healthy sexuality. Right now, there's this big lie that young heterosexual teenagers who are not bisexual and don't shoot drugs are at risk for AIDS. You won't find this disputed very much, but there is no scientific basis to this particular allegation. It comes from people who hate adolescent sexuality, and who are misrepresenting the scientific data.

So we have to literally change the way the young ones are being raised, change the kind of world they're having to grow up to, if we want the next crop of adults to be any different. And that's not an easy thing to do. It's a very demanding one.

But before we get all pessimistic about it, there is the finding of Reich's sex—economic discoveries which were vindicated by the Saharasian thesis. We don't *have* to be this way. It's not innate for *Homo sapiens* to be violent and sadistic and punitive. And the more people who know about this, the more potentials for change exist that did not exist beforehand.

In the last years I've lectured on these subjects in Japan, Greece, Germany and across the USA, and I see a revolution brewing in the sciences that has not yet started to surface. And I see more and more people, average people, willing to stand up for what they know is true. They will not tolerate their children being abused anymore. They will not tolerate their sexuality being put down or interpreted in a religious context as per the taboos of Christianity or Judaism or Islam.

So these are small things, but you have to start with them, with yourself and with your family, because that's where this all got started — and that's where the hope for any real change is. Remember, the more decentralized one's efforts are, the more potential for effective change exists. We just *have* to get away from this political plague of thinking that "if we can only elect so and so, *they* will solve the problem". That kind of thinking *is* the problem. We must begin to take personal responsibility by direct action within our personal spheres of influence, at home, workplace, and community.

Renewable Energy: Solar, Wind, Biomass, Conservation Equipment Suppliers (also check the telephone directory)
Active Technology, 4808 MacArthur Blvd., Oakland, CA 94619; 415/ 482-8025
Alternative Energy Engineering, PO Box 339, Redway, CA 95560; Catalog & Design Guide $3; 707/ 923-2277
Energy Conservation Services, 4110 SW 34th St., #15, Gainesville, FL 32608; 904/ 373-3220
Energy Store, PO Box 3507, Santa Cruz, CA 95063; Free Catalog, 800/ 288-1938
Integral Energy Systems, 105 Argall Way, Nevada City, CA 95959; Catalog & Design Manual $4; 800/ 735-6790
Jade Mountain, PO Box 4616, Boulder, CO 80306; Catalog $2; 800/ 442-1972
National Appropriate Technology Assistance Service, PO Box 2525, Butte, MT 59702;
        Free help and information, 800/ 428-2525; in Montana 800/ 428-1718
National Center for Appropriate Technology, PO Box 3657, Fayetteville, AR 72702.
Photocomm, 930 Idaho-Maryland Rd., Grass Valley, CA 95945; Catalog & Design Guide $3; 800/544-6466
Real Goods Trading Co., 966 Mazzoni St., Ukiah, CA 95482; Free Catalog; 800/ 762-7325
Resource Conservation Technology, 2633 N. Calvert St., Baltimore, MD 21218; Free Catalog; 301/ 366-1146
Solar Energy Industries Association, 777 N. Capitol St., NE, #805, Washington, DC 20002;
        Free help and information, 202/ 408-0660
Sunelco Sun Electric Company, PO Box 1499, Hamilton, MT 59840; Catalog & Planning Guide $4; 800/ 338-6844

Healthy Home & Ecological Products
Choice Healthy Home Products, 2290 S. Lipan, Denver, CO 80223; Free Catalog, 303/ 936-1181
Earth Care Paper, PO Box 14140, Dept. 617, Madison, WI 53714; Free Catalog, 608/ 277-2900
Ecco Bella, 6 Provost Sq., G-1090, #602, Caldwell, NJ 07006; Catalog $1, 800/ 888-5320
Eco-Choice, PO Box 281, Montvale, NJ 07645; Free Catalog, 800/ 535-6304
Eco-Source, 9051 Mill Station Rd., Bldg.E, Sebastopol, CA 95472; Free Catalog, 707/ 829-7957
Everybody, 1738 Pearl St., Boulder, CO 80302; Free Catalog, 800/748-5675
Livos Plant Chemistry, 1365 Rufina Circle, Santa Fe, NM 87501; Free Catalog, 505/ 438-3448
Natural Energy Works, PO Box 864, El Cerrito, CA 94530; Catalog $1, 415/ 526-5978
Needs, 527 Charles Ave., Syracuse, NY 13209; Free Catalog, 800/ 634-1380
Safe Environments, 207-B 16th St., Pacific Grove, CA 93950; Free Catalog, 408/ 372-8626
Seventh Generation, Dept MO5035, Calchester, VT 05446; Free Catalog, 800/ 441-2538
Urban Forest Packaging Products, 1222 W. Spring St., Brownstown, IN 47220; 812/ 358-3150

OROP ARIZONA 1989:
A Cloudbusting Experiment to Bring Rains in the Desert Southwest *

James DeMeo, Ph.D.**

I. Preliminary Report on Five Cloudbusting Expeditions into the Desert Southwest, May-September, 1989

Background

Several years ago, the increasing pace of desert expansion, drought, and general atmospheric chaos prompted me to propose the development of a Desert Research Center in the arid portion of the American Southwest, with the goal of demonstrating a long-term desert greening effect using the Reich cloudbuster. During the years 1952-1956, Wilhelm Reich investigated basic energetic-atmospheric functions related to drought and desert formation, and demonstrated clear desert greening effects during his cloudbuster operations near Tucson, Arizona (1, 2). His early experiments provided the basic guiding principles for this more recent desert research (3,4). As a preliminary step in the development of a long-term desert greening project in the American Southwest, it was important to determine how much influence separate short-term cloudbusting operations could have on the desert atmosphere in an exceedingly dry area. In the summer of 1989, this experimental program of desert cloudbusting was conducted. Five seperate field experiments were planned and announced in advance, undertaken by teams of workers using portable equipment, and recorded in extensive field observations. National Weather Service (NWS) data were used later to analyze the geographical distribution and intensity of atmospheric effects.

The operations proceeded as follows: A transportable cloudbuster was driven to a very dry part of the American Southwest and set up for periods lasting a few days to a maximum of one week. After each operation, the research team† with its

---

† A number of individuals physically assisted with these desert operations, taking all the risks associated with cloudbusting in a hostile desert environment. My thanks to Donald Bill, Argyro Collins, Dennis Collins, Jurgen Fisher, Alberto Foglia, Beate Friehold, Sybilla Heck, Stephan Muschenich, Vittorio Nicola, and John Trettin for their help and participation. Special thanks to Theirrie Cook, who also assisted in these operations, and to Richard Blasband, Robert Nunley, and Shafia Lave, who assisted with additional cloudbusters in the San Francisco Bay area and in Kansas on several occasions. Funding for these field expeditions was provided primarily by the American College of Orgonomy.

---

* A previous version of this article appeared in the *Journal of Orgonomy*, 23(2): 271-272, November 1989, 24(1):111-124, May 1990, and 24(2):252-258, November 1990.

** Director of Research, Orgone Biophysical Research Lab, demeo@orgonelab.org

cloudbuster left the area, completely suspending operations until the next expedition aproximately 3-1/2 weeks later. This allowed natural weather conditions to reestablish themselves. One experiment with this quick, mobile cloudbusting was previously undertaken in the desert Southwest in 1988, with much success (3). A site along the Colorado River, about 100 miles north of Mexico on the California/Arizona border was chosen for the experiments because of its easy access to abundant water and central location within one of the driest parts of North America. A second cloudbuster located in the San Francisco Bay area, operated by Richard Blasband, was used on two separate occasions to assist these desert operations, with a third, very simple cloudbuster located in Kansas for similar purposes. The Kansas portions of the operations were undertaken to enhance the downstream rainfall effects of any energetic potentials propagating toward the central U.S. out of the desert Southwest.

The five operations took place at approximately monthly intervals, beginning the first week of May, 1989. A letter detailing the dates and purposes of the desert cloudbusting expedition was sent to the National Oceanographic and Atmospheric Administration offices in Washington, DC, and telegrams for documentation were dispatched to those offices as well just prior to the onset of each individual operation.

Operation #1: May 8-10, 1989

This first operation began at a time when daytime temperatures were peaking at around 100-105°F. There was a forecast possibility of slightly lower temperatures, but certainly no rains or significant clouds were expected. (May and June are the driest months in the desert Colorado River basin region.) Plate 1 shows the thick atmospheric haze -- what Reich called DOR (for deadly atmospheric orgone) -- at the start of this particular desert cloudbusting operation. Reich was the first natural scientist to observe that such atmospheric haze possesses an energetic component which acts to block the development of clouds and rain.

After one day of operations, thunderstorms blossomed to the south of the cloudbuster, over northwest Mexico. After three days of hard work under the hot sun, streams of energy and moisture began to enter the dryland Colorado River basin area from both the west and the south. The westerly energy stream was composed of what classical meteorology calls the "polar jet"; after observing this shifting weather pattern, one puzzled TV meteorologist said, "It just can't do that" -- there was no "frontal boundary" or air mass with a sufficient temperature differential to "push" the jet stream southward into Arizona. And yet, the energy stream danced southward many thousands of miles from its original location near Alaska. The second, southerly energy stream was composed of the "sub-tropical jet," and it moved north into our area from the Gulf of California.

These streams of energy and moisture appeared as "fronts" on the weather map and were seen clearly as a long line of thin, wispy, high-altitude cirrus clouds standing out against the desert sky (Plates 2 and 3). When the energy streams approached, hazy, stagnant DOR conditions in the lower atmosphere diminished, even though the clouds demarcating the energy streams were many thousands of feet in elevation. When the energy streams moved away, stagnanant DOR conditions worsened -- a clear demonstration these wispy clouds represented much more than just high-altitude "jet streams" of mechanically moving air.

As these two energy streams moved inland, one from the northwest, one from the south, they gradually approached and superimposed far to the north in Montana, causing gusty winds and heavy rainstorms. Even though this superimposition and sequestration was happening far away, the spiraling motion "sucked" much of the DOR-laden weather out of the desert basin, toward the center part of the storm system. With this effect, the cloudbusting operation was terminated late in the evening of May 10. The next morning saw an absolutely crisp and sparkling sky, completely free of the previous stagnent DOR haze, as seen in Plate 4. Temperatures that day only reached about 70°F! Crisp, cool, sparkling conditions persisted over the Arizona/New Mexico region for weeks, bringing widespread rains in the surrounding mountain areas. This rain also spread northeast into the Great Plains and Midwest.

The precipitation resulting from this first operation is documented in Figure 1, which shows the NWS daily precipitation averages for April and May of 1989 from 424 different weather stations in the southern half of both California and Nevada, and from the entire state of Arizona (5). As shown in Figure 1, increased rainfall was observed across this very large and mostly arid geographical basin and range territory immediately following the cloudbusting operation of May 8-10. A previous pulse of moisture on April 21-27 appears in Figure 1, the result of a coastal storm that moved into California but lacked enough energy to move inland over the high mountains (4000'-8000' elevation). The cloudbusting operation of May 8-10, however, was followed not only by increases in moisture and rains in coastal and southern California but also in moisture that moved far inland and upslope, into the basin and plateau regions of Nevada and Arizona where significant rainfall developed. The lag time between the onset of cloudbusting operations and the time of first rainfall was shortest (around one day) for the wetter coastal regions of California, and longest (around four days) in the arid desert basin regions. These data do not reflect thunderstorms blossoming in Mexico only one day after operations.

This kind of atmospheric response to cloudbusting during May, one of the driest months in the Colorado River Basin region,

was remarkable, most gratifying, and in complete agreement with the results of previous desert cloudbusting operations undertaken in August of 1988 (3).

Operation #2: June 5-8, 1989

The second desert cloudbusting operation proceeded at a time of increased aridity and temperature. Widespread agricultural burning hampered visibility at the cloudbusting site and appeared to "choke up" and reduce the responsiveness of the atmosphere to the cloudbuster. In addition, a French nuclear bomb was tested in the western Pacific. The atmosphere was sluggish, immobilized, and had a "resigned" quality. After two frustrating days of work with the cloudbuster, the major observable change was a slight reduction in peak air temperatures. An assisting coordinated draw from a second cloudbuster located in the San Francisco Bay area was then begun, and this provided the additional energetic push needed to get things moving. Within a short time, a low-pressure storm system swept southeast from the San Francisco Bay area toward our location on the Colorado River, on a line directly between the two cloudbusters. This storm system dropped abundant precipitation on the southern Sierra Nevada Mountains, an unlikely occurrence for the season. Plate 5 shows the first outbreak of

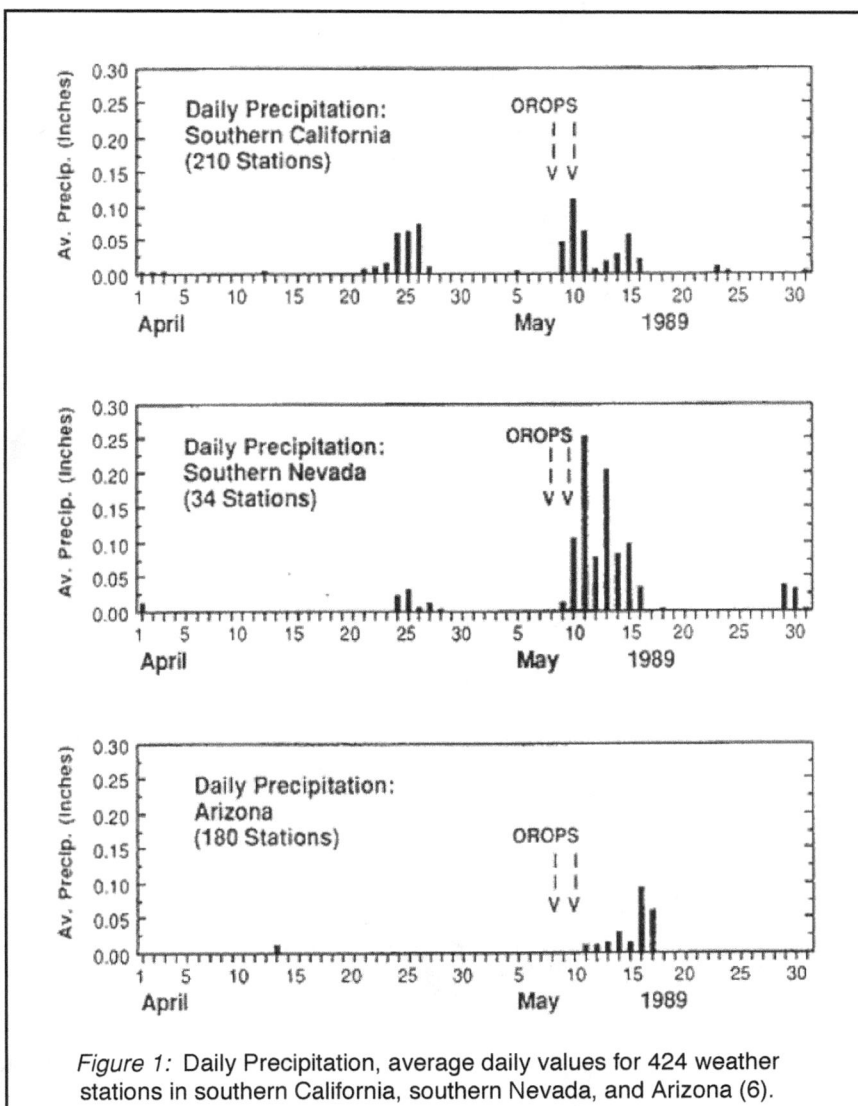

Figure 1: Daily Precipitation, average daily values for 424 weather stations in southern California, southern Nevada, and Arizona (6).

Plate 1: A harsh desert landscape prior to cloudbusting operations, near Blythe, California. Note the heavy DOR haze in the atmosphere, which obscures the nearby mountains. Shade temperatures at the time of this photo (11:00 AM) were around 99°F, with a 15% relative humidity, typical oppressive summertime conditions for the area.

Plate 2: Cloudbuster Icarus at work along the banks of the Colorado River, under broiling, DOR haze conditions.

Plate 3: Passage of the Galactic orgone energy stream (southerly "jet") overhead. Note the clearing of low-level DOR haze conditions with the passage of this energy stream, one of several indications that it is composed of more than just mechanically-driven, high altitude winds.

Plate 4: A moving, pulsing atmosphere in the desert, also close to Blythe, California, after completion of three days of cloudbusting operations in early May, one of the driest times of the year. Note the greater transparency of the atmosphere, and the developing cloud cover. Shade temperatures at the time of this photo (11:00 AM) were around 67°F, with a 50% relative humidity, very unusual crisp and clear conditions.

these mountain thunderstorms. Operations were terminated shortly after the onset of these rains, but clouds and showers continued to blossom in the same remote, dryland area of southern California over the following days, pushing even farther inland, into southern Nevada and Arizona.

Operation #3: July 3-5, 1989

This operation took place at a critical time. Two nuclear bombs were detonated underground in Nevada the previous week. Also, two of the three Palo Verde nuclear power plants, near Phoenix, were being loaded with nuclear fuel, while the third had a recent emergency shut-down. Nuclear refueling and emergency shut-down procedures are well-known as "dirty" procedures, releasing relatively large quantities of radioactivity into the local atmosphere. These events produced a severe *oranur* reaction in the Southwestern U.S., an overexpanded, overexcited, highly charged atmospheric energetic condition with virtually no cloud cover or rains (6). An unusual clockwise-rotating low pressure system dominated the area. In addition, this early July cloudbusting operation was scheduled to start just before the natural annual transition from the driest to the wettest period of the desert Southwest, the so-called Arizona "monsoon" season which is characterized by thunderstorms. We predicted our operations would result in an earlier onset of the monsoon with a greater than normal quantity of rain.

Operations began on a day with shade temperatures over 110°F, and relative humidities at or less than 15%. After one day of operations, some very large thunderstorms appeared to the south of the cloudbuster over northwest Mexico. A very rapid, high-altitude streaming movement from south to north also developed but failed to trigger any storms, until it reached into western Nebraska where thunderstorms blos-

somed. Conditions worsened as the draw continued. By the second day of operations, we recorded a shocking 123°F shade temperature, the hottest in the entire U.S. for that day. Oranur conditions intensified, making work near the cloudbuster almost unendurable. Satellite images continued to show no storms or rains in the vicinity, except in Mexico and southeast Arizona.

Cloudbusting operations were terminated after the third day of widespread high temperature, high-pressure conditions, which threatened to spread northeast, and trigger drought across the Great Plains and Midwest. However, a final short cloudbusting draw was made from both cloudbusters, on the Colorado River and in the San Francisco Bay area, to ensure a good westerly flow to push this dome of hot air eastward. This proved successful, and over the following week the dome of hot air moved eastward across the entire U.S. and the Atlantic Ocean, eventually appearing in Western Europe, where it "stuck" for a period before finally dissipating. As the hot, agitated air and energy moved eastward out of the desert Southwest region, clouds and rains filled in behind it, expanding northward from Mexico and southern Arizona, and increasing in tempo over subsequent weeks. The monsoon season had arrived.

Operation #4: July 31-August 3, 1989

This operation began at a time when the southerly flow of moisture into the desert Southwest had already been underway for several weeks, with widespread thunderstorms and rainshowers occurring every other day. The goal this time was to increase rainfall to above normal amounts. DOR levels were still markedly elevated in some regions despite the rains, and several forecasters indicated the really heavy "monsoon" rains had yet to appear. An easterly flow of air from out over the Pacific Ocean was also reported to be hindering rains in the area,

Plate 5: Thunderstorms develop over the southern Sierra Nevada Mountains following the cloudbusting operation of early June.

putting a "cap on instability"; this was contrary to expectation, given the air mass' origination over the moist ocean. More was learned about the stagnant DOR nature of some of these Pacific air currents and will be mentioned shortly.

After one day of operations, a large complex of thunderstorms developed just to the east of our site. The National Weather Service reported significant showers in a few surrounding areas, but no widespread rains were observed. In the eastern Pacific Ocean, Hurricane Gill began churning northward along the coast of Baja, and the counter-clockwise spiraling winds around this storm spread moisture inland up the Colorado River Valley. By the second day of operations, relative humidities were at 69%, although widespread rains failed to develop from this moisture. By the third day, Hurricane Gill completely fizzled out, and all clouds evaporated across the region.

We were very puzzled and frustrated by this turn of events, and acutely felt the need to better understand the nature of these Pacific air currents, which could sap the strength from approaching hurricanes and squelch rains far inland, even when high humidities and temperatures prevailed. This cloudbusting operation was terminated, and no assisting operations took place this time from the San Francisco Bay area. Moisture continued to stream into the desert Great Basin, however, and three days later, an incredible series of events occurred, evidencing a widespread break-up of the desert armor.

1.   A "mild" earthquake, magnitude 5.2 on the Richter scale, and some highly unusual "earthquake weather" occurred north of the San Francisco Bay area. Attention was drawn to the connection between the earthquake and the strange weather primarily because a local TV meteorologist spent considerable time denying the existence of such a connection. On moving satellite images, a unusual pattern of thunderstorms could be seen radiating outward from the earthquake zone, like spokes on a wheel, to the north, east, and south. Heavy rains and strong winds were produced from these storms in southern Oregon, Nevada, California, and Arizona. Rains, in fact, persisted for days in areas where, as one TV weatherman remarked, "It ought not to be raining."

2.   Yuma, Arizona, received two years' rainfall in one week, and other Arizona and southern California stations likewise experienced heavy rains, and some flooding of low-lying desert topography was reported. More rains broke out across the entire Great Basin region in subsequent days and continued through the month. By the end of August, the region near to the cloudbusting site had received a whopping 500% of its normal summertime rainfall, as seen in Figure 2.

Operation #5: August 28-September 1, 1989

This operation began under hazy, DOR-infested, cloud-free conditions. After one day of operations with the cloudbuster, partly cloudy conditions developed across the Colorado River Basin area, with scattered showers across Arizona, primarily in the mountains. A large complex of thunderstorms developed in Mexico, south of Yuma, Arizona, and also near the cloudbuster. Given this rather immediate response to the operations, it was decided to initiate a second draw from the Pacific coastal region just north of Los Angeles, hopefully to bring Pacific moisture more directly to the inland desert regions. The cloudbuster was moved to Moro Bay, on the coast, and grounded in the Pacific Ocean, drawing westward directly from a large fog bank. After one-half day of work at this site, the operation was terminated, and we returned home to observe subsequent events.

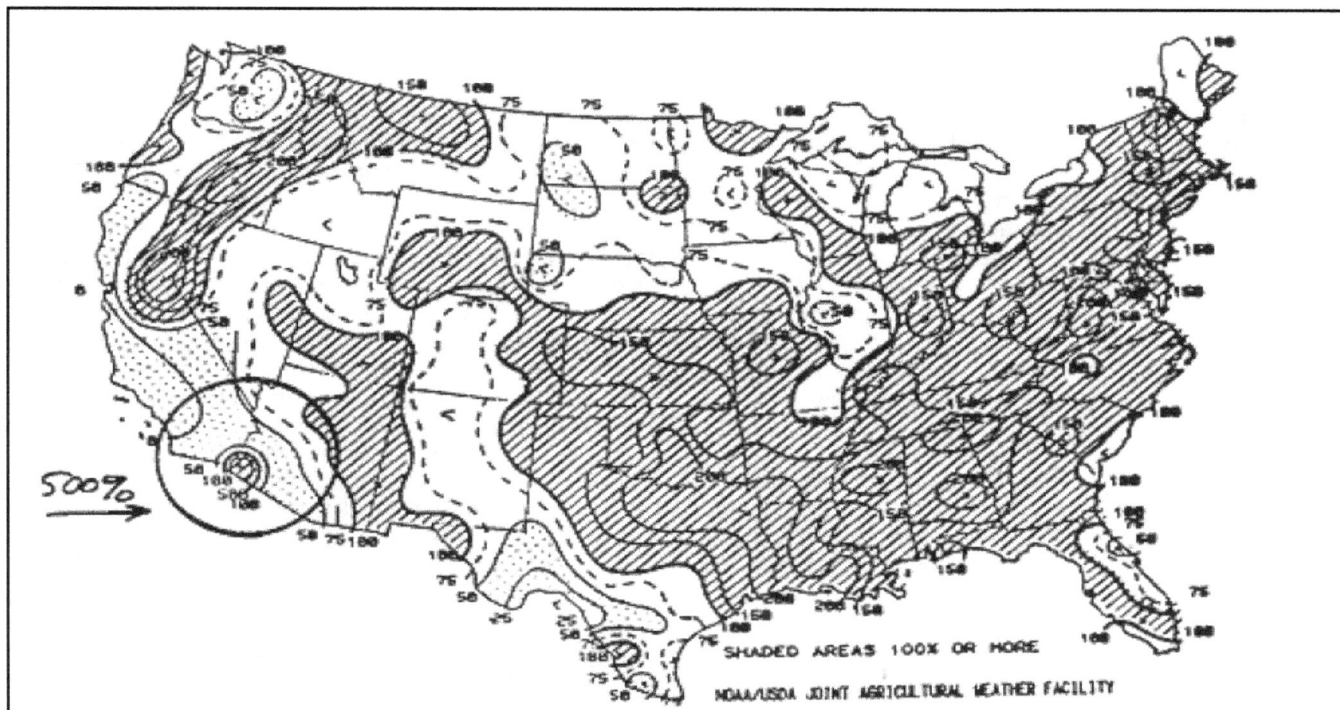

Figure 2: Map of Percentage of Normal Precipitation for the U.S., June through August, 1989. Shaded areas are over 100% to 250% of normal rains. Note the region of 500% of normal rainfall in the desert Southwest. The cloudbuster was located at the center of the large circle, close to that region of greatly increased rainfall.

The results were delayed, but significant. A tropical depression in the southern Pacific, about one thousand miles away, slowly moved northward off the coast of Mexico. Over a period of several days, this depression developed into Tropical Storm Octave and later became Hurricane Octave for a short period. This storm continued northward off the coast of Baja, but soon weakened and spread rains over a large area. There was enough momentum, however, to bring these rains ashore in southern California. This entire slow process took about 10 days. On September 15-17, highly unusual, unseasonal rainfalls moved inland across southern California. This was the first significant precipitation in two years for some parts of southern California. The atmosphere was refreshed and cleaned of its DOR-infested, hazy characteristics. The moisture was welcomed by firefighters and foresters, who were previously sounding an alarm about the dry conditions. A second major pulse of moisture occurred in the same area in late September.

Summary

With the exception of the first cloudbusting operation, NWS weather data have not been reviewed to provide "official" documentation for the field observations described here. Based on past experience, there is good reason to believe they will be fully corroborated and strengthened by such a review and data analysis (NOTE: These data are now presented in the report which follows on page 21). Certainly, these desert cloudbusting experiments have provided a wealth of observations, plus many new and important points worthy of future study and research:

1.   Preliminary indications suggest, as indicated in Figure 2, some areas of the desert southwest (southern California, southern Nevada, and Arizona) were unusually wet, receiving up to 500% of normal rains; other parts of the same region remained quite dry, however, receiving less than half their normal precipitation. This Figure suggests that DOR-laden desert air was sequestered in some areas but not in others. (Note: Figure 2 does not include data on the rains following the fifth cloudbusting operation, which was also followed by at least two major pulses of rainfall in southern California.)†

2.   Fire danger in the Western states dramatically lessened in 1989, particularly after rains in August and September when the major breakthrough occurred. For example, by September 15, 1988 (a record dry year in the U.S.), forest fires claimed nearly 4 1 million acres in the Western states, with a total of over 5 million acres burned for the year. As of September 15, 1989, following the series of desert cloudbusting operations, only 1.4 million acres were lost, with a very low fire danger for most parts of the U.S. (7).

3.   Stagnant, DOR-infested air was repeatedly observed moving into coastal southern California and the Southwestern desert Great Basin from off the Pacific Ocean itself. On several occasions, measurements of cool, low-humidity winds blowing in from the open ocean were made right on the Pacific shore. While normally attributed to "cool ocean currents," these dry ocean winds appear to involve other energetic factors. They were contaminated, for example, with a thick DOR haze which at times completely obscured the horizon, similar to the DOR

haze in the heart of the desert. Plate 6 shows DOR moving inland from the open ocean as seen from an altitude of several hundred feet. Usually identified as a "cooled marine layer" or "dry fog" with a characteristic temperature inversion, the DOR "inverted" air was literally followed as it blew inland up over mountain passes, into the dry inland valleys of California and Arizona. When such hazy DOR was first observed at our desert field station along the Colorado River, it was thought to be Los Angeles DOR air pollution; another time, "only" smoke from agricultural fires; then, DOR blown out to sea from the Mexican deserts, returning inland to the southern California coast. In the final analysis, however, this DOR probably originates far out over the Pacific Ocean, possibly from the deserts of Asia, as discussed in a prior paper on "Desert Expansion and Drought" (4).

4.   While the exact nature of the 1987-1988 drought crisis has yet to be understood, Figure 3 suggests there are at least two cyclical components. The figure first identifies a regular summertime increase in the area of severe or extreme drought within the U.S. This cyclical aspect is a result of the "normal," "dry season" climate affecting the California coast each year and appears to be related to the dry Pacific Ocean DOR layer, discussed above. However, in the summer of 1987, this "seasonal" component expanded from a yearly maximum of 10% to over 35% of the entire U.S. being affected by severe or extreme drought conditions. The reasons for the enormous expansion of drought during the late summer of 1987 remains a mystery. Part of the answer, however, may lie in the movement of a greater than "normal" quantity of DOR from its Pacific Ocean source into the U.S. This span of dry, DOR air, moving east-northeast from the Pacific Ocean inland across the Rocky Mountains into the Central U.S., can often be observed as a huge swath of cloud-free atmosphere on satellite images — one striking example of which was previously published in the *Journal of Orgonomy* (3:107).

5.   The desert Southwest exerts a controlling influence over rainfall in those wetter regions immediately adjacent to it. When DOR builds up and intensifies within the desert basin regions and "spills over" into adjacent wetter regions, droughts, heat waves, and forest fires are triggered. On the other hand, as rains break out within the desert region itself, the DOR desert air mass is reduced, and rains subsequently appear and are more easily maintained in those wetter borderlands. In this manner, the summer 1989 cloudbusting operations in the desert Southwest appear to have been helpful, perhaps crucial, in eliminating drought conditions in parts of the Great Plains and Midwest "downstream" from the Southwestern deserts. As seen in Figure 2 above, for the summer months of June, July, and August of 1989, most parts of the U.S. east of the Rocky Mountains experienced rainfall totals of over 100%, with many areas exceeding 200% of normal precipitation. The Great Plains and Midwest have clearly emerged from their prior drought condition, and rainfall totals in most of those farm belt states have increased above the deficit levels of prior drought years. Figure 4 emphasizes the unusual nature of the Summer 1989 rainfalls, which increased in magnitude and scope to levels not generally occurring in the U.S. since January of 1988. Indeed, the problem in some areas of the U.S. during the summer and fall of 1989 was too much rain. Although not an optimal situation, this was clearly better than the catastrophic drought and fire conditions prevailing in 1988.

---

† Recently obtained NWS rainfall data indicate the coastal and interior parts of central California received over 500% of its normal September rainfall (*Weekly Weather and Crop Bulletin*, October 11, 1989, pg. 12).

6.   Observations made during these experiments support the hypothesis relating earthquakes and weather. In particular, an earthquake of only moderate magnitude was followed by the geographically related genesis and movement of storms within a previously "stuck," immobilized atmosphere.

7.   The relationship between underground nuclear bomb testing and atmospheric phenomena needs further research and study. Soaring summertime temperatures were observed in the open desert shortly after a series of such tests. Additional systematic observations suggest a tendency for rainfall-diverting high pressures to build in the Pacific Ocean off the California coast following underground testing in Nevada. Similarly, an increasing buildup of DOR-laden air, followed by drought or severe storms, has been observed in the Great Plains also following testing in Nevada. These observations are in keeping with the recently published findings of Kato (8) and Whiteford (9), which can only be explained on the basis of a nonmechanical, energetic process.

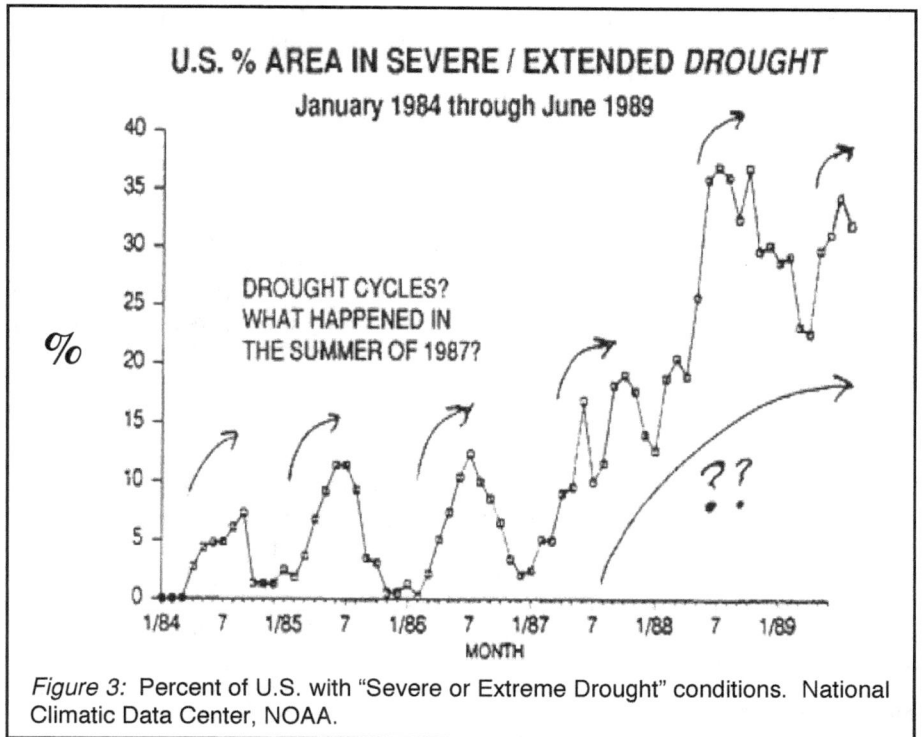

Figure 3: Percent of U.S. with "Severe or Extreme Drought" conditions. National Climatic Data Center, NOAA.

All the above observations and findings reinforce strongly the author's feeling that a more coherent and focused study of the atmosphere from an orgone-energetic viewpoint is vitally necessary. The connections between desert and drought and the global nature of various rain-inhibiting "inversion" phenomena need rigorous study (4). Careful documentation of the atmospheric effects of underground nuclear testing and practical solutions to the problem of desert expansion are

Plate 6: The Pacific DOR haze layer, mechanistically termed the "marine inversion" or "dry fog" layer. This low-humidity air is moving onshore from a dry zone over the eastern Pacific Ocean, into a largely coastal dryland region; it eventually merges with and feeds the DOR haze layer of the inland deserts.

absolute essentials for our collective survival. This cannot be left to mechanistic meteorologists, who generally deny any possible relationships, or that any problem exists.

REFERENCES

1. Reich, W.: "OROP Desert," *Cosmic Orgone Engineering*, VI(1-4):1-140, 1954.
2. Reich, W.: *Contact with Space, Oranur Second Report: Orop Desert Ea*, NY, Core Pilot Press, 1957.
3. DeMeo, J.: "CORE Progress Report #20: Breaking the Drought Barriers in the Southwest and Northwest U.S.," *Journal of Orgonomy*, 23(1):97-125, May, 1989.
4. DeMeo, J.: "Desert Expansion and Drought: Environmental Crisis," *Journal of Orgonomy*, 23(1):15-26, May, 1989.
5. *Climatological Data*, April and May 1989, for California, Nevada, and Arizona, USDOC, NOAA, Washington, DC.
6. Reich, W.: *The Oranur Experiment, First Report (1947-1952)*. Rangeley, Maine: Wilhelm Reich Foundation, 1951. Partially reprinted in *Selected Writings*, Farrar, Straus & Giroux, New York, 1960.
7. "Daily Situation Reports," mid-September, 1989, Boise Interagency Fire Center, Boise, Idaho.
8. Kato, Y.: "Recent Abnormal Phenomena on Earth and Atomic Power Tests," *Pulse of the Planet*, 1(1):4-9, Spring, 1989.
9. Whiteford, G.: "Earthquakes and Nuclear Testing: Dangerous Patterns and Trends," *Pulse of the Planet*, 1(2):10-21, Fall, 1989.

II. Data Analysis of Five 1989 Desert Cloudbusting Experiments

National Weather Service (NWS) precipitation data were obtained from a total of 424 different precipitation measuring gauges located in Southern California, Southern Nevada, and Arizona (1). Figure 5 shows the areas from which the new precipitation data were gathered, and the primary location of the cloudbuster during the experiments. The NWS data presented here indicate a significant increase in precipitation within the large dryland basin region following cloudbusting operations. It is recalled from the preceding report that one region, immediately adjacent to the site where the operating cloudbuster was located and about 100 miles in diameter, received up to 500% of the normal summer rainfall (see Figure 2.) The data presented here indicate that precipitation amounts following the cloudbusting operations generally increased for a much larger territory within the arid desert basin. This increased rainfall was not uniformly distributed, however, but concentrated according to the wind and cloud characteristics which developed after the operations were under way.

The first of the five 1989 cloudbusting operations, for example, took place in early May and was characterized by general west-to-east energy streaming ("jet stream") into the area and the inland movement of moist Pacific winds with frontal activity. These westerly winds brought moisture and rainfall first to the mountain areas of Southern California, including the Los Angeles basin region, and second to the arid Colorado River Valley and highlands of Arizona and Southern Nevada. Figure 6, which gives average daily rainfall data from 210 different NWS gauges in Southern California, shows this relationship. The first cloudbusting operation of early May, characterized by westerly winds, was accompanied and followed by a clear peak in southern California rains; the other four operations, which did not have as strong westerly winds, were not. Instead, the predominant energy streams and moist winds during and immediately following the last four 1989 cloudbusting operations of early June, July, August, and September were characterized by a generally south-to-north movement — from the equatorial Pacific Ocean, northeast across Baja and the Gulf of California, northward through the Colorado River Valley, and then north and east into Arizona and the Nevada highlands. In these latter four cases, coastal, frontal activity was minimal, and isolated inland thunderstorms predominated; moisture moving north through the Gulf of California, therefore, did not significantly affect Southern California, but rains did develop in Arizona and Southern Nevada. Figure 7 documents this observation and summarizes NWS data from 214 different rain gauges in those latter two states. Figure 7 shows a very clear peak in rainfall immediately following the early May, July, and September cloudbusting oeprations. The early August cloudbusting operation took place during the normal,

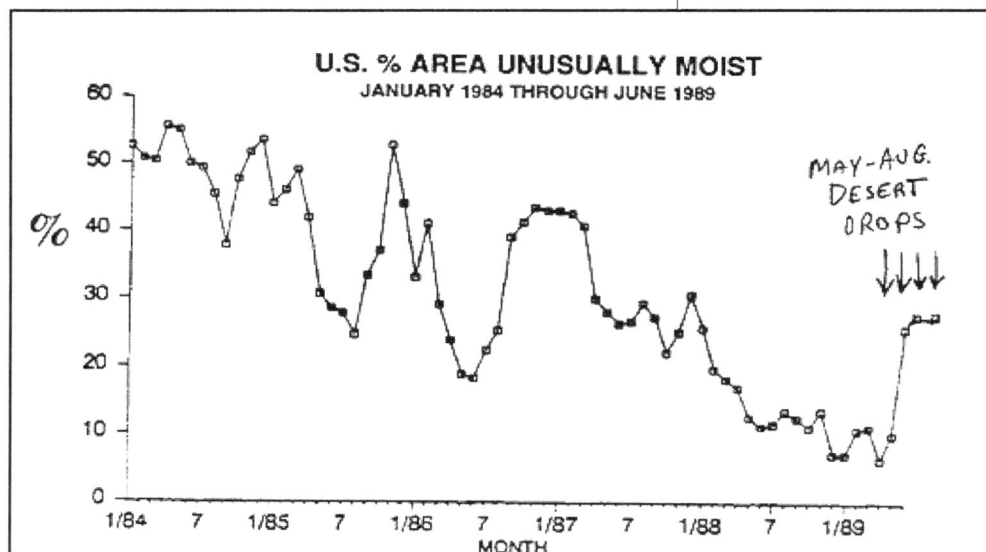

Figure 4: Percent of U.S. with "Unusually Moist" conditions. Note the dramatic increase in moist regions during the May-August cloudbusting operations.

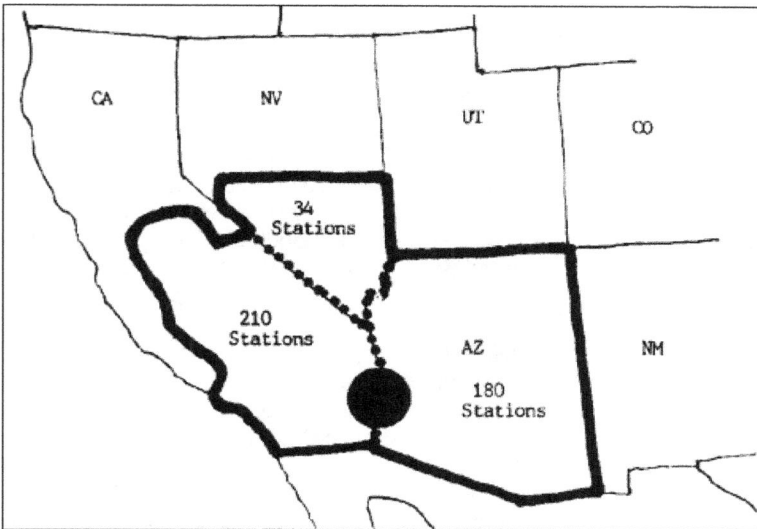

Figure 5: The area of the Southwestern USA used for the precipitation analysis is outlined, with the number of rainfall gauges given. The cloudbuster was operated from a site on the Colorado River, marked with a large dot.

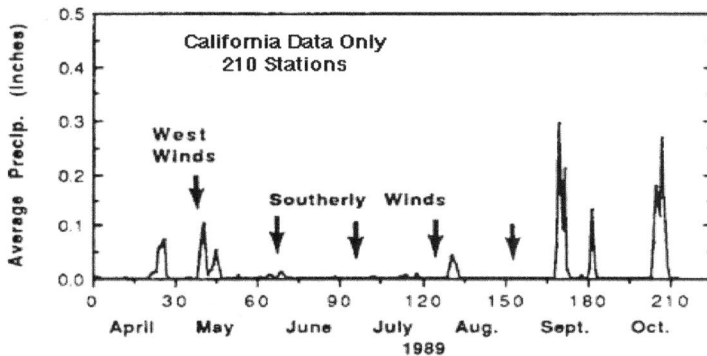

Figure 6: Averaged Daily Precipitation from 210 rainfall measuring stations in Southern California. The dates of the five different cloudbusting operations are marked with an arrow. Only the first of these operations was characterized by strong westerly winds and moisture from the Pacific Ocean, and only this first operation yielded significant rains in California.

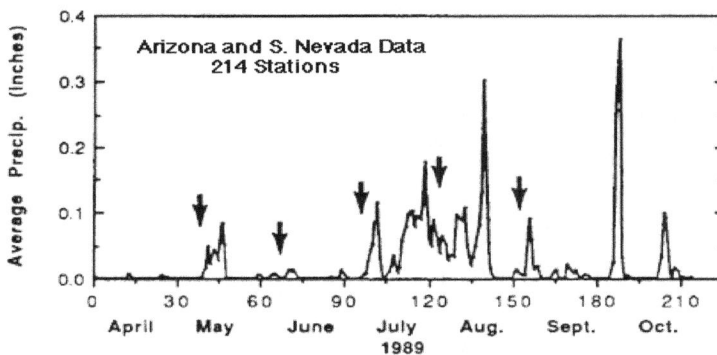

Figure 7: Averaged Daily Precipitation from 214 rainfall measuring stations in Arizona and Southern Nevada. Strong southerly winds and moisture brought rains to this area in the last four of the cloudbusting operations (marked with an arrow), even though rains did not develop at those times in Southern California.

rainy summer monsoon of the arid Colorado River Basin region, and so its influences were lost in the background chaos of the normal August rainfall. The early June operation was followed by only a very small rainfall episode.

Figure 8 shows the daily rainfall data from all 424 rain gauges in the three state regions combined. A periodicity, or atmospheric pulsation similar to that seen in wetland regions, was observed in the data for the months of September and October, and persisted for several weeks after the last of the cloudbusting operations. However, no attempt has been made to find out how often this pulsatory rainfall characteristic naturally develops in this large dryland basin region. These data indicate that four out of the five cloudbusting operations were followed by significant peaks in rainfall, which is similar to the success rate (80%) observed in prior cloudbusting operations (2).

The averaged daily precipitation values for all 424 rain gauges in the entire three-state region were converted to percentage-of-30-day maximum values. This was done for each of the five individual cloudbusting operations, centered on the first day of each operation, such that daily precipitation percentage values for two weeks before and after each operation were obtained. The percentage data preceding and following the five separate cloudbusting operations were then overlapped and averaged together. This approach demonstrated an even more solid connection between the five 1989 cloudbusting operations and rainfall in the arid Colorado River Basin. As seen in Figure 9, the rainfall occurring within the basin region after the cloudbusting operations was approximately twice that observed during the pre-operations period. This period of doubled rainfall developed about 48 hours after the onset of operations and persisted for approximately one week thereafter.

The implications of these atmospheric responses to the influence of a single large cloudbuster working for only a few days of each month are startling, and their importance during an epoch of expanding deserts is significant. While these positive results are very encouraging, much remains to be learned about the atmospheric-energetic processes involved in the creation and maintenance of this particular desert region. The lessons learned from this region may or may not be fully applicable to other desert regions. Also, many questions were raised about the potential ecological and social effects of greening a large desert region through use of the cloudbuster. Desert greening is *climate change*, after all, and this implies changes in flora, fauna, and human habitat. Decisions on any major undertaking to bring about climate change, even apparently beneficial change, must be made on a social level. However, unplanned climate change is already occurring in the reckless widespread assault on the Earth's forests and grasslands by nearly every human culture around the world and in the related artificial, human-forced expansion of the Earth's deserts. The drylands and deserts of the world do not exist in a stable equilibrium with respect to surrounding wetland regions; the deserts are getting larger by the year, while the green areas of the Earth are shrinking in size (3). It is doubtful that any rational collective social decisions can be made regarding these new scientific findings until at least the irrational, antisocial destruction of the Earth's living plant cover has been halted.

These important concerns aside, one thing is very clear: The data from this experiment clearly support Dr. Reich's findings from the 1950s that *the cloudbuster possesses a true desert greening capability* (4). Certainly, an extended period of field research is warranted.

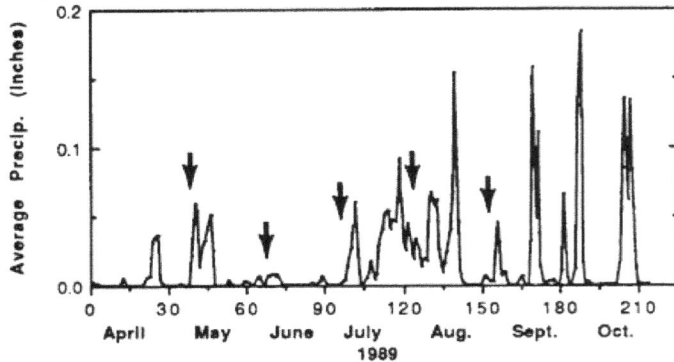

Daily Precipitation: 424 Stations in AZ, S. CA, S. NV

Figure 8: Averaged Daily Precipitation from 424 rainfall measuring stations in Arizona, Southern California, and Southern Nevada combined. Peaks in rainfall follow closely after the cloudbusting operations, marked with an arrow. A regular pulsation of rainfall is apparent in September and October.

REFERENCES

1. *Climatological Data*, NOAA, USDOC, Washington, DC, for Arizona, Nevada, and California, April-October, 1989.
2. DeMeo, J.: "Field Experiments with the Reich Cloudbuster: 1977-1983," *Journal of Orgonomy*, 19(1):57-79, 1985; DeMeo, J. and Morris, R.: "Preliminary Report on a Cloudbusting Experiment in the Southeastern Drought Zone, August 1986," *Southeastern Drought Symposium Proceedings*, March 4-5, 1987, South Carolina State Climatology Office Publication G-30, Columbia, SC, 1987; DeMeo, J.: "Nine Years of Field Experiments with a Reich Cloudbuster: Positive Evidence for a New Technique to Lessen Atmospheric Stagnation and Bring Rains in Droughty or Arid Atmospheres," *Abstracts of Papers*, Program of the Association of Arid Lands Studies, Western Social Science Association, El Paso, Texas, April 22-24, 1987, p. 6; DeMeo, J.: "CORE Progress Report #20: Breaking the Drought Barriers in the Southwest and Northwest U.S.," *Journal of Orgonomy*, 23(1):97-125, 1989.
3. DeMeo, J.: "Desert Expansion and Drought: Environmental Crisis (Part I)," *Journal of Orgonomy*, 23(1):15-26, 1989.
4. Reich, W.: "OROP Desert. Part I: Spaceships, DOR and Drought," *Cosmic Orgone Engineering*, VI(1-4):1-140, 1954; Reich, W.: *Contact With Space, Oranur Second Report: OROP Desert Ea.* New York: Core Pilot Press, 1957.

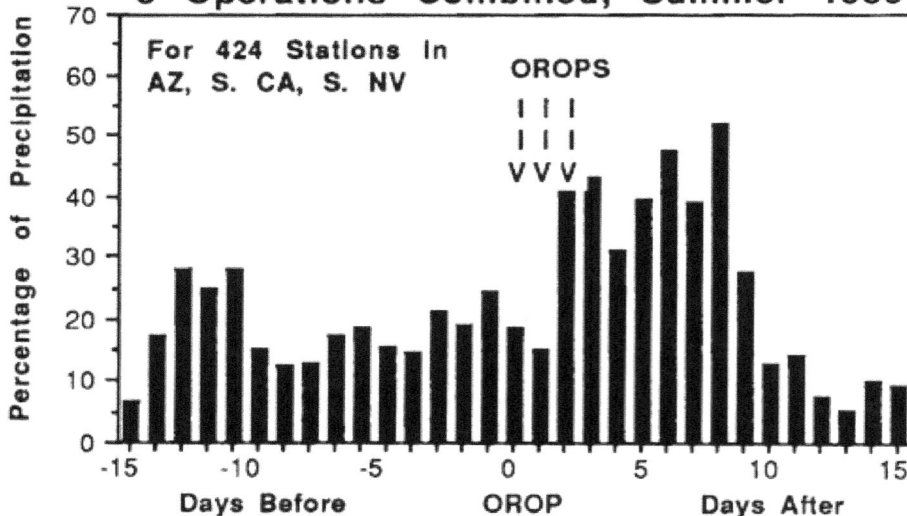

Daily Precipitation Percentages
5 Operations Combined, Summer 1989

Figure 9: Averaged Daily Precipitation Percentages for the five cloudbusting operations combined, as overlapped and centered on the first day of each operation. A clear and significant increase in precipitation begins about 48 hours after the onset of cloudbusting operations, lasting for approximately one week thereafter. This increase in rains is about double the amount observed during the two-week period prior to operations, or during the last week in the study period.

Thank You For This Honor*

Eva Reich, M.D.

Editor's Note: Over the last decade, Dr. Eva Reich, daughter and close co-worker of the late Dr. Wilhelm Reich, has traveled across Europe, giving lectures and seminars on the gentle birthing and handling of infants. Her apprentices have primarily been midwives, and she has been a prime mover in the establishment of a gentle-birth social movement in East Germany, Austria, and elsewhere. Thanks to her tireless work, Vienna now has four different gentle-birth midwifery clinics, and the City of Vienna recently gave her a public "Thank You" and recognition for her work, as well as a big party, complete with orchestra. This is certainly a first for anyone working in orgonomy, and few are as deserving of such an honor as Eva Reich. At her insistence, the award was broadened to encompass the work of both herself and her father, Wilhelm Reich.

Thank you for this honor. I am accustomed to emmigrate, to be a stranger, a minority in opposition. It's not easy for me to tell my biography in a short time, because it is varied and colorful. And its not easy for me to be so honored. This is also for me a reconcilliation that helps to heal the wounds and persecutions that I and my family, especially my father Wilhelm Reich, had to suffer.

I ask myself, what are the results of my 66 years on Earth. It is hard to tell what was important, and what will last. My grandfather, Alfred Pink, was a tradesman in Vienna, and he also was a municipal councillor. He was a kind-hearted man who invited orphans to a restaurant party at Christmas every year. When he was old he said to me "The work for a better world will not be finished for a long time". This gets to a main theme in my life.

My father, Wilhelm Reich, was always a fighter for the new, his whole life long. He was a psychoanalyst in Vienna — in Berggasse — and was a student of Sigmund Freud. He worked out how the neuroses is manifested in the human body, and dedicated his life to the uncovering of the energy that also expresses itself in the sexual impulse. For many decades he discovered more and more aspects of life energy, which he called orgone energy. The research on scientific orgonomy led him to the development of devices that are able to concentrate this energy. I want to point out the orgone accumulator, which I myself have used in my work as a practical doctor, the medical dor-buster, and the cloudbuster, with which my father could bring into movement the stagnated atmosphere, the so-called smog. I was present at many of his experiments. He liked to talk with me and often told me about his latest research. He said that "civilization had not yet started", and that the life force was being systematically suppressed and destroyed, again and again, in generations of new babies and little children, all in the name of

* Speech by Dr. Eva Reich at the award of the Silver Medal of Honor of the City of Vienna, 7 March 1991, presented by Vice-Mayor Smejkal, Vienna City Hall. Translation by Renate Purgathofer, from text published in *Bukamatula*, 2:5-8, 1991, newsletter of the Vienna Wilhelm Reich Institute.

the State, religion, school, and so forth. That's why he talked of the "battle for the human race", which for me — as a pacifist — changed to "humanizing humanity from birth on". He felt that this work to avoid neuroses was of essential importance. These ideas are contained in the book "The Murder of Christ", which for me became the most important of all his writings. In it he also defines all his ideas for the children of the future.

I've also worked on that, first in my practice as country doctor where — with the permission of the parents — I informed adolescents about birth control. As early as 1952, I made gentle homebirths. With that I learned that babies born in this manner, which are not separated from the mother, do not scream, have a warm rosy skin color, and can be satisfied. During my internship at Harlem Hospital, in 1951-1952, it became clear to me that premature infants are feeling human beings.

I developed my own kind of vegetotherapy — a gentle massage — to revitalize newborn babies whose mothers were routinely given drugs at birth. This led to my baby massage which relaxes the infants, avoids the so-called colic, helps autistic children come out of themselves again, leads to a physical connection between parents and child, and helps to restore the disturbed parent-child relationship. I want to call that *rebounding*.

Since 1975, I have been invited to give seminars and speeches in more than 30 countries. I had the opportunity to learn from various gentle bioenergetic works, and could also experience how useful these principles were on adults. And I also saw that in every country the newborn babies are hurt in different ways. I always think about what could be changed, in a way that wouldn't cost anything. Interventions in the beginning can avoid much suffering later on.

I wish to point out a few examples of the possibilities which exist to change awareness, that can help to avoid armoring and the development of neuroses: Babies and mothers must be seen as a *bonded pair*; they should not be separated, not even in the case of disease. That doesn't cost anything. Children should not be hit or humiliated; instead, they must be given full human rights in a democratic family. We are just at the beginning of a new era, in which we women, who are more than half of humankind, also take part in politics and governments, and when the world will no longer spend half of its money for weapons and destructive wars.

I wish to thank my friends Johanna Sengschmid, Ingebord Hildebrandt and Irene Hocher, who helped and supported me so much during these many months in Vienna. And I'm looking forward to more work together this autumn when I'm invited with others to cooperate in the Birth and Parentship Center.

The delicate beginnings of life are of the biggest importance. They are the foundations of our well-being for body and soul; I want to ask you to support these foundational efforts. We need peace on Earth, and it begins in the womb!

The Fifth International Orgonomic Conference*

Matthew Appleton**

The Fifth International Orgonomic Conference was held between 15th - 18th June 1990, at the Acropolis in Nice, organized by medical orgonomist Dr. Giuseppe Cammarella, under the auspices of the American College of Orgonomy. The first three orgonomic conferences were at Orgonon, while Reich was still alive. The last one had been in 1984, in Munich. Since Neill's death there had been no real link between Summerhill and orgonomy. So, as I set off from Leiston for Nice, I had the sense of retracing historical footsteps, as one might feel retracing the route of any great pioneer. For many years I had been interested in Reich's work, and the links between Summerhill and orgonomy seemed to me to be extremely valuable. I was somewhat daunted by the prospect of representing Summerhill, for my own sense of the school is that its meaning cannot be adequately conveyed in words, but is more readily perceived in the dynamics of everyday relationships; the fluidity of movement and depth of contact. Also daunting were the rumors I had heard about the state of orgonomy since Reich's death; that rigidity and reaction had set in like rigormortis on this most alive of all sciences.

All I can say is that during this Conference I was greeted with warmth, courtesy and a spirit of openness that moved me deeply. At the airport I was met by Dr. Cammarella, and his wife, Mme. Maria Gamaleia, along with Peter Robbins, another of the speakers. My first impressions put me at ease straight away. Throughout the Conference Dr. Cammarella and Mme. Gamaleia extended to me a heartwarming hospitality, and Peter Robbins proved to be a most pleasant and cheering of companions. Before the Conference started I had a free day, which I made the most of, exploring Old Nice, and swimming in the Mediterranean; a bit of a culture shock after the rough and ready raucousness of the House kids at Summerhill.

The Conference began with an overview of orgonomy; it's development, the trial of Reich, and an account of recent investigations into the Food and Drug Administration archives on the case. This set the context for the Conference. Over the next few days all the main areas of orgonomic work were to be given an airing biology, physics, astro-physics, psychiatry, education, and child rearing. Almost all the other speakers were either Doctors or Professors, and here was I armed only with a handful of stories about Summerhill and the strength of my own faith in the school. Yet, as I listened to one speaker after another I felt more comfortable, for although a vast array of subject matter was being presented here, in many different forms, it seemed to me that the underlying foundation formed a much

deeper, much broader totality than I had ever experienced before, and the stories of how a 'bunch of kids' ran their own school in Suffolk had as much a place in the scheme of things as the expansive lecture on the formation of galaxies.

It was clear from the Conference that the science Reich had given birth to had not died with him, but was still being developed and expanded upon. Professor James DeMeo, Director of the Orgone Biophysical Research Laboratory, presented the results of cloudbuster operations he has been conducting in America over the past years. In eighty percent of cases, the cloudbuster had achieved positive results over vast areas, at times bringing spectacular increases in rainfall in areas suffering from drought. In a further lecture he presented remarkable evidence to suggest that human armoring had originated as a response to sudden desertification, in and around the Middle East, brought about by climatic changes in the past. Specific dates and archaeological findings were quoted, along with anthropological evidence which indicated the pathways by which patriarchal, warlike attitudes, riddled with sex-taboos, and the subjugation of women and children, emanated from this area and period.

Reich's interest in unidentified flying objects was given voice by Peter Robbins, from America, who, following a 'close encounter' in his own youth, had embarked upon many years of serious study of the phenomenon. After Reich had pointed a cloudbuster at some UFO's over Orgonon, and they had disappeared, he had gone on to ask himself, could it be that these flying machines were propelled by orgone energy. Further evidence would suggest this is the case. Peter Robbins went on to support this hypothesis with a series of U.S. Government documents, which had recently been declassified by the America Freedom of Information Act. As with many of the speakers at the Conference, Peter Robbins presented what at first might seem to be wildly extravagant ideas in a matter of fact manner, devoid of sensationalism or outlandish speculation, but based on meticulous research, careful observation and backed up by personal integrity.

The Conference did not only engage the head but the solar plexus too. The team of professional translators, sitting in a glass booth to one side, were later to comment on the attentiveness of the audience, something they had never met before over such a prolonged period. The emotional charge in the lecture theatre was intense, as the many facets of orgonomy were given shape. Dr. Richard Blasband, President of the American College of Orgonomy, showed a film of the bions, the tiny vesicles of orgone energy discovered by Reich, which bridge the gap between non-life and the living. It was as beautiful as it was awe inspiring. He also lectured on the effect of DOR and oranur in the atmosphere today, a lecture which left me feeling drained and saddened. There was a workshop on

---

* Reprinted from *Wilhelm Reich, A.S. Neill, and Orgonomy,* Friends of the Summerhill Trust, Great Britain, 1990.

**A personal report by Matthew Appleton, Houseparent at Summerhill School, Leiston, Suffolk, 1P16 4HY England.

orgone therapy, with careful admonitions to avoid "Reichians", who were not properly trained and did not fully understand the forces they were dealing with. Armoring should be treated with respect, not simply smashed through to achieve a quick result. The effects of such dabbling were often disastrous. As well as orgone psychiatry, medical orgonomy was also discussed; the treatment of cancer, AIDS, and other degenerative diseases, each given their root in the disruption of the bio-energy.

The discussions on childhood and education began with a lecture by Dr. Michel Odent, a leading pioneer in new approaches to childbirth and obstetrics. Reich had written:

"There is no way of basically changing people once they have developed the wrong character structure. You cannot make a crooked tree straight again. Therefore, let's concentrate on the newborn ones and let's divert human attention away from evil politics, and toward the child."

Although Dr. Odent is not in fact working in the field of orgonomy, his work parallels Reich's observations in this area, and has led to many conclusions of interest to those involved in orgonomy. His comparisons between conventional childbirth and his own methods, drawn from years of experience and research, were startling. On the one hand, the needs of both mother and child are undermined and disrupted by the environment and mechanistic nature of childbirth as it "normally" proceeds in the hospital. On the other, an atmosphere of intimacy and non-intrusive support, as exists in the laboring mother's home or natural-birth center, fosters full contact between mother and infant, creating the conditions whereby the inherent expectations of both parties are met. The basis of future well being, argued Odent, as did Reich, lie in the establishment of this initial contact. Dr. Odent was followed by Dr. Cammarella, who spoke on the biological rhythms of childhood, the need for children to be able to express themselves in spontaneous movement, and exercise their autonomy in everyday life. Only from this bioenergetic basis of health could the child retain his or her basic integrity and creativity. This led naturally into my own talk on Summerhill School.

I began by introducing a twenty-five minute video tape, which was followed by a questions and answers session. The questions were provocative and thoughtful, showing a deep understanding of the dynamics of the school and the problems it has to face in a society that has not begun to think in such terms. "What contact does the school have with parents?" "How do the teachers respond to 'free children'? What is their reaction?" "How do adults at Summerhill deal with the problem of transference?" "Why is it that so few ex-Summerhillians send their children to Summerhill?" "Is society frightened by Summerhillians?" "Do children miss out on a home life being sent to Summerhill?" "What power does the Summerhill meeting have over the sexual lives of the pupils?" "How is Reich regarded at Summerhill these days?"

I answered the questions as best I could, drawing upon specific incidents to illustrate my observations. Later in the Conference, one of the medical orgonomists was to put very concisely my own feelings as I spoke. When asked by someone to describe a typical orgone therapy session, he declined, saying it was not simply a matter of mechanical procedure, but the way forward arose out of the contact between the therapist and the client. So it is with children; there is no procedure, no "right answer" to fall back on. The way forward arises, and can only arise, from the quality of contact between the child and the society they live in.

There were two further events I took part in; a workshop on "Orgonomy, Education and Society", and the closing discussion on "The Prevention of Medical and Social Biopathies". That Summerhill is not simply an "isolated" experiment, but has a place in a much broader movement of thought and development, the significance of which has not yet been grasped, became abundantly clear. Both participants and the audience were warm and enthusiastic in their response to the school. Many expressed their happiness that the school still existed. For many of the parents and educationalists who attended the Conference, Summerhill is seen as a beacon in a world that has yet to understand its children, and the function of childhood, an understanding that humanity has not merely failed to comprehend, but has consistently evaded in its relentless attempt to mould its children in its own armored image.

Speakers at the Fifth International Conference of Orgonomy held in Nice, June 1990. Starting on left: M. Appleton, R. Blasband, C. Haydon, G. Cammarella, M. Odent, P. Robbins, R. Schwartzman, C. Konia, J. DeMeo, J. Bell.

Acropolis Conference Center

Nice Harbor, France

Report on the Fifth International Conference on Orgonomy*

Gulio Gelibter**

Condemned by an American court and let to die in prison after a very strong complaint of deffamation, accused to be a madman and visionary, and boycotted by the official scientific community, Dr. Wilhelm Reich, assistant of Sigmund Freud and founder of the science of Orgonomy, 30 years after his death, is more alive than ever as demonstrated in this International Conference on Orgonomy, which has been these days in Nice. Here, the main people working in the Reichian movement both in Europe and the United States of America have presented their research and studies, with many astonishing revelations.

Especially of interest in the Conference — the first of its kind since 1984, and organized by the Italian analyst Dr. Guiseppe Cammarella under the auspices of the American College of Orgonomy — is the revelation of Dr. James DeMeo, Director of the Orgone Biophysical Research Laboratory, USA, on the so-called cloudbusting of the atmosphere with an "atmospheric cannon". This work is based on Reich's discovery of the vital energy of the cosmos: Orgone. The cloudbuster cannon, made from long metal tubes aimed toward the sky, would make the orgone energy present in the atmosphere move, and provoke the formation of clouds and rain. In spite of "official" science that doesn't accept even the existence of the orgone, Dr. DeMeo affirmed the results are real and they are astonishing. "I have done 35 expeditions with the cloudbuster", says the researcher with a degree in Geography. "We have gotten positive results in 80 percent of the cases. Last year, we were able to double the amount of rain in a large area covering Arizona, Southern California and parts of Nevada. In April of this year, we used the instrument to bring rains across a very large part of Greece." DeMeo said that the cannons would be an exceptioal tool to fight drought and the lack of rain in Africa, but any effort in this quest has been blocked by the total indifference of international organizations and the governments who should have been interested.

These cloudbusters are the reason why Reich was also interested in flying saucers, because during his experiments with the clouds in the 1950s in Arizona, this scientific man had, in fact, pointed the instrument at some UFO which appeared, and he provoked their disappearance. This fact made him think that those flying machines might be powered by an orgonomic kind of propulsion. At the International Conference in Nice, an apparent confirmation has come from an international expert on the UFO question: Mr. Peter Robbins, as a matter of fact, presented documents from the US Government which have been de-classified because of the Freedom of Information Act, documents which seem to confirm the hypothesis that UFOs are propelled by non-mechanical means which are unknown to our science, and which may confirm Reich's theories.

In addition to the research of DeMeo, other people in this field, using the orgone accumulator, are able to concentrate the vital energy for use in physical therapy. These accumulators are still the subject of a grand conspiracy of silence in the international scientific community. The reason for this conspiracy of silence is that the discovery goes against the beliefs of the big scientists. This is an example of how things are, says Dr. Cammarella. Dr. Cammarella also said that for this Conference he had invited hundreds of scientists working in other fields outside of orgonomy from around the world, but none of them came. The same point was raised by Dr. Richard Blasband, President of the American College of Orgonomy, who presented a film from the microscope which confirmed the existence of the very small organic particles that Reich called bions, but the scientific community did not care to see it. Dr. Cammarella said that Reich was a genius who has rarely been understood or seriously studied. Reich hoped that by 2007, when all his archives of about 100,000 pages would be released to the public, that the world would be ready to understand. "I ask myself", said Dr. Cammarella, "if he was not wrong by one thousand years, that it should have been 3007".

* Translated from the Italian weekly magazine *Panorama*, 21 June 1990

** ANSA corespondent, Nice.

Report on the 1990 Conference at Orgonon "The Voice of Wilhelm Reich" *

James Strick, Ph.D. **

This year's summer conference on "The Voice of Wilhelm Reich", sponsored by the Wilhelm Reich Infant Trust Fund at Orgonon on the 23-27 of July of 1990, was a rare opportunity for students of orgonomy to hear an entire week of tapes from the Reich Archives. These tapes were selected from among many made by Reich to record his lectures, seminars with physicians and social workers, discussions, business meetings, and other material which he wanted to document for posterity. Excerpts of this material have been used before at previous summer conferences, but this was the first time in 40 years in which an entire 5-day course was to be taught essentially by Wilhelm Reich. The meeting was attended by 30 physicians, teachers, social workers and others from all over the United States and Europe. Dr. Chester M. Raphael, a participant present at all the meetings, was able to consult on his recollections about the context of the proceedings, where the tapes were difficult to interpret, and so forth; this was another very fortunate and useful aspect of the class.

The first day was devoted to "The Importance of Method in Scientific Research and Reich's Discovery of the Armor". The tape was of a seminar led by Reich with orgonomically trained physicians. Reich had opened up by inviting the doctors to bring in material from specific cases for discussion at the seminar, but wished first at this meeting to go into some basic theory with them. He elaborated on the theory of scientific method beginning with Kant, who he said "was the first to investigate how we investigate." Referring to the Einsteinian school, Reich commented, "They work with reason, but they don't know how reason comes about." Reich described the beginnings of his method in his first years as a psychoanalyst. Psychoanalytic method was inadequate because the patients "were shut, like dead cats, corpses...", and that caused him to feel like pushing at the defenses, pushing repeatedly at the same spot until he got some feeling expressed. One student asked "So what got you going was that the patients' armor caused motor unrest in you and made you feel compelled to push at it?" Reich concurred. He went on: "In every activity it's like an ameba attracted to a bion. You see movement in the ameba first, then the bion begins to move." This interchange was typical of the lively give and take between Reich and his students, which we heard all week. It also reveals one of Reich's features most appealing to me: the use of graphic examples and metaphors in plain language which makes even his difficult technical works accessible.

Reich's direct manner of pointing out mistakes to students was also impressive. One frequently heard him respond "Why do you bring that up now? It's not related." Or "That's just talky-talk. It's evading the point". He did not hesitate to suggest to his

Wilhelm Reich

students that they needed to constantly observe their own motives in asking particular questions, since the nature of his work was emotionally disturbing to many, and the tendency to evade the central issues, e.g. the sexuality of children and adolescents, was great. To an example brought up by one student, Reich said that it illustrated:

> "...taking something completely in the wrong realm...Take pleasure in realizing your mistakes instead of being a 'fake knower.' I want you all to keep this term going, it's a very good one. A 'fake knower'; someone who doesn't know but who pretends to."

When discussing the armoring process, Reich referred much to a diagram (1) which he had drawn on the blackboard. The process begins with the expression of some need or drive: "...you're a child who wants to play but your father tells you to sit still, or who wants to investigate something and your mother slaps your hand." The organism reacts by withdrawing, becoming restless and/or fearful. The next response, said Reich, is to "try to get that damned obstacle out of the way!", i.e., rage.

---

* Republished with the permission of the Wilhelm Reich Infant Trust Fund, from the *Newsletter, Friends of the Wilhelm Reich Museum*, No.28, Fall 1990.

** Professor, History of Science, jstrick@fandm.edu

---

1. Reich, W.: *Ether, God and Devil*, Farrar, Straus & Giroux, NY, 1973, p.65.

> **"Reich's emphasis (was) that no mechanical prescription for handling these problems is possible...the essential element was *quality of contact*..**

"This sequence repeats itself again and again, and only if the organism cannot avoid it and cannot find another pleasure to replace the original, or get beyond the obstacle, does the original impulse become forced, violent or destructive. So far this process is all natural. It occurs all throughout nature in many different animals."

In the discussion Reich makes the point that if the child has a tantrum at this point in the process, it is not neurotic. Only later, when the child begins to hold in its anger, has armoring begun. "The obstacles outside (the organism) are natural, they're just not neurotic. The ameba has its obstacles too."

The second day of the conference was devoted to a taped meeting of the Orgone Infant Research Center sometime in 1950. It involved a talk with the parents of one of the children studied (2), an examination of the child, and a discussion of the case by Reich. In addition to the material later published on this case, a number of important points were made. For instance, referring to the mother's depression over feeling less than ideal, Reich noted:

"When she felt tense, the child nursed poorly and was not satisfied at the breast. The sucking process is not only a mechanical physical intake of fluid. We have always assumed that the quality of the milk must change when a woman is depressed... whether the nipple is sweet or sour. Now we don't know what that means, 'sour'. But I'm going to try to relate it to hyperacidity in the stomach in depression... When we say that somebody 'turns sour' emotionally, it refers to an actual physical reality in the body of the person."

This child eventually developed a bronchitis. Reich noted that most doctors would see this as nothing abnormal:

" 'Why shouldn't the child get a cold?' and so on. But we try to keep an open mind on these matters... If we assume orgone biophysical unity of child and mother, then how can we separate the cold from the mother's depression? ...I am inclined to understand a cold as a disturbance of the bioenergetic equilibrium of the child. Perhaps the lining of the tubula react to the emotional changes one way or another... It's important to ask such questions."

Perhaps the most important point in this meeting was Reich's emphasis that no mechanical prescription for handling these problems is possible; i.e. that the essential element was the quality of the contact on the part of the physician, social worker, parent, or whoever the intervening person might be. The child being examined was crying and in considerable distress. The chest was held rigidly high. Reich felt a need to help it. He just put his hands on the sides of the child's thorax, and it felt the warmth. He very gently stroked the intercostal

muscles, but did not 'massage' the chest.(3) Reich described the results:

"The child caved in. It peed. Felt relieved. I was surprised. I had never seen anything like that before. My daughter, Dr. Eva Reich, after 6 years in medical school and 2 years working in a hospital...asked 'What did you do there?! How did you do that?' And I said to her: 'What does a painter do when painting a landscape? Does he paint a blue spot 2 cm. by 2 cm., and then outside of it a green spot 1/2 cm in diameter, and then a streak of white 10 cm. long by 1 cm. wide? No. He just paints it.'...
"Of course I could see that the chest was high, but I wasn't going on a formula. I just came there, and out of my knowledge came what I did. I didn't massage or even tickle... If we could just teach this much to doctors and nurses, we'd be accomplishing a great deal."

The child soon began crying again, and it became clear that its chest was held rigidly high. Reich tried again to get it down. After awhile of no success, he said:

"I can't deal with that armor. You can't (turning to his students) do anything about it. The mother needs to work on this. (Then, to the child, as it cries) Yes, be angry! Be angry!"

The child responds by crying more vigorously and deeply, then after awhile calms a bit. Reich:

"Now what did I do?... It has nothing to do with quantity... It has only to do with having contact with this baby and how it feels... Now I'm sorry to have to repeat this again, but what I don't want is for social workers, psychologists, etc. to go out and press down babies' chests [mechanically]..."

Yet he feels anxious about the danger that the professions would turn it into this, to make money. He says it could be important for parents to be able to do this kind of first aid. Dr. Singer says that in the community he works in, however, out of 100 mothers he had interviewed, only two had the contactfulness to be able to do this. Reich: "This cannot be ignored. It must be taken into account. It's clearly a major obstacle."

At this point in the meeting, Dr. Raphael had presented his report on orgone treatment during labor, later published in 1951 in *Orgone Energy Bulletin*.(4) Since he was present at our meeting this summer, Dr. Raphael told us about these cases in person. He emphasized that in modern medicine:

2 . Reich, W.: *Children of the Future*, Farrar, Straus & Giroux, NY, 1986, pp.89-113.

3. Reich, W.: *Children of the Future*, Farrar, Straus & Giroux, NY, 1986, p.107.

4. Raphael, C.M.: "Orgone Treatment During Labor", *Orgone Energy Bulletin*, III(2):90-98, 1951.

"The tendency is to direct attention only to where the pain is, rather than to the total organism. I hadn't the slightest idea what to do in this situation (having been called to help out a woman having difficulty in labor with some kind of orgonomic 'first aid'), but in observing her, I felt what I could try... You have the feeling that you can do something about it. It seemed so clear and effective, and after all these years nothing more has been done with it yet..."(5)

Day three of the conference featured a tape from 1950 on "The Relationship Between Work and Organization". In this meeting with his students, Reich discussed how the Wilhelm Reich Foundation, its daughter departments like the Orgonomic Infant Research Center, the Orgone Institute Press, and the Orgone Institute Research Fund, as well as the separate, independent Orgone Institute were related to one another, and why he had set up the organizations in this way. He also went into some depth on his attitude about how co-workers must behave in order to remain associated with him, and his reasons for this. Since many writers have described Reich's attitude on this as rigid and authoritarian, this material is of great importance. I cannot emphasize enough in this context the importance for the participants of actually hearing Reich's voice on this subject, as well as his own words. The kind of bossy, inflexible tone of voice and manner of dealing with people conjured up by those writers as their image of Reich was simply not there in what we heard. We heard the highly animated voice of a man convinced of the seriousness of his work and completely rational about his reasons:

"As long as you are connected with the Orgone Institute, you are not free agents, you cannot do as you please. If you want to be on your own and do as you please, you must disconnect yourself from the Orgone Institute. Why? Because if you're on an army staff at war, with lives at stake, self-regulation is not valid there... There is a basic natural law which I described in the paper on orgonometric equations, and that law binds you down... I would like to ask you to help me kill the license ideology in our midst...the main necessity is that of fighting the plague. That means that we are organized not like an organization of free human beings, but more like a university staff or the staff of a military service... I am at war for 30 years. That's not my personal choice, but I am at war... "

Reich went on to emphasize to his students that the Orgone Institute was a legally registered name synonomous with him himself, and kept separate from all the organizations in which his students and co-workers participated, i.e. that only he was the Orgone Institute:

"No one here has the right to say anything in the Orgone Institute. You don't like that, eh? The Orgone Institute is I. If I hadn't done it that way, we would have gone down many times already... A worker in a clinic or the laboratory is an assistant, a helper in this task..."

5. Dr. Raphael discusses the issue of labor in more depth, and the non-mechanical nature of orgonomic treatment in his pamphlet *Some Questions and Answers About Orgone Therapy*, 1977, $6.50 from the WR Museum Bookstore, PO Box 687, Rangeley, ME 04970.

Reich was very concerned about the standard American notions of democracy and voting. He felt that both could be very destructive and went to some length to set up a structure in which voting by people who had not actually done the work could not take over something created by the real work of others like himself. He cited the takeover of the Hamilton School, and throwing out of Hamilton, the man who had actually created it:

"I know from experience over a long period of years that the little man, just with his vote, could take over anything. It happened to Hamilton. The discoverer of television dynamics is out, and a few little, clever politicking fellows took it over... You vote by taking responsibility for what you're doing, by working. We can vote with or without raising our hands. Raising of hands is not a responsible democracy — it lacks responsibility and toll and worrying... All formal democracy went down because of it... Dr. MacDonald, for example, works in the lab here as my assistant. If she wants to go and set up a research lab elsewhere, and develops in accordance with our line, we'll have the development of a new function. But if she makes a discovery which goes against me, then she's got to fight for it on her own. She's got to support such a thing based on her own work, not just on a vote. We cannot go and vote about whether to freeze a tube of a bion preparation. You can't vote about it: that's work! You have to do it! You have to watch out for the one who, knowing the least, wants the most of honor. Those are the Stalins, the destroyers of human society."

A bit earlier, Reich had described the process of the development of a new function:

"When a new organization starts, we watch it for a couple of years to see if its capable of functioning independently along our line. For instance, I started the Infant Research Center last year, and Dr. Singer and Dr. Duvall have taken it over and are working more and more independently, and they seem to be doing this with no regard for public opinion and focused only on the interest of the child. It may be a little too soon to say, but so far they seem to be able to do it."

Reich also discussed how the various organizations were funded: the Orgone Institute from his own personal income, since it was him personally; the Wilhelm Reich Foundation and the non-profit public service agencies under its umbrella by private donations, fees for service at clinics, rental fees for accumulators, etc. He firmly stated that "No contributions go to the Orgone Institute. Any money like that goes to the daughter institutions which are under the Foundation. Now, does anyone have any questions about this? Don't be embarrassed to ask because I'm talking about money..." Nobody asked any questions, which is significant because later, toward the end of the meeting, Reich brought up for discussion a rumor he had heard that contributions the physicians were making to the Orgone Research Fund were actually kickbacks, going into his pocket. Reich had discovered that many of the physicians knew about the rumor and none had done anything about it, implying that they believed it might be true. He was very direct in his response to this behavior:

"We will not accept any checks from any of you for a long time to come. I will not accept any money which is not freely given. I want you all to understand that we are completely independent of your checks. The income from the accumulators is the majority of our income now. But I don't want your contributions unless all this mud and shit is cleared out of the work."

Questioning the physicians one by one, he asked "What did you think? Why didn't you do anything about the rumor? Why were you giving the money? What did you think it was going for? Was it of your own free will?" After much hemming and hawing (one could imagine the squirming), the main feeling which seemed to emerge from most of the doctors was "I felt guilty about riding on your back" and "I thought you deserved the money, that it might go for lab supplies and such." In other words, none of them knew clearly their own motives or had felt responsible to find out what their contributions were going to support. One participant at this year's conference pointed out that Reich was trying to show them: "Your giving money is just like the way you vote. You're thinking that just giving money, like voting, is what's necessary to be working. But empty voting without knowing what it's going for, is like empty money giving." Most of us present felt this a very apt analysis. The unity of Reich's approach was clear, and his tenacity in going after emotional "dirt" behind the scenes was admirable. This was an impressive demonstration of what he called doing practical social psychiatry. After the air was cleared, the taped meeting ended on a note of "I feel much better now that all this is out", echoed by all present.

The fourth day featured a tape made by Reich in March 1952, to document the aftermath of the Oranur Experiment. It began on March 8th with Reich using a Geiger counter to demonstrate the usually high rates of activity in the observatory building, especially near the large stone fireplace whose rocks had been disintegrating noticeably for the previous two months. Many of us had read Reich's account of this in "The Blackening Rocks" (6), but to actually hear the clicking of the Geiger counter increase from 50 or 100 counts per minute to 2000 cpm as he approached the fireplace, and to 20,000 cpm over the red tile top of one worktable in the observatory, was a spine-tingling experience. One felt the danger, the anxiety of working in a contaminated environment, in a way that books alone could not communicate. Reich:

"The rock gives off secondary nuclear radiation. The primary orgone energy in the atmosphere fights with the secondary nuclear radiation being released from the disintegrating rock. This produces disease symptoms... The processes are self-regulatory. They cannot be stopped... I am taking the personal responsibility of staying here in the observatory despite the hardships...to investigate the effects on the living organism. All the mice in the students' lab have died... Who will or will not be left alive, we shall see... I believe a boy of 18, 19 or 20 fighting in Korea is doing about the same, fighting for freedom."

A number of other workers on the tape also verified observing the position of the Geiger tube, the high readings at the fireplace and tabletop, the crumbling of the fireplace stones, etc.

At a meeting on Saturday 29 March 1952, all of them were present to discuss the situation; Reich, Simeon and Helen Tropp, Lois Wyvell, Michael Silvert, Myron Sharaf, Tom Ross and Ilse Ollendorf. Reich opened the meeting by summing up:

"I personally believe that everything will come out all right. I have exposed myself to extra doses of oranur since January 5th, 1951, have lived in it continuously, and I'm still alive..."

He described the general effects on organisms exposed: initial shock and paralysis, followed by:

"...a ferocious fighting back against the disturbance. This we call the DOR reaction. Third comes adjustment to the situation and a slow adaptation to the requirements of a higher energy level of functioning. The level keeps getting higher and higher here. Will it ever stop? I don't know."

The final day's session was a tape of the final lecture to the 1950 International Orgonomic Conference, Reich's talk of 26 August on "Man's Roots in Nature." This was later expanded upon in Chapters 2 and 8 of Cosmic Superimposition. The full text of the talk is to be transcribed and published in the forthcoming volume of the new journal from the Wilhelm Reich Infant Trust Fund, Orgonomic Functionalism.(7) In this talk, Reich presented his findings on the significance of the "ring" of the aurora borealis, and the common functioning principle with formation of galaxies and hurricanes. Reich spoke powerfully, and my attention was riveted, even during highly technical astronomical sections. During the question and answer session, he was lively, working with profound ideas, yet he still always showed the intensely personal nature of superimposing orgone energy streams. His closing remarks were deeply moving as well, and at no time during the week did we feel more privileged, as if we were personally at a talk by Wilhelm Reich, and as if his remarks were to us personally. Congratulations are due to Mary Boyd Higgins for an excellent job of organization and selection of material. Next year's conference will have a truly tough act to follow!

I wish to let Reich's comments close this report as they did the 1950 and the 1990 conferences:

"Laws which we find in psychiatry guide us right through into the universe... we must start seeing human beings as cosmic concentrations of energy, with cosmic laws. Otherwise we shall get lost in talky talk... We must see children here, not as psychological beings, not even biological beings, but as representations, concentrations of two orgone streams, superimposing, merging energetically so that further creation can emerge."

And finally:

"You will all go home now... I hope you got one thing from here. And that is a bit of courage and conviction. We are not alone... What we need is a bit more FIST... Hardness, rational hardness in our approach to our problems... if you maintain your position, if you are a

6. Reich, W.: "The Blackening Rocks: Melanor", Orgone Energy Bulletin, V(1-2):28-59, 1953.

7. Orgonomic Functionalism, available from the Wilhelm Reich Museum and Trust, www.wilhelmreichtrust.org

learning human being and modest, very modest, then you'll gain respect, very much among the people, and every single one of you will be a center. And this is not a very pleasant position to have. The responsibilities are tremendous, quite tremendous. But, it can be done."

An Additional Perspective
Myron Sharaf, Ph.D.*

Some Comments on the 1990 Conference at Orgonon "The Voice of Wilhelm Reich"

We are indebted to James Strick for his report on "The Voice of Wilhelm Reich". His selection of quotes from Reich are apt, and his comments are generally informative and illuminating. Strick's excitement listening to Reich's tapes are in accord with my own experience hearing some of Reich's recordings. They are extraordinarily alive and vivid. They convey nuances of emphases that cannot be carried by his written words alone.

My only major differences with Strick concern the tape on "Organizations as Beings — Clearing the Air". We do not have the complete transcript of this recording. However, it is not my impression that Strick quotes Reich out of context. Rather, I believe Reich himself at times spoke out of the context of the totality of his work, and his relationships with his students, assistants, and colleagues.

To illustrate: Strick quotes Reich as saying: "No one had the right to say anything in the Orgone Institute... A worker in a clinic or the laboratory is an assistant, a helper in the work". It needs to be stressed, as Reich did on the tape, that one of his big concerns at the time was: "I would like to ask you to help me kill the license ideology in our midst". Reich was very anguished over emotional plague utterances and actions over undisciplined behavior, over mind-numbing irrelevancies that broke his contact with his thought and work.

It was typical of Reich to focus completely on an important challenge facing him at a given time. However, neither Reich's utterances nor Strick's comments call attention to the fact that during this period Reich was giving tremendous freedom to several researchers working with him. Robert McCullough has well described this aspect of Reich, an aspect not at all conveyed by Reich's emphasis on people being "assistants". He wanted someone like McCullough specifically not to be an "assistant" only, but to pursue independent research, which McCullough did.(1) It was only when McCullough violated important orgonomic principles that Reich sharply interfered.

It is true that he expected someone directly assisting him, e.g., a secretary, to do what he asked her to do. He did not want her to start arguing with him about the existence of the orgone energy. However, even in this context the statement "No one here has the right to say anything" suggests that Reich wanted mute automatons around him. That was not the case. Reich certainly wanted a secretary to tell him when he had misspelt a word and to make other rational suggestions.

Strick devotes some space to the section of the tape which involved Reich's interrogating those present about the rumor that he received "kickbacks" from physicians practicing orgonomic therapy. Strick expresses unreserved admiration for the way Reich handled the matter (and relates some amusement "One could imagine the squirming" on the part of the doctors present).

Having participated in several "confessional" meetings of this kind led by Reich, I find Reich's concern admirable, but not the way he pursued it. His method seems to me to have been authoritarian, as I remember it, and as it is recalled by Strick's report. I find the group atmosphere chilling. Nor is it my experience that people learn well when they "squirm" and feel humiliated.

And here I join my central issue with Strick. Strick defends Reich against certain unnamed authors: "The kind of bossy, inflexible tone of voice and manner of dealing with people conjured up by those writers as their image of Reich was simply not there in what we heard. We heard the highly animated voice of a man convinced of the seriousness of his work and completely rational about his reasons".

Many people seem to have difficulty grasping the idea that Reich may at times have been "bossy and inflexible" and at other times anything but. With them it is either-or. However, it was not either-or. It was both-and. Reich was trying hard and admirably to cut out the tumor of irrationality in the people he dealt with. In the process, he sometimes also took the healthy tissue of rational inquiry and dissent. This kind of authoritarianism rarely happened in clinical or scientific realms. It happened more frequently in social issues, both microcosmic and macrocosmic. To cite one example, I would mention Reich's insistence to the point of dogmatism during the late 1920s and early 1930s that psychoanalysis and the Marxist political parties were inextricably linked. Reich corrected this error. He did not have the chance to correct his possible mistakes from the period the tapes were made.

We need to honor Reich's extraordinary achievements of this time made under extraordinary stress. However, we do neither him nor orgonomy a service by embracing his every word and action as having been "completely rational".

* Lecturer at Harvard Medical School, Department of Psychiatry (18 Duncklee St., Newton Highlands, MA 02161). Author of *Fury on Earth: A Biography of Wilhelm Reich.*

1. Sharaf, M., *Fury on Earth: A Biography of Wilhelm Reich*, St. Martin's Press, NY, 1983, pp.407-409.

CONFERENCE REPORTS (Continued)

Reviewed by James DeMeo, Ph.D.

The First International Symposium on Circumcision
Anaheim, California, March 1-3, 1989: NOCIRC (National Organization of Circumcision Information Resource Centers) PO Box 2512, San Anselmo, CA 94960.

NOCIRC, the National Organization of Circumcision Information Resource Centers, sponsored this First International Symposium on Circumcision, bringing together scholars and clinicians from around the world who have studied various "practices" that affect infant health. Dr. George Denniston educated the Symposium participants about the natural history of the foreskin, that the foreskin of the young boy is fused to the head of the glans penis until around age 10-14, by which time the two tissues naturally separate from each other. Dr. Paul Fleiss reviewed what is known about the care of the intact boy's penis, which can be easily summarized: "Leave it alone". This phrase became the unofficial motto of the entire Symposium, and was often repeated by speakers and audience participants following the other presentations, which dealt with the pathology and upsetting consequences of the circumcision mutilation.

During "routine" circumcision, the sensitive foreskin must literally be ripped and torn away from the equally sensitive glans penis, a process that is very painful and quite bloody. Color photos of the mutilation process, presented by some of the speakers, made this quite clear. Dr. James Snyder and Richard Steiner presented heartbreaking evidence of severe mutilations of boys who had lost small or large portions of their penis due to botched circumcisions, and the various reconstructive surgeries used to help some of the infant boys. From their presentations, it was clear that much of the true pathology associated with "routine" circumcision of baby boys in hospitals, and in private religious ceremonies, is not being studied or reported by the medical community.

Clinical cases of men with deformed penises, either due to accidental removal too much foreskin, or removal of parts of the glans, were presented. Young men, and even mothers spoke out at the Conference about the tragedies of painful erections, bent-penis erections, and genital anesthesia, due to excessive scar tissue from botched circumcisions. I must confess to having been rather ignorant of some these issues, even though only a few years before I had reviewed nearly 60 different published papers and books on the circumcision issue; it was absolutely clear that the true pathology associated with circumcision is much greater than has been reported.

Several of the papers presented at the Symposium addressed the typical "reasons" given for circumcision, demonstrating that there is no legitimate scientific or medical justification to perform it. For example, none of the claims for health benefits from circumcision have survived careful scientific scrutiny. Only cultural and religious reasons could be identified, and the rationality of these were challenged also by many speakers. Dr. James Prescott's paper on "Genital Pain vs. Genital Pleasure; Why One and Note the Other?" and my own paper on "Attacks on the Sexuality of Children: The Geography of Genital Mutilations" addressed the cross-cultural and historical evidence on genital mutilations — this evidence demonstrates that circumcision is a bodily mutilation, performed for nearly identical reasons in both "civilized" and subsistence-level cultures; they

provide an outlet for sadistic aggression of adults towards children, with an emphasis upon pain and sex-negation in childrearing, and certainly do not benefit the child. Fran Hosken's presentation identified similar underlying motives at work in female genital mutilations, which today are mostly confined to Africa. Marilyn Milos, Symposium organizer and tireless champion of the rights of infants to an intact body, spoke about the tremendous opposition to reform that has been given by obstetrician/gynecologists, who profit from infant circumcision, which constitutes the most widely-practiced surgery in America today. Dr. Michel Odent, author of books on natural childbirth, spoke with puzzlement about circumcision: "It is strange for me to even speak about this practice, as we do not do it in France; there is no need."

A few of the doctors at the Conference became upset because no speaker made a case for the supposed "benefits" of circumcision. Some openly expressed dismay that "equal time" was not being afforded to the proponents of circumcision, and others were upset about perceived "doctor bashing" over the practices. Conference organizer Milos pointed out that several of the proponents of circumcision had been asked to speak at the Symposium, and respond to open criticism from the floor; but they had declined the invitation, knowing that they would be called, some for the first time, to defend their data and conclusions in a very public manner. Regarding the "doctor bashing" charge, it was true that the many midwives attending the Symposium were, to a person, against the practice, and horrified about the evidence being presented. Also, early in the second day of the Symposium, a man in the audience had openly shouted "butchers! butchers!" during Dr. Snyder's presentation, which recounted cases of penile loss or extreme deformity from botched circumcisions. Many persons in the audience were openly crying after his moving presentation.

From many speakers it became quite clear that the medical community, notably the obstetricians and pediatricians, was engaged in a big evasion and cover-up about the true effects of the ritual circumcision mutilation, and some very sharp and penetrating questions about the role of the medical community in continuing with the practice came from the floor. Some of the doctors attending (but not presenting) at the Symposium, and who at first openly expressed support for circumcision, became understandably uncomfortable at the way things were proceeding. They could not marshal any scientific or clinical defense for the practice, which they had been doing for years, and the hidden antisexual, antichild psychological motives for the practices were being laid bare by various speakers. Emotional reactions from some individuals was inevitable, especially when one could hear about and even see, in large color photo slides at the front of the hall, the horrid and disastrous life-long consequences of what some apologist-physi-

International Symposia on Circumcision

cians had been calling a "mild, routine practice". As Dr. Denniston stated, a primary rule of every physician is: *Primum non cere! First, Do No Harm!* And circumcision of infant boys (and girls in Africa) was a clear and unacceptable violation of that rule. This Symposium put the issue into sharp focus: one could be on the side of "popular culture" and "medical tradition", or one could be on the side of the infant. The evidence presented at the Symposium spoke heavily in favor of the infant's biological needs to have its genitals left alone, and its rights to an intact and whole body. Will the world listen?

The 17th Annual Cancer Control Society Convention

Pasadena, California, July 28-30, 1989: Cancer Control Society, 2043 N. Berendo St., Los Angeles, CA 90027.

This Convention has been an annual affair in California for many years, and brings together speakers on a variety of unorthodox and "unofficial" or "unaccepted" treatments for cancer and other degenerative diseases. Dr's. Eva Reich and Richard Blasband have presented information on Reich's Cancer Biopathy at past conventions, and the Director of the Cancer Control Society, Elaine Rosenthal, is the daughter of a woman who worked with Reich for many years: Dr. Helen MacDonald (co-editor of *Orgonomic Diagnosis of Cancer Biopathy,* xerox available from the *Reich Museum,* PO Box 687, Rangeley, ME 04970). I accepted an invitation to speak at this convention with some reservation, as I am not a medical practitioner, and could not speak with the authority of someone who was experienced in the actual treatment of such diseases. However, my new book, *The Orgone Accumulator Handbook* was just published (www.naturalenergyworks.net), and it discussed the medical effects of the orgone energy accumulator — I had also returned from a Conference in Germany with doctors who had presented their observations regarding the accumulator's health benefits. Hence, my own presentation to this Convention, on "Life Energy and Its Role in Health and Disease", was in the manner of a "report from overseas", mixed with observations from my own accumulator experiments on plant life, water evaporation, and other physical factors. My talk was well received, and stimulated a number of interesting discussions. Many other interesting speakers presented at this Conference, some of whom were known around the world for their pioneering works. The various presentations suggested a major common theme: There are numerous, reasonably effective therapies and approaches for degenerative illness, but virtually all of them are suffering from repressive attacks by government agencies and big medical organizations. Here are samples of a few of those presentations:

1. Charlotte Gerson, daughter of Dr. Max Gerson, presented on the Gerson Therapy, a dietary regimen for degenerative disease that is today used around the world (see *A Cancer Therapy, Results of 50 Cases*, by M. Gerson, available from the *Gerson Clinic*, PO Box 430, Bonita, CA 92002, or NEW, PO Box 864, El Cerrito, CA 94530). Max Gerson lived in Forest Hills around the same time that Wilhelm Reich did, though it does not seem that the two men met each other. Gerson's dietary, detoxification therapy is often recommended by orgonomic physicians as an adjunct for use with the orgone energy accumulator, and is considered quite effective. The Gerson Clinic is located in Tijuana, as the technique is nearly banned in the USA.

2. Dr. Virginia Livingston presented on her innovative vaccine and dietary therapy against cancer. Livingston also claims the discovery of a specific cancer "microbe" that closely parallels Reich's findings on the bions and t-bacilli. Her clinic in southern California (3232 Duke St., #114, San Diego, CA 92110) was under intense critical scrutiny by establishment doctors, who have been trying to shut her down for years.

3. Dr. Vicki Hufnagel, author of the book *No More Hysterectomies* (New American Library, NY 1988; order from the *Institutie for Reproductive Health,* 8721 Beverly Blvd., Los Angeles, CA 90048), and pioneer of alternative surgical methods for saving the sexual organs of women, instead of merely cutting them out. In her book, Hufnagel pointed out that some 650,000 hysterectomies are performed each year, the vast majority of which were completely unnecessary. For publicly exposing this lucrative but dirty business in female castration, and pushing for legal reforms, various doctors with political power attacked Hufnagel with a vengeance; she lost her hospital privileges, and is now fighting to save her medical license.

4. Dr. James Privitera spoke about live blood examinations using the dark field microscope, in a manner quite reminiscent of Reich's own test on live blood. Privitera is currently under assault by medical boards for using a dark-field microscope in his medical office. Apparently, only "pathologists" are allowed to use dark-field microscopes, and so Dr. Privitera's microscope had been recently confiscated in a raid on his office by medical police agents!

5. Peter Barry Chowka spoke generally about the growing medical tyranny and repression in the USA, and specifically about the case of Harry Hoxsey and the Hoxsey herbal therapy. Chowka produced a special videotape documentary of the Hoxsey treatment, and the repression of Hoxsey by a political collusion and conspiracy between the American Medical Association and Food and Drug Administration. Hoxsey is dead now, but the *BioMedical Clinic* (PO Box 727, 615 General Ferreira/Colonia Juarez, Tijuana, B.C. Mexico) continues with his herbal treatment methods.

Readers with an interest in holistic, non-toxic treatments for degenerative disease are encouraged to contact these various institutes, and to also consult the listing given on page 48 in this issue of the *Pulse*.

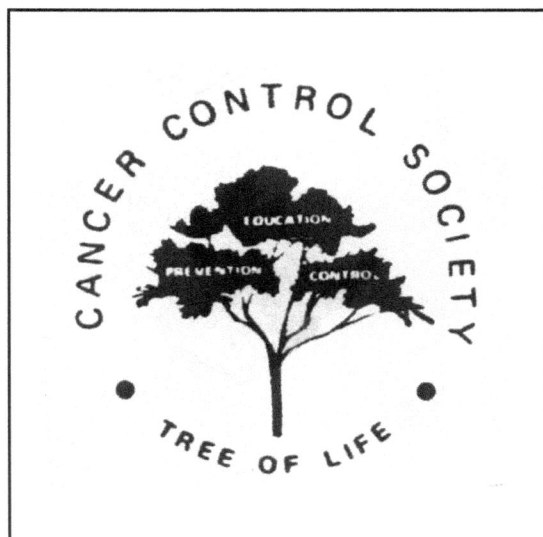

Wilhelm Reich Conference

Berlin, Germany, October 21-22, 1989; Wilhelm Reich Society, ~~Karlsbergallee 25-E, D-1000 Berlin 22~~, Germany.

This Conference took place at a time when West Berlin was still a walled city, surrounded by contingents of soldiers and tanks from America, Britain, France, and both West and East Germany. Traffic signs in streets gave indications for tanks and military vehicles, as with "cars go this way, tanks go that way". It appeared to me, at first, as a puzzlement and contradiction that a group of physicians, social workers, professors, and natural scientists could have undertaken an extended program of research into physical orgonomy and even the medical applications of the orgone accumulator in this apparently restricted environment. However, I learned that Berlin was much like a "San Francisco" of Europe, that is, a European refuge, haven, and breeding ground for all sorts of new and unorthodox ideas (some healthy, some not). It was additionally a completely separate political entity from the rest of Germany, and so capable of going its own way. Young men wishing to escape the compulsory military draft of West Germany came to West Berlin, which had its own ethnic minority neighborhoods (rare in Germany), plus a system of independent passports and citizenship. Most people could not stand the contrast of freedom within the very large cage of West Berlin, while others enjoyed life and thrived there.

The workers in Berlin with interests in Reich had formed a very lose, work-democratic association, with shared responsibilities for educational activities, such as workshops and conferences, and for the publication of a research journal, titled *Emotion*, now into its 10th annual volume (available from ~~Karlsbergallee 25-E, D-1000 Berlin 22~~, Germany). Individuals and small groups of workers were also engaged in the evaluation or application of various aspects of Reich's work. For example, Dr. Heiko Lassek's paper is presented in this issue of the *Pulse*, and so the reader will gain a impression of some of the clinical work being undertaking with the orgone accumulator. Lassek is one of a group of perhaps 6 or more physicians in

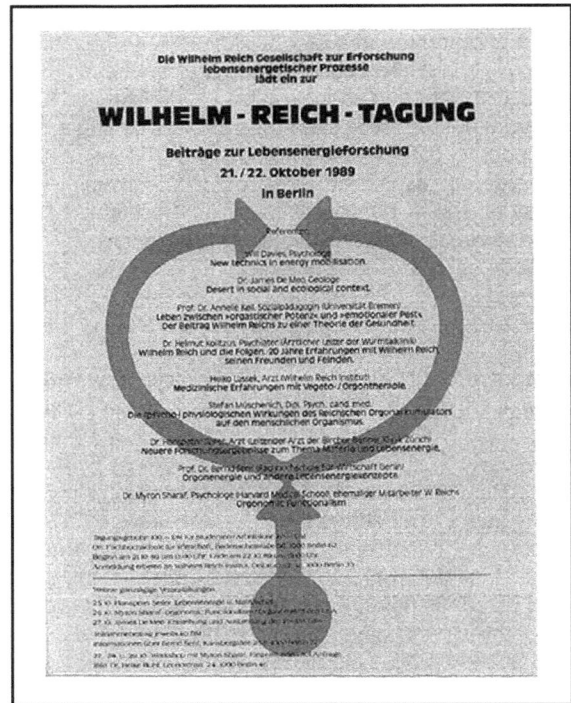

Berlin who are now gaining practical, hands-on experience with the clinical use of the orgone energy accumulator. Dr. Bernd Senf is a professor of political economy at the *Fachhochschule für Wirtschaft*, where for the last 15 years he has been teaching a course on the *Life and Work of Wilhelm Reich*, and another on *Life Energy Research*. He regularly draws from 100 to 300 students to his classes, with guest speakers such as Dr. Eva Reich, Lassek, and others. I was able to attend one such lecture by Dr. Eva Reich, who now spends much time lecturing and traveling in Europe, stimulating great interest in her father's findings. She has worked toward establishing a natural childbirth program in East Berlin, initiated even before the Berlin Wall was torn down. On one occasion, I traveled with her to East Berlin to meet several of the childbirth physicians and midwives. She lectured there in a church, to overflow crowds, on natural childbirth and infant care methods. Her lecture in West Berlin, "Personal Recollections about Wilhelm Reich" was also attended by several hundred people, a startling contrast to the situation in America, where attendance for orgonomic events is not nearly so great.

During the Conference, Lassek presented a paper on "Medical Experiences with Vegeto/Orgone Therapy", covering his experiences with orgone accumulator treatment of diseases, and the Reich energetic blood test. Senf discussed "Orgone Energy and Other Life Energy Concepts", integrating the findings of Reich with several German workers such as Shauberger, whom I previously knew little about. Other presenters from Germany included Stefan Müschenich, a psychologist and medical student who, with Rainer Gebauer, had recently completed a double-blind, controlled study on "The Psycho-Physiological Effects of the Orgone Energy Accumulator."† Dr. Hanspeter Seiler, of the Bircher-Benner Clinic in Zürich, Switzerland, also presented a cross-cultural study on spiral formations in the

---

† Published in German as *Der Reichsche Orgonakkumulator* , Nexus Press, ~~Fichardstr.38, 6000 Frankfurt 1, Germany, 1987 (available from NEW, PO Box 864, El Cerrito, CA 94530~~). Abstract published in the *Pulse of the Planet*, 2:22-23, Fall 1989.

artwork of ancient and contemporary peoples, finding that the spiral form was linked with the relatively unarmored, matristic peoples of Earth. Seiler argued that spiral-form art was the product of direct observation of the spiraling orgone energy units in nature, something only an unarmored, or relatively unarmored life system could directly perceive. By contrast, more armored, patrist peoples gave up the spiral form in their artwork. Seiler was also very much interested in the physics of the orgone energy; he has engaged in studies on the accumulator To-T effect, and has made some new findings on gravitational functions that have yet to appear in English. In fact, I found myself quite frustrated during this Conference, as meine Deutsche ist schrecklish, and these presentations were rich in new findings.

My own presentation was about "Deserts in Social and Ecological Context", on Saharasia and the origins of armor. Other English presentations were made by Will Davis, an American psychologist living in France with an interest in "New Techniques in Energy Mobilization", and by Dr. Myron Sharaf, biographer of Reich, on the topic of "Orgonomic Functionalism". Davis presented on new gentle-touch methods for therapeutic release of pent-up emotional tension and armor, while Sharaf's talk emphasized the need for workers in orgonomy to be open to new approaches while still maintaining clarity and focus on central, demonstrated orgonomic principles. Lack of central focus, he pointed out, led to compromising away the essentials of orgonomy, while a too-rigid approach, of censoring all new approaches, led to dogma and immobilization. He argued for a functional combination of the apparent opposite aspects, in a deeper common functioning principle.

Other speakers presented at this Conference, but I was either not in attendance at the time, or did not get enough translations to make meaningful comment. I also must elaborate a bit about other workers in the region who did not present at the Conference, but who have been making significant and important contributions to orgonomy in Germany (and my listing here is by no means comprehensive). Berlin is also the home of John Trettin, who with Beate Friehold and others, established an orgone accumulator factory in the 1980s. Trettin and associates have published a small book on the accumulator, and have constructed and sold around 400 full-sized accumula-

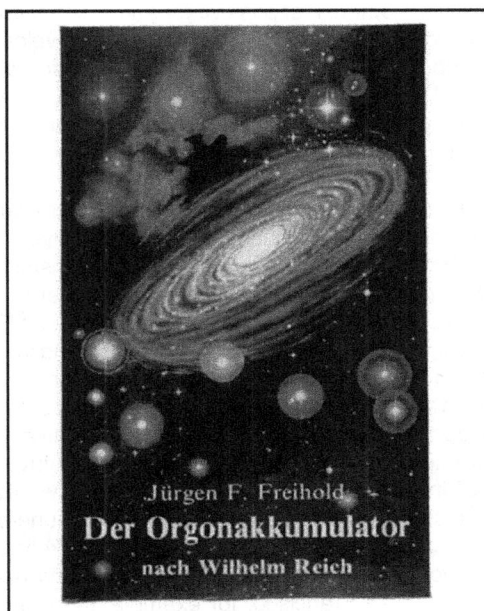

tors in Germany; they are well made, and priced economically. Trettin was brought into contact with Reich by the late Dr. Walter Hoppe, who worked with Reich in the 1950s and translated many of his articles from the American period into German. Trettin has additionally developed an ambitious video archive, which holds a large number of interview and lecture videotapes from various American and European researchers with an interest in Reich and orgonomy. (*Orgonomisches Video Archiv und Orgontechnik*, ~~Mullerstr.145, 1000 Berlin 65~~, Germany)

The *Zentrum für Orgonomy* in Eberbach is directed by Dr. Manfred Fuckert and Dr. Dorothy Opferman-Fuckert. Students Hoppe also, the Fuckerts have engaged in significant clinical and biophysical research with the orgone energy accumulator and Reich energetic blood test. Dr. Opferman-Fuckert's clinical report on the medical use of the accumulator was only recently translated into English and published in the *Annals, Institute for Orgonomic Science* (Volume 6(1):33-52, September 1989, available from ~~PO Box 304, Gwynedd Valley, PA 19437~~). The *Zentrum* also recently began publishing the journal *Lebensenergie* (~~Memelstr.3, Eberrbach 6930~~, Germany).

My general impression, backed up by numerous facts, is that the physical research aspects of orgonomy are flourishing in Germany in a manner that probably would have taken place in the USA in the 1960s, if Reich had not been imprisoned and orgonomy repressed. The Germans, mostly younger people in their 30s, were born after the collapse of the Third Reich, and have been able to develop their work free of political restraint. The 1950s FDA repression and government crackdown against orgonomy has acted as a continuing restraint and impediment to basic scientific research and progress in the USA. The Germans I met were relatively free of such concerns, using the accumulator quite openly, even in medical practices and clinics, to the point that many could not even comprehend the reality of the situation in America. During the entire Conference, for example, a large human-sized orgone accumulator was in the Conference hall, available to be examined and tried out by anyone who wished to do so. Hopefully, English translations of the German accumulator studies will help the situation of neglect in America to change for the better.

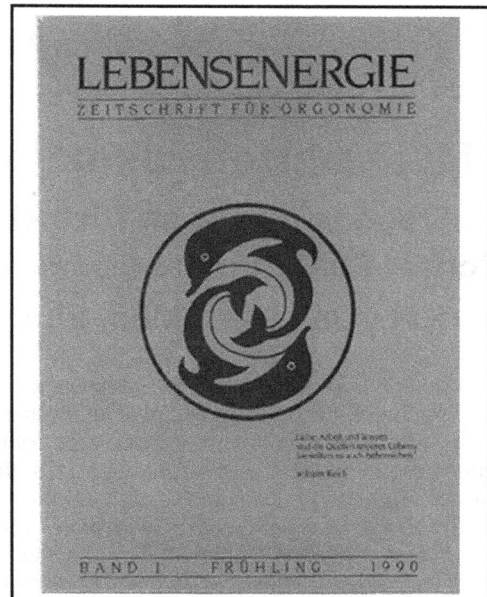

The Myth of Heterosexual AIDS:

How a Tragedy has been Distorted by Media and Partisan Politics*

by Michael Fumento (Basic Books, NY, 1990, 411 pages, $22.95)

Reviewed by James DeMeo, Ph.D.

Michael Fumento's book, in this reviewer's opinion, is a concise and thorough source of accurate information on the problem of AIDS (Acquired Immune Deficiency Syndrome). This includes the biomedical and social aspects of the AIDS disease which, according to Fumento, is currently the subject of widespread propaganda regarding its epidemiology. Fumento's book also reveals the political shadow that has hidden many facts about this disease, the risks of which, to people outside of major risk factor groups, have been wildly exaggerated. Below is a review of some of the salient points from this important book, which find support from other sources which are cited at the end of the review.

Like most people, the issue of AIDS came to my awareness rather suddenly in the early 1980s. What first was presented as a blood-borne agent afflicting primarily promiscuous homosexuals and intravenous drug users who shared needles, suddenly became a major concern for heterosexuals, even teenagers. This was so, even in the rural hinterlands of Illinois and Iowa, where I was teaching to large classes of young, energetic college students. In my classes, we often discussed research on human behavior, to include human family life and sexual behavior in different cultures. My research, and the work of others, had indicated that cultures with a high degree of sexual freedom had low levels of social violence (see the articles by DeMeo and Prescott in this issue of the Pulse). In these discussions, serious question were often raised: "What would our society be like if a true sexual freedom were to exist". "Would social violence decline?" "Would people be monogamous or polygamous?" "Would we become a nation of pacifists?" "Would people become more or less responsible sexually?" "Would there be more or fewer unwanted babies?" One new question increasingly began to be asked: "What about AIDS?"

This latter question about AIDS became foremost, overshadowing all others, precisely because it was so much less theoretical than the other questions about social violence and distant cultures. After all, every television and radio station was constantly saying that everyone was at risk for AIDS, including or even especially teenage heterosexuals in the rural Midwest. Sexual abstinence, or so-called "safe sex" practices involving devices that were guaranteed to reduce or kill sexual feeling, were promoted as the only fail-safe proofs against AIDS. These statements from the press, and even from the Centers for Disease Control, backed up by slick flyers and publications out of Washington, D.C., were all quite clear — and yet, I had suspicions that something was not right with it all. These suspicions were heightened as various antisexual church groups

and clergymen joined into the chorus, expounding upon AIDS as "god's plan" for separating the saved from the sinner, or for punishing the "wicked" homosexual. This was hardly science, and yet, though the words were different, the same tone of voice emanated from the Centers for Disease Control and other scientific institutes.

By the late 1980s, it was clear that a "heterosexual AIDS epidemic" had not taken place in the USA, or Europe, and that some of the statistics had been cooked. While there certainly was a very serious epidemic among male homosexuals, and among intraveneous drug users, with some infection spreading to a much smaller number of individuals who were the sexual partners of those two groups, the epidemic had not spread significantly into heterosexual populations. What had taken place was, instead, a widespread misrepresenting of data, for purely political reasons. In the USA, the syndrome was still largely confined to the two major groups engaging in high-risk practices associated with blood transmission of the AIDS virus, namely male homosexuals and bisexuals, and intraveneous drug users. The overwhelming majority of cases of "heterosexual" transmission — which in any case never constituted more than a small percentage of the total — were mostly confined to the sexual partners of persons in those high risk groups. A smaller number of AIDS infected infants appeared, having contracted the syndrome from their infected mothers, and other individuals who contracted the syndrome via infected blood transfusions. But even here, it was noted, the AIDS virus is very difficult to transfer. Only a minority of babies born to AIDS-positive mothers test positive for AIDS, and many people who received AIDS infected blood have never developed infections, antibody response, or symptoms. The so-called "epidemic" of AIDS has not entered the heterosexual sphere to any significant extent, mainly because it was not, and still is not, easily transmittable through ordinary means, sexual or otherwise.

The actual epidemiological evidence challenged the official position of "abstinence as the only prevention", and reaffirmed the primary significance of blood transmission factors. Gay males and intravenous drug users were most seriously at risk, particularly in certain cities such as San Francisco or New York, where large numbers of gay men and intravenous drug users were concentrated. If someone had a regular sexual partner who was in one of the two major high risk categories — that is, a partner who engaged in active or receptive promiscuous anal intercourse, or intravenous drug use involving the sharing of injection needles — then their risk would slightly increase also. If an individual also used immune-depleting drugs or antibiotics, then it appeared more likely an AIDS exposure would take root with AIDS infection and later, AIDS symptoms. In San Francisco, for example, the promiscuous,

* For more information on this subject, consult this website: www.orgonelab.org/aidscrit.htm

bath-house, anonymous sex-party lifestyle prevalent in some gay communities, where a single man might have several hundred different, mostly anonymous sexual partners per year, also has suffered from high levels of drug-taking, as well as repeated doses of antibiotics to stave off multi-spectral infections. Gonorrhea, syphilis, hepatitis and other infectious diseases were deeply rooted in the gay community long before AIDS appeared, solely because of the highly promiscuous, anonymous, and unsanitary nature of many gay sex practices. These infectious diseases also exist among heterosexual populations, and are often transmitted via heterosexual genital contact. But not AIDS -- to any significant degree -- which in nearly all cases requires a blood-borne route of infection. For this reason, AIDS is a relatively difficult disease to catch. Certainly, casual contact will not transmit AIDS, and even heterosexual male-female, penile-vaginal intercourse will not allow any easy or quick transmission of infection, even when one of the two individuals has clearly tested positive for AIDS. What is needed is not simply sexual relations, but a direct route for blood-to-blood exposure.

For example, a woman with a promiscuous bisexual male partner is at a much greater risk than a woman with a promiscuous heterosexual male partner, because the bisexual partner might be engaging in active or passive anal intercourse with other bisexual or strictly homosexual men. A heterosexual man and woman, by contrast, would have an exceedingly low risk for AIDS, even if one or both were somewhat promiscuous with other heterosexuals. However, if one of the heterosexual individuals was using intravenous drugs, or if the man was bisexual, their risk would increase. Studies fail to indicate that the absolute number of sexual partners has anything to do with the risk of AIDS infection. In select brothels in Nevada and Amsterdam, for example, where anal intercourse and intravenous drug use are forbidden, the risk of AIDS through vaginal or oral intercourse is demonstrated to be exceedingly small, even nonexistent. Street prostitutes in New York and elsewhere have tested positive in high percentages mainly due to their use of shared intravenous drugs. This also appears to be a major factor in the current increase of AIDS among inner-city populations of teenagers, where such drug use is more common. The lesson from the data on AIDS is therefore not "abstinence", or even questionably effective "safe sex" practices. The lesson is, instead, *know your partner!*

Some examples can be given regarding problems with the AIDS data. It is known that people often lie about their sexual practices. How would the average man or woman respond to the question "How many times have you had anal intercourse in the last year?", or "Are you homosexual or bisexual?" "Is your husband or boyfriend bisexual?" "Do you shoot drugs?" Obtaining the facts from someone found to be infected with AIDS becomes all the more difficult. It is far simpler to identify "heterosexual transmission" as the means of infection than to risk what is still to many a tremendous personal and family disgrace — "I must have caught it from a prostitute" is more acceptable to many people than to say "I am gay", or "I shoot drugs". In the case of immigrants, drug use is a deportable offense, and some major increases in the AIDS data have appeared when entire blocks of immigrants were changed, for purely political reasons, from "unknown" classifications into the "heterosexual" category. The fact is that in the USA, AIDS has not broken out into the heterosexual population in any major way, or even in a minor way, except within those groups still engaged in high risk practices. If it were transmissible through heterosexual means, then by now AIDS would truly be at epidemic proportions, following patterns similar for syphilis and gonorrhea. This is not the case.

But what about Africa, where AIDS is truly more widespread? AIDS in Africa affects entire villages, both men and women, and in these areas, it truly takes on nightmarish proportions. But the fact is, in Africa, the situation is dramatically different than in the USA or Europe. AIDS spread widely in Africa because of social and hygenic factors which do not significantly exist in the USA or Europe. Fumento does not elaborate much on the African AIDS question, but some examples of these social and hygenic factors can be listed: 1) female genital mutilations are widespread in parts of Africa, and these lead to bleeding during heterosexual, vaginal intercourse; 2) genital ulcers and sores are also widespread in Africa, providing additional heterosexual blood transmission routes; 3) AIDS infections are often masked by tuberculosis or syphillis infections, or vice-versa, in a manner that is not well understood -- and these latter two diseases are common in parts of Africa; 4) anal intercourse may be preferred as a population control measure in some areas; 5) cultural institutions such as child betrothal exist in Africa, where grown men have forced sexual intercourse with small girls following arranged marriages, leading to more genital bleeding; 6) unsanitary conditions exist in many hospitals, and injection needles may be recycled repeatedly, without sterilization; 6) male bisexuality may be at high levels in some areas, but due to stronger homosexual taboos, such information is very difficult to obtain. In short, the data from Africa cannot be extrapolated meaningfully to either the USA or Europe. And yet, the African data are constantly being raised as a frightening example of what's in store for the rest of the world.

Fumento also failed to seriously review a number of important penetrating criticisms which have been raised regarding the special status accorded the AIDS virus, rendering it different from all other viruses. For example, a large number of people who have tested positive for the AIDS antibody, and who have been classified as "AIDS-carriers", have never had the AIDS virus cultured or observed in their blood or tissues, nor have AIDS symptoms been observed in such individuals — only the AIDS *antibody* has been observed, but not the virus that prompted the antibody. There is a rational opinion which says these people are not at risk, and should not be classified as "carriers". Dr. Peter Deusberg, a biologist at the University of California at Berkeley, has strongly criticized the whole fabric of the "AIDS infection" argument, given that in the USA there are a growing number of such people who test positive for AIDS antibodies, but in whom *no AIDS virus or AIDS symptoms can be found.* He asserts that these people were exposed to, but successfully *defeated* the virus, developing the well-known antibody response. He says these people should celebrate, as they do not have the virus, and most will never develop symptoms. The fact is, this latter group of "exposed, asymptomatic" individuals grows larger year by year.

Medical people with a vested interest in perpetuating the heterosexual AIDS epidemic mythology are also making several fallacious claims: 1) The AIDS latency period is getting longer each year, growing by about one year for each year that passes, and 2) Many who test positive for AIDS antibodies but who remain symptom-free owe their lives to better treatments for AIDS. The first claim becomes more tenuous with each passing year, while the second has no empirical support, and appears designed to promote continued widespread use of untested medicines and vaccines. Indeed, one of the more popular medicines for AIDS, called AZT, causes a severe shock to the immune system, and may itself be responsible for creating immune system disorders and AIDS-like symptoms among antibody-positive groups who, without the "medicine", would not have developed such symptoms in the first place! Today, there is also much talk of AIDS vaccinations, with the hint that once a vaccine is developed, it will become mandatory for every young child, to "protect" them from horrible sexual death — like holy water, to fend off demons.

The prevailing antisexual hysteria and evasion of essential facts about AIDS has grown such that the sexual lives and happiness of many people have needlessly suffered, and very few scientists have dared to raise an objection. When they do, a variety of social and economic forces are marshalled against them. "Official science" says that abstinence or condoms are the only rational solutions, and no scientists employed in a Federal position can challenge this "Truth" without risking attack and unemployment. Evidence also suggests that "safe sex" practices are relatively ineffective for curbing the very real epidemic among the high-risk gay groups, but you would not know this from the widely publicized propaganda for condoms. While properly used condoms will prevent pregnancy, and halt the spread of a variety of sexually transmitted diseases, they do reduce or kill sexual feeling, and so a great deal of compromising takes place during their use — they can slip off a man's penis during intercourse, and can tear or break, particularly the more sensitive varieties made from thinner materials. The condom companies certainly don't object to the "official" pro-condom scare-tactics and propaganda, but the effectiveness of condoms in preventing the spread of AIDS is another question entirely. Among heterosexual teens in New York City, clean injection needles and effective drug treatment programs to reduce intraveneous drug use, not condoms, are needed to stem the problem.

Abstinence enforcement has never worked to suppress or stamp out adolescent sexuality — it only works to inflict confused feelings of sexual fear, anxiety and guilt. Heterosexual activity in teenagers may be repressed by increasing their sexual anxiety, but by denying young people their natural biological rights to physical love and affection, distrust, guilt, and confusion is created about their own sexuality, and many are driven into a state of intense frustration and misery. Abstinence education can thereby push young people toward increased drug use, and toward homosexuality, the two main areas of increased risk for AIDS. In the final analysis, this major thrust of the "AIDS education" program, namely the "abstinence" propaganda aimed at schoolchildren, may actually work to *increase* the incidence of AIDS. But this fact is hardly acknowledged outside of sex-economic circles.

Gay-rights organizations also have their own agenda regarding AIDS, and they have often raised a strong protest towards any program aimed at changing sexual practices. Fumento cites numerous examples of harassment of scientists and authors by the various gay-rights groups, who have tried to "heterosexualize" AIDS, and divert attention away fromunsanitary, high-risk gay sex practices. Bath houses in San Francisco, for example, were closed quite late in the very real homosexual AIDS epidemic, mainly due to charges that city officials were being "homophobic"; this delay, and the connected evasion of the facts of AIDS epidemiology, led to the deaths of many thousands of gay men who might otherwise have avoided infection. The various Christian moralists have also been disinterested in the facts about AIDS transmission, and have spread false or misleading information about the risks for casual infection, leading to many regrettable incidents. The AIDS mythology conveniently supports the anti-sexual agenda of fundamentalists. AIDS was and is a sex-hater's dream come true. AIDS is also big business, and millions of dollars are now going into various big-time medical research programs, education programs, and social programs, few of which emphasize directly to high risk populations just what their actual risk factors are — these continue to be the blood transmission routes of promiscuous anal intercourse and the sharing of infected drug needles.

Michael Fumento has done a marvelous job in bringing together materials and data from a variety of sources on the AIDS problem, touching nearly all the points given above in a thorough and well-documented manner. The points it raises have not been undermined by time or new findings. One major telling point is that his book was reviewed in the scholarly journal *Science*. Instead of itemizing and refuting the major points in his book, however, the reviewer instead *trashed* the book in a most emotional, condemnatory manner. This was, in itself, telling.

"Everyone is at risk for AIDS", we are told, and "sex can kill you". "Sleeping with someone is like sleeping with all their partners for the previous seven years" goes one television commercial for condoms. While the AIDS epidemic has failed to materialize among the general heterosexual population, AIDS symptoms have actually increased among homosexuals and drug users (and their sex partners), and among homeless teens and other low-income, distressed populations where not enough was done to curb actual high-risk practices. And why should we expect the high-risk practices to quickly decline, when the information from the medical community and news media for years has failed to emphasize in plain language the real risk factors involved. After all, "Isn't AIDS really a problem for everybody?"

Additional References
1. Gould, R. E.: "Reassuring News About AIDS", *Cosmopolitan*, 204, January 1988.
2. Harman, R. A.: "An Evaluation of the Risks of AIDS Transmission", and "The Emotional Plague and the AIDS Hysteria", *Journal of Orgonomy*, November 1988.
3. Deusberg, P.: "HIV is Not the Cause of AIDS", *Science*, 241:514, (29 July 1988).

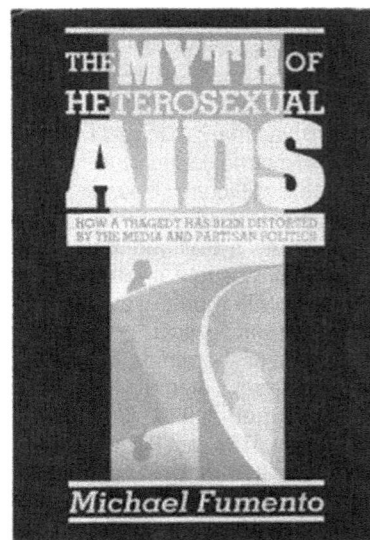

OTHER BOOKS OF INTEREST

* The German-Language 4th Edition of Thure von Uexkull's **Psychosomatische Medizin** (Urban & Schwarzienberg, Munich, 1990) has a new chapter on "Body Oriented Psychotherapy" which is generous in its favorable discussion of Reich's medical contributions. The chapter was authored by Han Mueller-Braunschweig, and is notable because Uexkill's volume (updated and republished every year or two) is considered to be the definitive European work on psychosomatic research. It is not known if this work is available in English.

* Robert Anton Wilson recently published two books in which Reich, and the irrational attacks against him, are central themes. In **The New Inquisition: Irrational Rationalism and the Citadel of Science** (Falcon Press, 1987, ISBN 0-941404-49-8), Wilson lampoons various "quackbusters" like Martin Gardner, and "skeptics" organizations such as CSICOP, as well as the various major scientific institutions that feed from their propaganda. His work **Wilhelm Reich in Hell** (Falcon Press, 1987, ISBN 0-941404-47-1) is a marvelous musical comedy in two acts (recently performed, for the first time in the USA, at the University of California at Santa Cruz) with a cast that includes Reich, the Marquis de Sade, Sacher-Masoch, the American Medical Association, Marilyn Monroe, the Anti-Sex League and other notables. The reader will keep laughing for hours. He is merciless with the critics of Reich, showing through humor the insanity of the "normal" folks who are so deeply set against him. Wilson also edits a newsletter, *Trajectories*, that keeps the CSICOP folks in the cross-hairs.

* **The Cancer Microbe**, by Alan Cantwell, Jr., summarizes this author's years of biomedical research, in which he independently observed the bionous, pleiomorphic nature of cancer cells. Cantwell is sympathetic to Reich, and discusses how Reich's pioneering cancer research and discovery of the *bions* and *t-bacilli* parallel the more recent works of other innovate cancer specialists. The questions of bacterial-viral *pleiomorphism,* and the antiscientific forces at work in the modern Cancer Industry are addressed. The work contains full citations for the reader who wishes to follow the tracks of the research in more detail. (Aires Rising Press, 1990, ISBN 0-017211-01-4, Address: PO Box 29532, Los Angeles, CA 90029).

* **Geo-Cosmic Relations, the Earth and its Macro-Environment** is the published Proceedings (in English) from the First International Congress on Geo-Cosmic Relations, held in Amsterdam, Netherlands in April of 1989 (Pudoc Press,1990, ISBN 90-220-1006-6, available from NEW, PO Box 864, El Cerrito, CA 94530). A prior issue of the *Pulse* carried a report on this Congress (Volume1(2):37, Fall 1989), in which research findings from dozens of Europe's best scientists are presented, with a focus upon unusual cosmic forces at work in ordinary environments. Little of this subject material ever makes its way into the "mainstream" journals, mainly due to reasons of prejudice and censorship.

* The **Flatland 1990 Catalog of Unusual Lore**, compiled by author, satirist, and Reich enthusiast Jim Martin, was recently released upon an unsuspecting world. It contains a wide assortment of hard-to-get and, truly unusual lore, plus various interviews. *"The publications herein are...the creative gifts of unpaid, unwanted castoffs in a society gone mad".* ($2, from Flatland Press, PO Box 2420, Ft. Bragg, CA 95437)

* The German book **Ein Überblick über die Grauzone in der Wissenschaft**, by Lars Jörgenson, was recently published by WDB-Verlag, in Berlin (ISBN: 3-9801452-0-4). The author, a scientist working in the Max Planck Institute (using a pseudonym for protection against attack), devotes nearly 100 pages of the 360 page work to Reich and orgonomy. The works of other authors who discovered a similar life energy principle are also discussed, such as Reichenbach and Schauberger, and the work is rich in its coverage of anomalies in the natural sciences. Hopefully, this work will be translated into English.

* A doctoral dissertation examining Reich's early years in the Communist Party has recently been published in Germany. According to our reviewer Stefan Müschenich, the book **Wilhelm Reich in Wien - Psychoanalyse und Politik,** by Karl Fallend (Geyer-Edition, 1988, ISBN 3-85090-129-7) suggests that Reich was more deeply engaged in Party activities than is apparent from his writings during the American Period, and also that Reich was prepared to oppose the Nazis in a very forceful manner. One portion of the book indicates that Reich proposed, during a C.P. meeting, that guns be obtained to oppose the growing Nazi attacks upon clincs, libraries, and other party facilities. As Reich wrote in later years, the C.P. never took the Nazi threat seriously, and Reich proved to be too much of a thorn in the side of the party politicos — he was eventually expelled by the German communists. Additionally, a short correspondence between Reich and Leo Trotsky was discovered by the author.

* **Above Top Secret, The Worldwide UFO Cover-Up**, by Timothy Good, (ISBN 0-688-09202-0, Quill/Wm. Morrow, NY, 1988) provides a wealth of information and documentation from official sources and government documents (obtained through the Freedom of Information Act) demonstrating that the USA, Canadian, and various European governments knew for years about the reality of the extra-terrestrial nature of the UFO, and conspired to hide the facts from the public. This volume explains the offensive silent-treatment given to Reich by the US Air Force, to whom he had regularly sent detailed reports about his UFO observations in the 1950s.

Other Books of Interest:
* **Cosmic Life Force**, by Fred Hoyle & Chandra Wickramasinghe, Paragon Press, NY, 1990.
* **Cosmic Magnetism**, by Percy Seymour, Adam Hilger, London,1986.
* **Astrology: The Evidence of Science**, by Percy Seymour, Lennard Publishing, London, 1988.
* **The Big Bang Never Happened**, by Eric Lerner, Times/ Random House, NY, 1991.
* **Advances in Biomagnetism**, Samuel J. Williamson, et al., Plenum, NY 1989.

THE DESERT GREENING PROGRAM

The third and final part of *OROP Arizona 1989*, presenting a final data analysis of five separate cloudbusting experiments in the deserts of the Southwestern USA, appeared in the November 1990 issue of the *Journal of Orgonomy*. Parts I and II appeared in the November 1989 and May 1990 issues, and many color plates were reproduced. Part III presented National Weather Service data showing an approximate *doubling* of rainfall over Arizona, Southern California, and Southern Nevada, for a full one week period which began about 48 hours after the onset of cloudbusting. The 1989 experiments confirmed, in a very powerful and significant manner, the prior findings of Wilhelm Reich on the desert greening potentials of the cloudbuster. These articles have been reprinted in this issues of the *Pulse* as the article: *OROP Arizona 1989: A Cloudbusting Experiment to Bring Rains in the Desert Southwest*. A special Prospectus describing the financial and logistical requirements for the Desert Greening Project will become available later in 1991, as described below. Continued field research is planned for the American Southwest, with outreach efforts to other nations suffering from desert spreading and drought. Additional desert greening field research is planned for the deserts of Israel and the Mideast.

CALL FOR VOLUNTEERS AND DONATIONS

Volunteer workers and financial donors are being sought for the Desert Greening and Drought Abatement Outreach Programs. Drought continues in the Western USA, and field work will possibly become necessary for the remaining months of 1991 and 1992. Field research also continues in the deserts of the American Southwest and Mediterranean/Mideast regions. Financial resources to continue with this work is a continuing difficulty. Additional help is needed for several different library research and data analysis projects, as well as for field research activities. Contact Dr. James DeMeo at the Orgone Biophysical Research Lab for details, and a list of current funding requirements.

PROSPECTUS AVAILABLE

A new Prospectus, *About the Drought Abatement Outreach Program and Desert Greening Programs*, will become available in late Summer 1991. This document describes in some detail the two major non-profit cloudbusting outreach projects of the Orgone Biophysical Research Laboratory. It is intended for individuals and organizations contemplating use of the Lab's cloudbusting services, for the purposes of experimental drought relief or desert greening. An Appendix to the document contains reprints of scholarly research articles on the subject of cloudbusting. To obtain a copy, contact the Orgone Biophysical Research Laboratory, www.orgonelab.org.

CORE OPERATIONS IN THE WESTERN USA

The following Cosmic Orgone Engineering (CORE) operations should be viewed against the background of five years of progressive drought in the Western USA. While some progress has been made, with at least one substantial breakthrough of significant rains, there is no clear indication that the drought has ended. Normal climate cycles have been badly disturbed in North America, since even before the severe drought disaster of 1988. Conditions during 1989 were better for parts of the Eastern USA, but remained quite dry in most western states. In 1990, several cloudbusting operations were undertaken, aimed at bringing relief to various drought areas, as described below. A more detailed report on these various operations will appear in the November 1991 issue of the *Journal of Orgonomy*.

Cloudbusting in the High Sierras, Central California:
Two cloudbusting experiments were undertaken in the central Sierras of California during late March and early April of 1990, at the invitation of a California grower. April rainfall amounts of over 400% of normal fell in an area just to the east of the cloudbusting site, in western Nevada, while amounts of over 150% of normal (less significant) occurred just to the south of the operations site. These same areas continued to receive fairly regular pulses of rainfall for several months, with heavy rains of 200% to 400% of normal in May (Figures 1 and 2). Local farmers commented on the unusual manner in which clouds continued to build over the same place where the cloudbusting operations had taken place. However, by June of 1990, rainfall

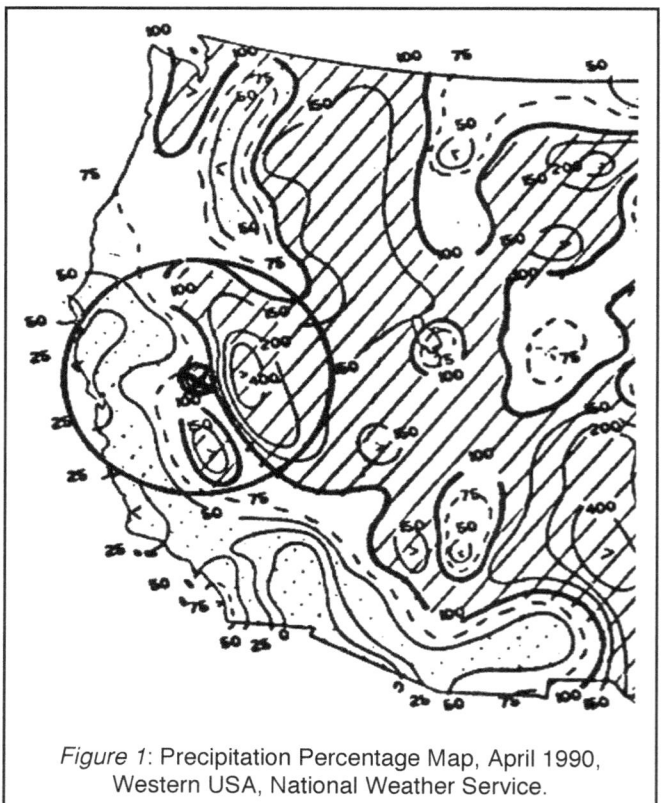

Figure 1: Precipitation Percentage Map, April 1990, Western USA, National Weather Service.

diminished significantly, with the onset of the normal summertime dry period of California. In April of 1990, most of the area stretching from Oregon to Mexico received 50% or less of normal rainfall. May rains were significantly better for this same broad region, including California. These operations were not effective in finally ending the overall drought, but did assist symptomatically, and much was learned regarding operational techniques in the West.

Cloudbusting in Northeast Montana:

In late May of 1990, a cloudbusting expedition was undertaken in Northeast Montana, at the invitation of a local group of farmers. The operations got underway in late May, and continued for approximately one week at different sites across Washington state, Idaho, Montana, and North Dakota, with a goal to restore an inland flow of moisture into the region. The following preliminary indications, taken from National Weather Service records, can be reported here:

1. Within 48 hours after the onset of operations in Northeast Montana, a major storm system swept across the region, releasing scattered but significant rainfall across the drought region. In some areas, it was the first rain, or heaviest rain, in nearly two years.

2. Scattered light rains and cool temperatures continued across the area for nearly two weeks after the onset of operations.

3. By mid-July, rains had nearly ceased and drought conditions intensified. Atmospheric pulsation was only temporarily restored to the drought area, and conditions worsened over the summer months. Funds were not sufficient to allow a second expedition into the area.

4. For Northeast Montana, rains received during the weeks ending on May 26 and June 2 (which was just after the May 23 start of the cloudbusting operations) constituted the second and third largest episodes of weekly rainfall to occur during the months of April, May, June and July. Only one other weekly period, ending on June 30, received more rain than the weeks of the cloudbusting operations. Expressed as a percentage of the total, the week ending on June 2, shortly after the cloudbusting, constituted the largest weekly percentage of the entire four month period, April through July.

5. A corridor of dry, arid air was identified on climate maps as extending from the Baha/Colorado River Basin, northeast into Eastern Montana and North Dakota. Figure 2 identifies this corridor of dry air, which was only temporarily interrupted by the cloudbusting operations. A small island of 100% of normal rains can be seen

6. After-the-fact, it was learned that a series of underground nuclear bomb tests got underway just after late May. In particular, French underground nuclear tests in the Pacific Ocean and American tests in Nevada strongly correlated with anomalous increases in the 500 mb pressure surface elevation over Southern Nevada. The pressure increases over Nevada were followed shortly afterward by similar pressure increases over Northeast Montana. Figure 3 presents some of these data, which suggest that underground nuclear tests reinforced the drought tendencies affecting the entire Western USA. More research is obviously called for to establish this relationship, but it is in complete agreement with the observations of Reich, Eden, and others on the drought-producing affects of underground nuclear explosions. Other findings by Kato and Whiteford, previously published in the *Pulse*, support the possibility of such atmospheric effects.

CORE RESEARCH AND FIELD OPERATIONS OVERSEAS

Field work and some limited cloudbusting operations were undertaken overseas, at the invitation of the Berlin Wilhelm Reich Society, and the Hellenic Orgonomic Association. These operations were directed by James DeMeo, with much help from the various members of the organizations. Given the difficulties of language translations, and access to overseas weather data, there will be a longer delay in the reporting of results from these overseas operations, but the following preliminary observations can be reported:

Greece, 1990:

At the invitation of Dr. Theodota Hassapi and the Hellenic Orgonomic Association, assistance with cloudbusting operations in Greece was given in 1990, following upon prior contact and assistance given by the College of Orgonomy in this same area. In 1990, severe, persisting drought conditions were present and widespread over Greece and much of Italy, Turkey, and the Balkan states also. The city of Athens was reduced to a 40-days supply of water by late Spring, when the first operation was undertaken. That operation, lasting two days and employing a newly-constructed, hydraulically controlled cloudbuster, was followed by a good week of rains, with storms that covered all of Greece, and dry parts of Italy and Turkey as well. The rains did not persist for long, possibly due to the onset of the normal

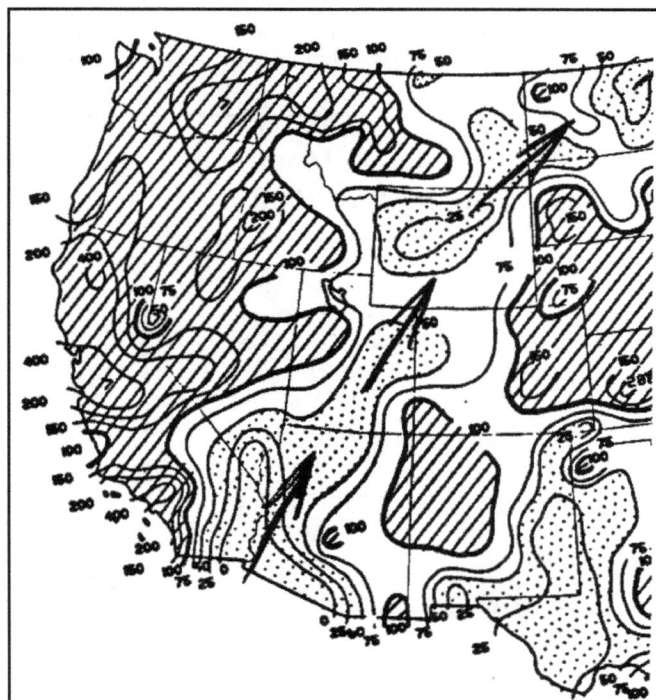

Figure 2: National Weather Service Precipitation Percentage Map for May 1990, showing a rain-free corridor extending from the American Southwest north into NE Montana and the western Dakotas. Observations suggest this corridor has been a semi-permanent feature characterizing dry episodes in the Northern Plains. Generalized air flow patterns are drawn in, showing the connection to the larger deserts of the American Southwest.

summertime dry period in the Mediterranean. A subsequent cloudbusting operation undertaken in summer, at the height of the dry season, produced no significant results of any kind, except a slight reduction in air temperatures and dor levels. As the fall and winter rainy season approached, however, no early onset of rains was observed, and it appeared that drought might persist for another year. With little water in reservoirs or as snowpack on the mountains, a major disaster appeared in the making. A third cloudbusting operation was therefore undertaken in November, this time employing several different cloudbusters stationed in different parts of Greece, ranging from Thessaloniki in the north to Crete in the south. The use of these instruments were coordinated by telephone instructions from the station in Thessaloniki, where the largest of the instruments was located. After three days of operations, a major breakthrough in rains developed. Storm systems began moving west to east, from as far west as the Gibraltar opening to the Atlantic Ocean, in a manner that experienced meteorologists said they had "not seen in 10 years". Heavy, soaking rains broke out in parts of the northern Mediterranean that had been rain-deficient for years, and heavy snowfalls began piling up in the Alps, and Balkan Range. By early 1991, Greek reservoirs were relatively full, and heavy snow accumulations were apparent in the highland mountains. Dr. Hassapi and her assistants have continued with this work independently, with excellent results that have provided a continuing flow of moisture across the northeast Mediterranean. This research program continues.

Germany, 1989-1990:

In 1989, the Orgonomische Waldheilung Projekt (Orgonomic Forest-Healing Project) was initiated, at the invitation of a group of German and Swiss workers, to investigate the

Cloudbuster in Greece

problem of widespread forest death (waldsterben) in Germany. The objective was two-fold: to investigate the possibility of an orgone-energetic component to the continuing atmospheric stagnation and dying of trees, and the possibility that cloudbusting operations might provide some change in atmospheric energetics and chemistry which could help the trees. Several preliminary cloudbusting operations were undertaken in the Berlin region, with a goal to affect the atmospheric chemistry in regions near to the operations site. Workers in Berlin reviewed data on rainfall, humidity, cloud cover, ozone, nitrate and sulfate concentrations, as well as background radioactivity (routinely measured by the German weather service since the Chernobyl

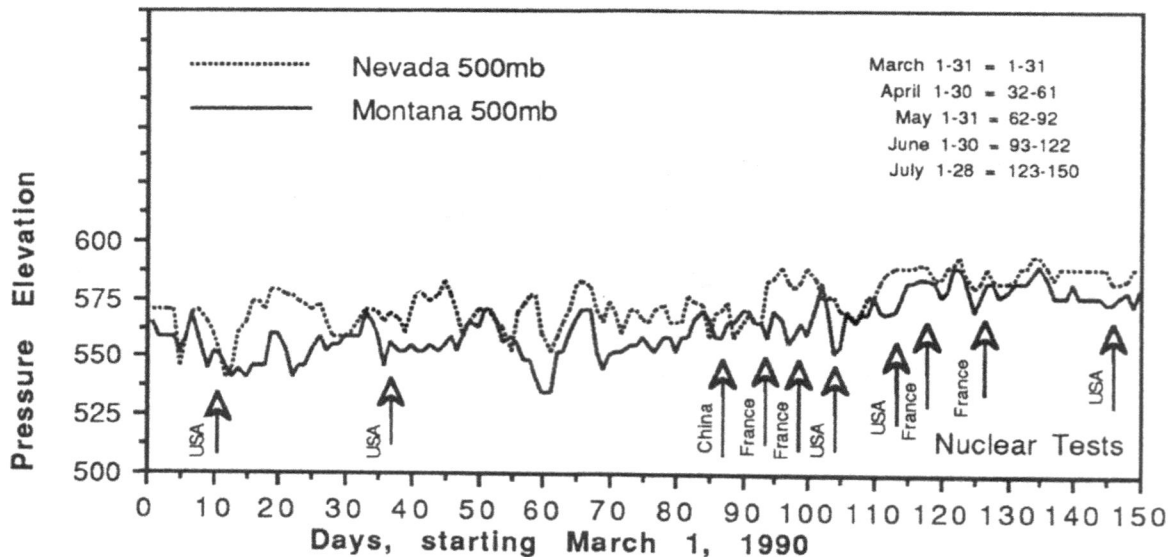

Upper Atmospheric Pressure (500 mb, Nevada and NE Montana) and Underground Nuclear Bomb Tests, March-July 1990

Figure 3: Upper atmospheric pressure patterns over Nevada and NE Montana. A similarity exists between the two pressure curves. Nuclear tests in the USA are shortly followed by dramatic increases in pressure over Nevada, and, to a lesser extent, in NE Montana. Nuclear tests in French Polynesia appear to have a similar correlation to pressure increases in the Western USA, though the relationship is not so clear as with USA tests. The single Chinese test shows no clear influence at all.

accident), and other measures of air pollution. In several of these initial experiments, very clear decreases in air pollutants were observed — other experiments did not show such clear effects, but the field research is clearly in a preliminary stage of development. As with the experiments in Greece, final publication of results is delayed until such time that the sponsoring groups gather and organize the data, and translations are made. The following report can be made public at the present time, extracted from Dr. DeMeo's "Survey of German Forests":

MEMORANDUM   16 November 1989

FROM: James DeMeo, Ph.D.
TO: All Participants and Supporters of the Orgonomisches
    Projekt Waldheilung (Orgonomic Forest-Healing Project)

Greeting from the USA. I hope you all survived the social earthquake and big party when the Berlin Wall recently fell down. It appears I left Berlin about one week too soon, and so missed all the fun.

After the two day meeting in Berlin with the Waldheilung group, I spent about a week traveling through Germany, visiting with several professional scientists studying the problem of waldsterben, and seeing first hand those areas most badly affected. My meetings and travel arrangements were scheduled and implemented with the help of Stefan Müschenich, Sibylle Heck, and Ralf Borgardt, and so I must firstly thank them for undertaking to guide me on this very educational adventure. The trip took place during stagnant, drizzly weather, and so exemplified some of the worst conditions for trees. The weather was mostly miserable, being cold and wet, but a lot was learned which both confirmed and extended the many things we have previously discussed. It is unfortunate that I could not have made this trip prior to our past meeting, as many questions were answered about the nature of the problem. During this same period, I also visited several individuals with a working interest in the forest death problem, including Dr. E. Wedler at the Berlin Meteorological Institute. Here is a more detailed report.

SURVEY OF GERMAN FORESTS, October 1989

1)   Mr. Rainer Fischer, of Sinzheim, spent half a day with us, showing various areas of tree damage near his home. He has been engaged in both remineralization experiments in regions of dying forest, as well as in theoretical work on the causation of waldsterben (forest death). His remineralization experiments, like those done in the USA, show great promise for re-invigorating small plots of trees, but the procedure is too costly and intensive for saving large forest regions, and has no guarantees for being a long-term solution. His theoretical work is about as functional as any classical approach could be, however, and he basically argues the following:
      a) There is a relationship between large forested regions of Earth and the strength of the planetary magnetic field. As the tropics are deforested, and as the Earth's ecosystems are polluted and decimated, the geomagnetic field strength, and the protective magnetosphere of the Earth, is becoming weakened. He also points to high altitude and atmospheric nuclear bomb testing as partly responsible here, linking together the deterioration of both the upper and lower atmosphere, via an energetic means.

      b) The weakening of the Earth's magnetic field has allowed more harsh, penetrating radiation from space to enter the lower atmosphere, notably affecting trees on the higher mountains, even in regions with little air pollution.
      c) The next several years will be critical, given that the sun reaches its sunspot maxima, and solar wind will also be at maximum. This will allow an increased entrapment of charged particles within the weakened geomagnetic field, affecting trees at high altitudes all the more.
      Mr. Fischer's arguments parallel those which I and others have developed via orgonomic reasoning: we may be facing a general weakening of the orgone energy envelope of the entire Earth, and not merely a localized problem due solely to air pollution and stagnation related to nearby deserts. His theory attempts to explain a common contradictory observation made by many orthodox scientists: there is greater tree death at higher altitudes, where in many cases there is less or little air pollution. At some lower altitude regions, air pollution is greater, but there is little tree death. The weak point in his theory, from the classical perspective, is the link between "geomagnetism" and forests, and the question of a weakening of the Earth's geomagnetic field. Neither of these latter points have any strong classical support, to my knowledge, though they have a good functional relationship to orgonomic theorems if we substitute "orgone energy envelope" for "geomagnetic field". I stress that my knowledge of Mr. Fischer's views were gained via on-the-spot translations, and while I believe I got his main points, I may not be 100% accurate in my presentation of his theory.

2)   Dr. Helmut Klein, who is a professional forester working for the BUND (Bund für Umwelt und Naturschutz Deutschland), gave us a tour of some high altitude forests that are dying, in the regions south of Munich, especially near Andechs-Erling. From him, I was able to confirm the following general observations that I had previously known only from textbooks:
      a) Approximately 80% of all trees in the Alpine regions are sick and dying. In other areas, the situation ranges from better or worse than this figure. This number is valid for all age groups of trees, and for all species, though tanne (fir) is the worst off. Through deceptive use of statistical measuring techniques, a lower figure of 52% has been concocted by government officials. For example, when trees reach a bad state, they are often cut down for firewood. This artificially reduces the number of sick and dying trees in a given region. Also, the healthiest tree in a given forest is often used as the standard by which all other trees are compared; that tree is given the most "healthy" rating, and it does not matter if this "healthy" tree is about to collapse. If all other trees are in a similar state of near-collapse, then the entire stand or forest will be classified as "healthy" also. "Normalcy" is now the standard of health, for both people and trees! Likewise, if half of the trees in a forest are cut down, because they are nearly dead, then there will be no "nearly dead" trees to be counted! And so, there has been a "leveling off" of the number or percentage of dead and dying trees in the national data base, allowing many politicians to proclaim publicly that the forests are "getting better". Other politicians in ski areas refuse to acknowledge the death of the forest, as it might scare off tourists! The mayor of Dr. Klein's home town, where we saw numerous sick or dead trees, has publicly said that "there are no dying trees in town".
      b) Given the widespread weakening of the vitality of forests, trees are putting down much more shallow root systems than in prior decades. Also, existing deep roots are weakening,

due to loss of symbiotic fungi and root hairs. Soil erosion with consequent downstream siltation has increased in all affected areas, and there have been an increasing number of highly unusual snow/mud avalanches starting *within the forested slopes*. Avalanches usually start on the snow-packed tops of mountains, where there is no tree cover to hold it in place. Avalanches that start *within* the tree-covered regions of a slope are rare, because the tree roots usually hold the snow, moisture, and soil in place. However, a record of 300 such within-forest avalanches occurred in Germany in 1980, increasing to over 1200 in 1987.

c) As the trees die off, a landscape consisting of pasty grey, nutrient-poor podzolic soils remains behind. These soils can be easily eroded, and are replaced by underbrush and scrub. Even with replanting, it may take centuries to regain a significant forested cover.

d) Dr. Klein is firmly convinced that there is air pollution in the high altitude regions, primarily ozone and acid rains. Measurements are made by many recording stations in the affected regions. The high levels of ozone pollution are attributed to the classically-described pathways of photochemistry in the atmosphere, as an end product of photochemical smog which is transported in from polluted regions. Likewise the acid problem with the rains. However, I point out that there are problems with this theory. The photochemistry of air pollution and ozone production at low altitudes is exceedingly complex, and is largely a theoretically top-heavy body of knowledge. Given a lack of clearly measured evidence for the pollution-ozone pathway, and the general absence of other kinds of evidence for the photochemical smog effects (such as thick atmospheric haze and the presence of other necessary pollutants, such as hydrocarbons, sulfates, and nitrates), I am personally not convinced that 100% of the ozone and acid rains in those areas come from air pollution, or that those two constituents are even the primary agents responsible for the tree death. I was a bit befuddled when supposedly "acidic" streams where fish had died off registered a pH of 8.2 on my pH meter. When I asked him about it, Klein admitted that neither he nor his assistants made such measurements themselves, but instead relied upon the measurements of the meteorologists and hydrologists. In fact, only one of the streams I measured ever registered a pH value below around 6.0. It was a stream which ran through a previously mined area, and likely was suffering from the long-term effects of acid-mine drainage, registering a pH of 4.5. The rainfall itself was around 4.5 to 5.0, but virtually everything the rainwater touched increased its pH value. These measured values lead me to feel that the whole pH question needs some deeper investigation, from an orgone-energetic point of view, and should not be left up to those who are uncritical about the measurements. The pH meter itself is, after all, a very sensitive, high impedance millivoltmeter, and the arguments of Reich about the emotional-energetic components of bioelectricity may well be true of the atmospheric-energetic components of pH readings!

e) Dr. Klein expressed to us his exasperation regarding the absence of any viable method to save the forests, save for massive pollution controls over all of Europe. Efforts to replant trees are tiny compared to the numbers of trees that are dying, and he even feels that government funding for replanting by students are primarily aimed at diverting the energy of young people, who might otherwise engage in more effective protests and political action. He feels that political action is the only effective way to end the problem.

f) Another BUND official we met with, Mr. Hubert Weinzierl, informed us that the German government has spent 250 million marks on 590 different studies on the forest death problem, and pointed out that the only certain thing which has been learned from all of this is that "air pollution is a factor in forest death". Beyond that, nothing else was certain. Mr. Weinzierl is looking into the possibility of finding funds for this project, but this is what we call "long shot".

g) A friend and co-worker of Dr. Klein gave us a tour of regions near to the German/Czechoslovakian border. These areas were very bad off, with the majority of trees in some areas already dead, cut down, and gone. Whole mountains, and large blocks of land hundreds of square kilometers in area, are now almost completely deforested, from dead and cut trees. The stark nature of these areas reminded me of the short-grass prairie grassland of the USA. It was the barren steppe of Central Asia moved westward. This shocking impression is significant: in the context of desertification of the planet, first the trees vanish, to be replaced by grassland. Next, the grasslands vanish and are replaced by deserts.

h) From Klein and his associates, we obtained general confirmation that heavier rainfall amounts are better for the forests than the lighter rains. The heavier raindrops wash the pollutants off the tree leaves and needles, preventing an accumulation, and are indeed of a higher pH value. Good moisture supplies keep the trees in better health also. This is significant, as the cloudbusting efforts can be aimed at gently mobilizing the atmosphere to change lighter drizzles into heavier showers.

3) On a train ride across central Germany, I nearly choked to death on the air pollution which appeared widespread and quite visible across the open rural countryside. I also read about

Waldsterben: skimpy treetops due to loss of needles.

the death of the flowers in Amsterdam in early September, when air pollution killed, in one single night, some 80% of all the flowers grown by commercial growers near the city. On my trip through this area, I learned that a significant percentage of young children in The Netherlands and Germany suffer from "non-specific croup", which is a heavy mucous flow with a low grade infection. Again, sickness is considered "normal" for both humans and trees! These factors underscore that air pollution is taking a heavy toll, irrespective of the stagnated, inversion-type weather that must also be present for pollutants to concentrate. Air pollution is indeed a "significant factor". Regardless of the role of energetic stagnation in trapping air pollutants, or the problems with air pollution theory in high altitude waldsterben regions, Dr. Klein is absolutely correct about the necessity for concerted political action to stop the pollution (and the nuclear bomb testing).

4)   These observations above leave me with the feeling that the problem of forest death is enormous, no less than that

Waldsterben: down-drooping branches and needles.

of deserts. Like the problem of desert spreading, the social aspects of the problem are far more difficult to deal with than the question of atmospheric stagnation.

5)   Dr. E. Wedler, of the Meteorological Institute of the Freie U. of Berlin, has provided a lot of materials on the question of atmospheric dynamics over Europe. Other members of his Institute provided samples of residues remaining from evaporated rainwater. The residues are blackish in coloration, as are various tree limbs and bark samples taken from the dying forest areas. These materials are reminiscent of the "melanor" which Reich observed and described in the 1950s, during a phase of localized tree death in Maine. I predict that they will reveal a lot of t-bacilli, but presently do not have a microscope capable of resolving to the necessary high magnifications. Therefore, I must rely upon those of you who have such equipment and time to make the observations. If any of you can undertake this kind of analysis, please try to obtain some sample tree needles, bark and limbs from dying trees, or soils from affected areas, and compare them microscopically to healthy trees and soils. I have a few of such samples here, and can mail them to you if necessary, but as you are all closer to the problem there, such samples should be fairly easy to come by.

CORE BREAKTHROUGH IN CALIFORNIA RAINS, MARCH 1991

In February 1991, two major cloudbusting expeditions were undertaken in California, employing techniques of mobile operations, and several different cloudbusters across a large geographical area ranging from southern Oregon to extreme southern California. Prior cloudbusting operations, undertaken from a single station near San Francisco were followed by some good rainfalls, but these rains were mostly confined to areas of Northern California, or they were not persistent, even when stimulated during the normal California rainy season. Efforts were therefore undertaken in February to broaden the scope of operations, to break up concentrations of stagnant dor over a wider geographical area. These February operations were followed by the most widespread and heaviest rains to arrive on the West Coast since the beginning of the year, and possibly for many years past. Specifically, an early February operation stimulated rainfall increases across the entire West Coast, but those rains did not persist. A second, late February cloudbusting operation was undertaken on the 18th of the month; this was followed within 24 hours by a major southerly shift of jet stream flow, and a "lining-up" of storm potentials, over the central and eastern Pacific Ocean. Widespread, persisting rains arrived in coastal California a little more than a week later. Rains were persisting and cyclical for most of March, being stimulated by additional operations from two cloudbusters stationed north of San Francisco. March rains in California were from 200% to over 600% of normal, as shown in Figure 4 (next page). These rains greatly benefited the region's forests and farms, providing sufficient water supplies for the subsequent dry season, and for many additional months (assuming natural rainfall does not spontaneously return in the winter). More detailed reports on these operations will appear in the forthcoming May and November 1991 issues of the *Journal of Orgonomy*.

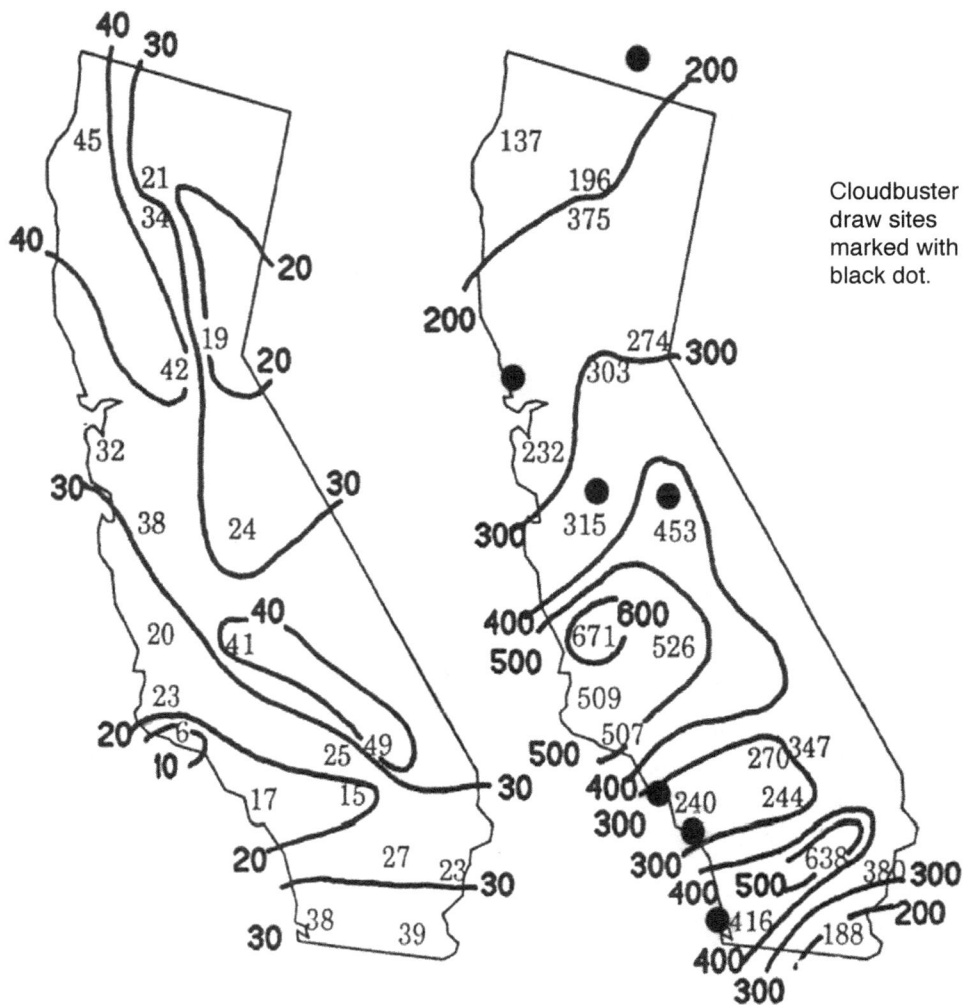

PERCENT OF NORMAL PRECIPITATION
October 1, 1990–February 26, 1991

PERCENT OF NORMAL PRECIPITATION
February 27, 1991–March 31, 1991

Cloudbuster draw sites marked with black dot.

Figure 4: Precipitation Percentages for California, for the months Before (left) and After (right) onset of cloudbusting operations, which began in late February, and continued on and off through March.

Nicasio Reservoir, central California, at record low levels.
Before cloudbusting, February 1991.

Nicasio Reservoir filled to near capacity,
After cloudbusting, April 1991.

Cloudbuster *Icarus.*
What's in the name?

Some readers of the *Pulse* have asked why the experimental cloudbuster used in our field research is named *Icarus*. One questioner wonders why we do not use the name *Daedalus*, from the same Greek legend. In the legend, Daedalus and his son, Icarus, escaped from the isle of Crete by making wings of wax and feathers. Daedalus warned his son not to fly too close to the Sun, but his son ignored the warning. Icarus soared higher and higher until, finally, his wings melted and he fell into the sea. Daedalus, who safely stayed at a lower altitude, and who did not soar to the heights, made it back to the mainland. Daedalus was also credited with the invention of the Labyrinth, a complicated maze in which the half-bull, half-man *Minotaur* was confined. On the surface, these myths seem to condemn Icarus for his foolhardiness, in ignoring the wise words of his father and flying too close to the Sun. But upon analysis, the myth indicates something else. Daedalus played it safe, and never experienced either the ecstasy of soaring, nor of touching the Sun, nor of the fall (surrender). The young Icarus yields to feeling and soars; he melts; he surrenders and drowns in the sea, which is a metaphor for oceanic feeling, or sea of emotion. Daedalus, instead, traps his bull-like rage inside the complex Labyrinth of armored structure, and cannot soar, cannot melt, or feel deeply, except for the sorrow at the loss of his untamed son, Icarus. In this sense, Daedalus is Reich's *Homo normalis*, afraid to make any deep contact with nature or self, while Icarus is the *Child of the Future.* The essence of this interpretation of the Icarus myth was captured in a large bronze sculpture by Charles Umlauf. I first observed that bronze statue of Icarus in 1977, in front of Nichols Hall at the University of Kansas Space Techology Center, when I was a student there. It captured the ecstasy of Icarus in the fall, or surrender, which also appears as an embrace. Hopefully, in my drawing, I have captured some of Umlauf's portrayal. I was so deeply impressed with Umlauf's sculpture, and this interpretation of the legend, that, in 1977, the newly-constructed cloudbuster was named: *Icarus.*   J.D.

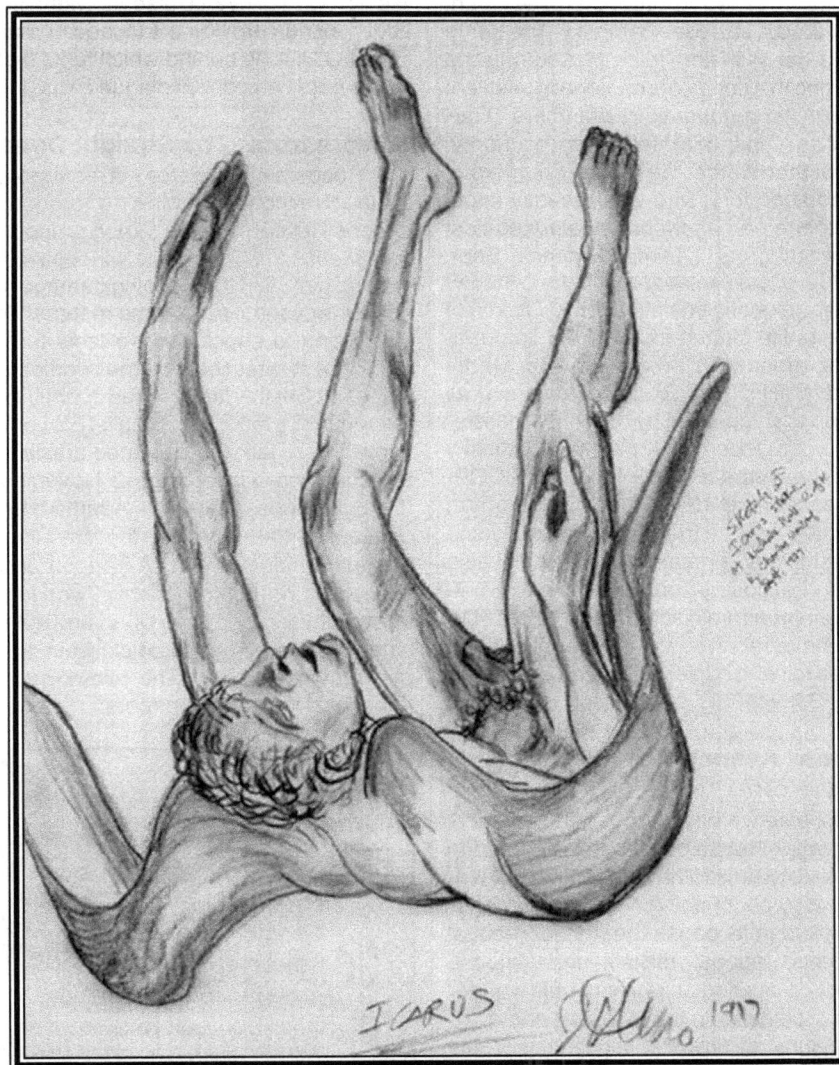

by James DeMeo, Ph.D.

Over the last two years, tremendous progress has been made in orgonomic research, confirming Wilhelm Reich's sex-economic and orgone biophysical findings repeatedly, and extending them in significant ways. It is therefore not surprising that certain "critics" of orgonomy would increase the tempo of their hostile actions, which are designed to thwart and kill this important body of research. The following are but a few recent examples which have grown to proportions requiring exposure to the light of day.

Recent Smear Letters, and Other "Activities" of Mr. Joel Carlinsky

In recent months, several smear letters were circulated in the US and overseas, purporting to present "facts" about James DeMeo, and other persons engaged in orgonomic research. The author of these letters was Mr. Joel Carlinsky, the same individual who burglarized the Wilhelm Reich Museum in the 1970s, and who has engaged in a series of emotionally pestilent and threatening actions against orgonomic researchers. Carlinksy has masked himself as a "friend" of Reich and orgonomy, but his actions speak louder than words. He has sent numerous threatening letters to this Laboratory, and was recently implicated as a possible FBI stooge and *agent provacateur* against the Earth First! environmental group (*Animals Agenda*, Sept. 1989, p.20). He has a history of cozy relations with the CSICOP network (Committee for the Scientific Investigation of Claims of the Paranormal), and boasted to us that he provided "information" on Reich and other orgonomic researchers to Martin Gardner, for a nasty smear article in the *Skeptical Inquirer* (Fall 1988), which was picked up and repeated by other "journalists" nationwide (see "The Straight Dope" note, below). Carlinsky personally authored one recent smear article for the CSICOP-affiliated *New Jersey Skeptics Newsletter*. Titled "Orgonomy in New Jersey", the article called orgone therapy "medical quackery" and cloudbusting "fraudulent rainmaking". Carlinsky also urged the "investigation and exposure" of orgonomy "to whatever degree possible" in a manner reminiscent of the 1950s FDA campaign. Persons who have received the above-mentioned smear letters are requested to send copies to Dr. James DeMeo (PO Box 1395, El Cerrito, CA 94530).

About those "Scientists" Associated with the CSICOP Organization

In prior issues of the *Pulse* we have exposed a few facts about the CSICOP organization, which publishes the *Skeptical Inquirer* magazine. The group claims to have a scientific orientation, but has an official policy of not doing research itself; nor does it allow persons attacked in its pages the right of rebuttal and fair response. As research, rational criticism, and response to criticism are major activities of science, one can only wonder about the use of the word "Scientific" in the CSICOP name. To give the organization an "aura" of scientific legitimacy, it has appointed many scientific people to the status of "Fellow". These individuals include Carl Sagan, Isaac Asimov, Paul MacReady, Stephen Jay Gould, and other scientists not so well-known, but

with powerful positions in academic circles. Other organizations, such as the National Center for Science Education, with its "Committees of Correspondence", are dominated by CSICOP Fellows at the executive level. Together with various affiliated "skeptics clubs" around the USA, they have a growing influence over university appointments and tenure decisions, editorial policies of science journals and major book publishers, funding decisions within science institutions, and so on. As such, CSICOP has a network of tentacles that stretch deep within the scientific-academic institutions, and truly constitutes a highly organized form of emotional plague. Do these "Fellows" of CSICOP see any contradiction between their formal obligations to scientific research, and their high-status in an organization which blindly attacks, rather than examines and criticizes unorthodox research subjects, and which in any event does not allow rebuttal by persons so assaulted? How do they feel about the growing association between their organization and persons like Mr. Carlinsky? These individuals, whether they acknowledge it or not, provide the facade or mask of "scholarly objectivity" for CSICOP, behind which lurks a great deal of nasty, stab-in-the-back, emotional plague behavior.

Smears from "The Straight Dope"

Readers of the *Pulse* will recall an article from the Fall 1990 issue, "Response to Martin Gardner's Attack on Reich and Orgone Research in the *Skeptical Inquirer*". Gardner's article in the *Skeptical Inquirer* was widely circulated, and continues to crop up here and there, doing damage to this work. In one case, a reporter used it as "source material" for his own article, never attempting to check out the facts from independent sources. The case in question was the syndicated column "The Straight Dope" by Cecil Adams. Adams' untitled article was printed as a response to a letter by journalist Steven Stocker (who previously wrote a very fair and balanced article on Reich and myself for the *Baltimore City Paper*, 2 November 1990, pp.8-10) The "Straight Dope" article by Adams was anything but fair, and condensed the worst of the smears and slime from Martin Gardner's prior article. A cartoon accompanying the Adams article showed a mad scientist working a strange gizmo which threw lightning bolts at a cloud and at the genitals of an ameba-type creature. Adam's article went out to 20 different newspapers around the USA! The following letter was drafted, and sent to each of the various papers:

To the Editor:

As an independent scholar with nearly 20 years of research and fieldwork expertise investigating the biophysical research findings of the late Dr. Wilhelm Reich, I was outraged to read the incredibly derogatory and twisted statements about Reich, and about myself, in "The Straight Dope" by Cecil Adams. Adams Relied upon a previously published attack upon Reich and myself by Martin Gardner (Skeptical Inquirer, Fall 1988), but it is a big mistake to elevate Gardner's unstudied criticisms to the level of fact, while so completely ignoring and trivializing the decades of solid published experimental evidence on the orgone energy question by Reich, his co-workers, and by younger researchers like myself who have replicated his controversial experiments in more recent times.

For the record, SI magazine refused to publish my rebuttal to Gardner's attack on Reich, and I later discovered that the parent organization of that magazine, CSICOP (Committee for the Scientific Investigation of Claims of the Paranormal), has a long-standing policy of not doing research, and also of not allowing persons attacked in its pages the right of rebuttal. It is not a scientific organization in the least, and most of its members are professional or amateur magicians. Gardner, himself a card-trick and math game expert, has made a career out of smearing Reich, and his articles are notable for their nasty tone, the highly selective lifting of materials out of context, and most importantly from the perspective of the sciences, the total ignoring of published positive research evidence supporting Reich's claims. Adams successfully condensed the worst of the journalistic slime from Gardner's earlier article into his own. Given this fact, I will respond to "Adams" article by enclosing a copy of my article "Response to Martin Gardner's Attack on Reich and Orgone Research in the Skeptical Inquirer", which I will send free of charge to anyone who mails me a self-addressed, stamped envelope.

Given the police-state nature of the continuing nasty assault upon unorthodox scientific and medical research findings in the USA, which has resulted in the burning of scientific books, the harassment and imprisonment of medical pioneers and natural scientists, and the continuing repression and neglect of numerous workable solutions to major social and environmental problems, I demand that your paper reprint my complete letter. I am sick and tired of seeing Reich's name, and now my name, dragged through the mud by "journalists" or other self-proclaimed "science experts" who are too prejudiced, hostile, and lazy to get their facts straight.

James DeMeo, Ph.D.
Director of Research
Orgone Biophysical Research Lab

My letter was reprinted by a number of national newspapers, and in turn, I received quite a collection of interesting and generally supportive responses. Following are selected quotes from a few of those letters:

I've read a little of Reich, and I have several of his books on the shelf (waiting), and it has always seemed to me as if the criticisms of his work went far beyond mere scientific disagreement. S.P., Chicago

I've never been able to really figure out why people are so afraid of [Reich's] work. So much of his bioenergetic ideas are now totally accepted in psychiatric-new-age massage-self-help regimens — but never acknowledging their debt or source. R.P., Chicago

I want to build an orgone accumulator, but I lack the building skills. Please send advice or who to contact for help. S.N., Phoenix

I was very pleased to see your letter in defense of Reich. Where the hell did Adams pull that stuff from anyway? I was pretty upset by it. B., Berkeley

I think you deserve a chance to explain things your way. See, the way I figure it, in all likelihood Orgone is a crock and you're a crackpot. If you're not, I'll be able to tell by your response... I keep hearing stories about unjustly ignored scientists, and I want one to be true. Additionally, I'm fascinated by mental illnesses that cause delusions. M.S., Baltimore

...I wanted to express my support and solidarity with you in your struggle to transcend the tyrannization of the human imagination and soul by those who have closed their minds in blind homage to the god of 'scientific' rationality. P.F. Oakland

Frankly, I am not interested in slanderous tripe towards you or Reich..., P.O., Berkeley

...I crave more information on this fascinating figure, Reich, whose fundamental ideas form the basis of my own philosophy; unfortunately, I know very little of his work... Do people actually practice a Reichian therapy as a Freudian alternative? H.D., New York

Although I am relatively ignorant with regard to Reich's theories, it seems like he's been unfairly blackballed..., M.A., Berkeley

Cecil Adams is, as you must realize, using his mind as armor; he has made a nice thing out of being a professional know-it-all, or smart-ass... Adam's article [is] carried locally in an ego-trip yuppie-oral-sadistic local free weekly full of otherworld ads... B.T., New York

Censorship and a Death at the International Society for Biometeorology

Readers of the Pulse will recall that, in prior issues, Conference Reports were given regarding the International Society for Biometeorology (ISB), which has a study group on "Physico-Chemical and Biological Fluctuating Phenomena", and the International Committee for Research and Study of Environmental Factors, which sponsored the First International Congress on Geo-Cosmic Relations in Amsterdam. One of the leaders of these European groups was Dr. Erich Wedler, a man who had worked for years as laboratory assistant to the late Dr. Giorgio Piccardi. It was Piccardi who opened up this new territory of research, which is widely known in Europe, but hardly known in the USA. Dr. Wedler had on two occasions provided opportunities for orgonomic research findings to be presented, at the ISB group in the USA, and to audiences in Europe at the Amsterdam Congress. Additionally, Wedler was working as a Professor of Biometeorology at the Free University of Berlin; he had contributed to the Orgonomische Projekt Waldheilung (see page 112), and several of his students were openly pursuing research interest in orgonomy. We therefore deeply regret the recent passing away of Dr. Wedler, doubly so given the awful circumstances of his death — a stroke and heart attack during his paper session at the ISB Conference in Vienna. It appears that his frail medical condition was brought to a crisis as a consequence of the censorship of various papers scheduled by Wedler for presentation at that same Vienna Conference.

Firstly, the ISB organization is international, with a fairly large and powerful American contingent. Additionally, the study group led by Wedler, on "fluctuating phenomenon", has been looked down upon by many influential American ISB members such that, only a few weeks before this big international Conference, Wedler was informed that the Americans had *vetoed* the participation of the various persons he had invited to speak! Without any prior notification, discussion, or debate, an entire list of speakers from around the world were effectively *censored from participation*. This censorship was apparently undertaken as the unilateral decision of Professor Dennis M. Driscoll, a meteorologist at Texas A. & M. University, who had assumed the title of "Scientific Program Chairman" for the Conference (though no "Scientific Committee" as such even existed). After reviewing the list of speakers, he selected those paper topics which he personally objected to, and sent each of those individuals a letter informing them that their papers were unacceptable, because they were "not within the purview of biometeorology". The various international speakers who were censored and their topics included the following:

A. Ansaloni, et al., "Analysis of Water Samples by Means of the Gold Test: Water From Marian Sanctuaries Show Peculiar Characteristics"

N. Bodrova, "Macroscopic Quantum Effect: Discrete Distributions of Physicochemical Parameters"

G. Bonacina, "New Features of Solar Activity Revealed by the Solar Maximum Mission"

C. Capel-Boute, et al., "Variability in the Surface Tension of Distilled Water Due to Usually Uncontrolled Environmental Factors"

J. DeMeo, "Preliminary Results of a Desert Greening Experiment in the American Southwest, Summer 1989"

D. K. Dhal, "Climate and Human Welfare"

P. Faraone, "The Frequency of Colonies with Sector Differentiated as an Indicative Test of Periodical Fluctuations of Biological Phenomena, Influenced by Environmental Factors Such as Those of Cosmic Origin"

A. Gmitrova, "Permanent Magnetic Field Hypotensive Effect Correlation with Geomagnetic Activity"

V. P. Kaznacheyev, et al., "Phenomenon of Heliogephysical Imprinting and Individual Sensitivity of Human Organism to Magnetic Fields and Weather"

L. Krivsky, "Forbush Decreases on the Cosmic Ray Level in the Course of Sunspot Cycles"

N. I. Mzalevskaya, "Interphase Region as a Primary System of Weak Electromagnetic Field Reception"

A. M. Opalinskaya, et al., "Fluctuations in Physico-Chemical and Biological Systems, Correlations with Cosmogeophysical Factors"

V. Uritskiy, et al. "Cosmic Noise Controlled Structures Forming in Non-Equilibrium Systems"

N. Yago, "A Mathematical Formulation of Compartmental Model with Fluctuations"

T. Zeithammer, "Dynamics of Synergetic Systems in Solar-Terrestrial Physics, From the Viewpoint of Biometeorology"

In case it is not apparent to the reader, the above papers constituted the strongest evidence to be presented at the Conference on the question of a cosmic energy in space, which influences both weather and organism. When the form letter went out informing the above speakers that they were censored, there was a protest from some ISB members, but to no avail — the decision of the American meteorologist was cast into stone. As an attempt at compromise, the Austrian Society of Medical Meteorology, which was hosting the larger ISB affair, invited each of the speakers to present their papers at a "special session", to take place during a slack period of the larger ISB Conference. This slack period, however, turned out to be the same time that the rest of the Conference was scheduled to be "out on the town", or on field trips. Most of the censored speakers declined to attend under such circumstances. Apparently, the emotional blow of these events were too much for Wedler, who was frail and aging, and whose study group was gutted by the censorship. He collapsed in the middle of his paper session, and died a few days later in a Vienna hospital. Here are quotes from two letters that followed in the wake of the cenorship:

Concerning my paper, I was surprised to read that you did not consider it "within the purview of biometeorology", as this research has been sponsored by Dr. Tromp himself [one of the founders of the ISB], and the paper intended for the permanent working group on "physico-chemical and biological fluctuating phenomena". It has been accepted by its actual President Dipl. Ing. E. Wedler. Fluctuating physico-chemical processes and specially properties of water itself are, in fact, basic for the understanding of all biological processes, and their relations with the environment. This matter has been discussed and settled in the frame of the ISB since its birth! ... I had to face similar attempts to suppress our working session at the 4th ISB Congress (1966, Rutgers University, USA), at the 6th (1972 Noordwijk-Netherlands), and at the 8th (1979, Shefayim-Israel), usually by the influence of some American scientists or local organizers, but we have always reacted rather successfully. The problem of "uncontrolled" environmental factors is still very "disturbing" or even "unacceptable" for many members of the scientific community. M. Capel-Boute, Past ISB President

My presentation to the Physico-Chemical and Biological Fluctuating Phenomena study group was proposed in the way of continuing a scientific dialogue and exchange that began at the Purdue Congress several years ago, at the invitation of European members of the ISB. My presentation scheduled for Vienna very clearly impinges upon the question of energetic exchange within the biosphere, as a regulatory mechanism in both plants and clouds, and likely also at work in solar-terrestrial exchanges. The findings in question, as undertaken in both Europe and the USA, are precisely focused upon the link between organisms and environment, and contrary to your assertion, are within the "purview of biometeorology"... You refer me to the American Meteorological Society, of which I am a full member, and already know a good deal about. In my professional judgement, the vast majority of members of the AMS, and especially those dealing with weather modification issues, have not worked through the various conceptual and experimental prerequisites by which a rational judgement of this work could be based. Very few have read either Reich or Piccardi, much less my own published works. They, like you and me, are not equipped to engage in peer review of issues they have not personally studied. However, some individuals working within the community of biometeorologists, in particular those ISB members working within the Physico-Chemical and Biological Fluctuating Phenomena study group, have such a conceptual and experimental basis, and that is why I was glad to initiate the dialogue and exchange with them. Why do you now wish to intercede in these research matters, which do not appear to either interest or involve you? Yours is an unwarranted and unethical intrusion. Your decision to cancel my paper (and the papers of numerous other scholars) has no basis or validity from a scientific perspective. As you made this decision unilaterally, by yourself, in isolation, without consultation or overview by others with research expertise in the subject areas being reviewed, the political motivations are laid bare. You have transformed your narrow personal viewpoints into official ISB policy, but without any attempts to solicit the views of the broader ISB Membership, or to legitimize your decision according to ISB protocols. Indeed, you insist on titling yourself as "Chairman" of a "Scientific Committee" which does not appear to factually exist!... Like most scientists, I am fully prepared to defend my research once it is presented for review. But the issue here is one of censorship, the irrational erecting of barriers to the public presentation of findings. I take very seriously all the talk about scientific freedom and freedom of inquiry, but I have little patience for those who would claim this freedom for themselves while denying it to others... I demand your resignation from these key decision-making positions within the ISB.

J. DeMeo, Orgone Biophysical Research Lab

Watercolor by Deborah Carrino

* "Extraterrestrial Exposure" can land you in a "quarantine", surrounded by armed guards, with a suspension of all your civil liberties, including complete isolation from all friends and relatives. An obscure set of National Aeronautics and Space Administration Regulations, dating from 1969, provides for the immediate seizure and complete quarantine of any *"person, property, animal or other form of life or matter who or which has been extraterrestrial exposed"* (Title 14, Chapter V, part 1211). While originally drafted to govern situations where a person was exposed to a returning NASA or military satellite, which might harbor germs from the Moon, Venus, or Mars, or possibly radioactive materials or harsh rocket chemicals, NASA officials more recently ruled that the same regulations would apply to extraterrestrial visitors from outer space, as well as to any humans who had been witness to such a visitation. (NY Times, 14 Sept. 1982) No hearings would have to be held, and the forced quarantine could not be broken even by court order. The regulations authorize participation of other Federal agencies to likewise participate in such quarantine measures. This obscure law could pass as a mere curiosity, unless one reads the documentation contained in Timothy Good's book *Above Top Secret: The Worldwide UFO Cover-Up* (Quill/William Morrow, NY 1988). This book documents that the US, Canadian, and other governments have *for years* known about, and had solid evidence to prove the extra-terrestrial nature of the UFO; and worse, that a planned conspiracy exists to prevent the general public from learning this truth.

* Why did the Hubble Space Telescope malfunction? The people who were in charge of the malfunctioning satellite say there is a flaw in the telescope mirror, a flaw which would have been detected were it not for the fact that money-managers, instead of engineers, were running the program. However another possibility exists. Did it fail because *starlight is not as bright in space as it is seen from the surface of the Earth?* In a prior issue of the *Pulse* (Vol.1, No.1, Spring 1989, p.48), we discussed the observation that there are no, or only a very few photographs taken from space on which

stars can be directly seen. That's right, few or nearly no star photographs taken from space. Even in photos where the camera would have been set for a relatively long exposure, one rarely will see stars, and even when you do, they are only the brightest of stars, such as those defining constellations. There are lots of photos with bits of floating space debris in them, and one may even be fooled by a dust particle or two on the photographs. We have reviewed many different large-format photo album books about the astronauts and cosmonauts in space, including several video-tape presentations on the space program, including shots of lunar walks, space walks, and so forth — out of *hundreds* of cases, on only one or two of these photos could a few stars be seen in the black background of space! Moreover, the quotations from the astronauts and cosmonauts are revealing: they comment on the overwhelming *blackness* of space, and say nothing that would indicate that their view of the stars and galaxy was anything spectacular or extraordinary. Statements by NASA scientists about why the telescope failed do not mention *diminished starlight* as a possible explanation, but some of their own theories about the failure appear to be in agreement with this hypothesis. For example, the explanations given for the failure include "poor focus due to improperly constructed lenses" and "improper index of refraction in the telescope glass"; these explanations do account for the fact that less light is arriving at the focal point of the telescope than was originally expected. But the explanation of diminished starlight would also account for this lessening of light intensity. In fact, one computation suggests that only 20% of the expected light is actually being detected at the telescope focal point. This figure appears reasonable considering that photographs from space made with hand-held cameras rarely show even a single star! (*Science,* 1 June 1990; *Science News,* 7 July 1990). We again point to Wilhelm Reich's observations and findings on lumination between two objects possessing an orgone energy field:

"The phenomena of light, then, would be a local lumination of the atmospheric orgone, determined by the atmospheric orgone's existence, its capacity for lumination and thick-

ness. Further, we must distinguish the phenomenon 'light' from 'electromagnetic' light waves and say: what are 'transmitted with the speed of light through space' are only the invisible electromagnetic waves. They swing in a cosmic orgone ocean and excite the Earth's orgone envelope at the point where it is dense enough for lumination. ...light [is] a local lumination." (Reich, W.: Orgonotic Light Functions 1, *Orgone Energy Bulletin,* 1(1):5-6, 1949)

Blue Blobs from Space!!

* Astronomers have for years been investigating the question of the so-called "dark matter" of space, the invisible "stuff" that supposedly comprises at least 90% of the mass in the universe. So far, it has only been detected by its gravitational effects on the visible stars and galaxies. More recently, this dark "matter" has been photographed by making very long telescopic exposures. According to a recent report in *Science* (8 June 1990), this "invisible ectoplasm...will eventually reveal a dense patchwork of faint, irregular blue blobs..." Now, in addition to contending with "blue fuzzies" on their photographic plates (reported in two prior issues of the *Pulse*), astronomers must deal with the new "blue blobs". One wonders when these astronomical hard-heads will take a look at Reich's blue orgone energy, which also exists in high vacuum and, presumably, also in space itself.

Bions from Outer Space!!

* Astronomers keep finding substances in meteorites that resemble the membranes of modern day cells. As Reich demonstrated, *bions* -- observable microscopical vesicles transitory between living and non living matter -- can form from sterile rock or metal which has been heated to incandescence. As meteorites have certainly been heated to exceptionally hot temperatures during their transit through the Earth's atmosphere, the possibility exists that the astronomers are in fact identifying the existence of bionous processes which have been initiated within the meteoritic material itself.

* Is "empty space" an endangered species? Even some physicists are now challenging the notion of "empty space", as evidenced by the increasing study of "vacuum energy" by various scientists. *Omni* magazine recently ran a small item, which read as follows: *"There may be more out there in space than emptiness. That seemingly endless vacuum may actually be a sea of energy. 'Its like a strange fluid that's everywhere', says cosmologist Joshua Frieman of the Stanford Linear Accelerator Center."* Reich's pioneering study of orgone-charged vacuum tubes was not mentioned, and it was clear that the "vacuum energy" under discussion was not like Reich's orgone. *"It's not tangible in any way. The only thing we can feel is its gravitational effects,"* said Frieman, *"Vacuum energy has negative pressure; it forces matter apart...if a huge amount were around, the universe would blow up very rapidly. Things like galaxies could never even have formed."* From this, it is interesting to note how modern physics is so close to, and yet so far from Reich's very tangible and negatively entropic orgone energy. (*OMNI*, October 1988)

* "Cold Fusion" has been widely discussed in science journals since the phenomenon was first discovered by S. Pons and M. Fleischmann of the University of Utah. Not a small number of physics professors lined up to declare that the experimental results were "impossible", but so far, the experiment has survived the assaults of even the few critics who actually undertook some laboratory investigation of the matter. The experiment demonstrates very small temperature increases, on an order of magnitude similar to the to-t experiment developed by Reich. In fact, it is plausible that the Cold Fusion experiment actually demonstrates some kind of orgonotic heating effect. Reich's to-t experiment was also a hard pill for the physics community to swallow, even though it has been widely replicated with generally positive results; even the late Albert Einstein replicated the to-t experiment, declaring the phenomenon — which violates the second law of thermodynamics — to be "a bomb in physics". The Cold Fusion experiment was, initially at least, a topic for open discussion. Soon, however, the subject was greeted with the same aggressive hostility which greeted Reich's accumulator experiments. From one source, we recently learned that Japan is spending around $25 million to research the cold fusion phenomenon; in the USA, only derision and potential loss of employment greet such curious scientists. There are three new books on the subject: *Cold Fusion: The Making of a Controversy* (by David Peat, Contemporary Books, 1989) and *Fire From Ice: Searching for Truth Behind the Cold Fusion Furor* (by Eugene Mallove, John Wiley & Sons, 1991) take a balanced look, pro and con, while *Too Hot to Handle* (by Frank Close, Princeton U. Press, 1991) is distorted trash which fails to seriously review the positive evidence supporting the phenomenon and, worse, accuses advocates of cold fusion of fraud.

* An anti-gravity effect has been observed by two scientists, Hideo Hayasaka and Sakae Takeuchi, of the engineering faculty at Tohoku University, in Sendai, Japan. Their paper reporting the effect "Anomalous Weight Reduction on a Gyroscope's Right Rotations around the Vertical Axis on the Earth", appeared in the December 18th issue of *Physics Review Letters*, 63(5):2701-2704. Weight changes were found for three metal gyroscopes with rotor masses of around 150g, spinning inside a glass vacuum chamber on a chemical balance. The gyros were electrically driven to various rotation ranges, from 3,000 to 13,000 rpm, and then weighed under inertial rotation. Measurements were repeated 10 times for given rotational frequencies. Right rotations, with vertical axis, caused weight reductions of from 3 to 11 mg., with greater weight losses for higher frequencies. Left rotations showed no observable weight changes. Various control experiments did not eliminate the asymmetrical nature of the weight changes. The experiment violates conventional theory, which predicts no change in weight.

* Following the Chernobyl nuclear accident, which vented large quantities of radioactive pollution into the atmosphere, there was an observed increase in the frequency of lightning in downwind areas. The increase, observed by Swedish researchers at the University of Uppsala, lasted for approximately one month. (*J. Geophysical Research*, 20 Sept 1989)

* An American scientist, E. W. Silvertooth, recently published a paper on "Experimental Detection of the Ether" (*Speculations in Science and Technology*, 10:3-7, 1987). Silvertooth's experiment with laser beams and interference patterns shows the wavelength of light varying with its direction. From his results, Silvertooth was able to calculate the speed of the Earth relative to an absolute "ether frame of reference". The result: Earth is moving at "378 kilometers per second towards the Leo-Aquarius axis, aligned in that direction". This result is in agreement with prior studies on the anisotropy (unevenness) of the 3-degree Kelvin background radiation, which is 377 kilometers per second in the same direction. Another American scientist, Dayton Miller, worked for years on the same problem, following in the footsteps of his teacher, Albert Michelson. Michelson and his associate Morley, of the early Michelson-Morley experiments, detected only a very weak ether influence with their experiments. (See Dayton Miller, "The Ether-Drift Experiment and the Determination of the Absolute Motion of the Earth", *Reviews of Modern Physics*, 5:203-242, July 1933) Miller's work predates that of Silvertooth, and he also made a very solid detection of the ether. The problem about the "ether" has never been an absence of evidence, but rather, and absence of fairness among the orthodox empty-space physics theorists, who have consistently misrepresented, attacked, or ignored the positive evidence.

* The *Center for Frontier Science* at Temple University (Ritter Hall 003-00, Philadelphia, PA 19122) has become a major forum for discussion of research findings which have previously suffered from repression and censorship within American and European academic circles. Directed by biophysicist Beverly Rubik, the Center holds regular public lectures by innovative scientists from around the world, offering an important exposure to unorthodox research findings and scholars. The latest issue of *Frontier Perspectives*, the Center's newsletter, carried articles by the following researchers: Ernest Sternglass, M.D., who for years has sounded an alarm regarding the health effects of low-level atomic radiation; Beverly Rubik, Ph.D., and others on new evidence in support of homeopathy; Adolph Smith, Ph.D., physicist investigating Reich's bion discoveries, on the issue of self-organization in materials science; Robert O. Becker, M.D., on bioelectromagnetics; O. Edson Wagner, Ph.D., on the interaction between plants and gravity.

Population Growth

*   A population growth of 1% per year, which on the surface appears small, is actually sufficient to *double* a given population within only 70 years. A population which grows at 2% annually will double in 35 years, while a 3% population growth means a doubling time of only 23.3 years. Some areas of the world have growth rates even higher than this, such as east Africa, where populations grow annually by an estimated 3.5%, and double every 20 years. In areas such as this, women invariably are treated like chattel, and not given equal access to education, economic opportuntity, nor contraceptive education and devices. Every woman of child-bearing age is expected to become pregnant on a regular basis, and it is not unusual to see 30 year old women with 6 or 7 children. Of course, it is impossible for economic growth and social development programs to keep up with such expanding populations; medical care, schools, jobs, housing, and food supply fall far behind the demands of the ever-increasing numbers of people, leading to social chaos and catastrophe. Malthus discovered this "Principle of Population" decades ago, and the discovery helped to spur Europe through a sexual revolution, where the status of women was markedly improved and contraceptives made freely available. Non-Hispanic North America also participated in this change, which is still unfolding. Population growths drastically declined and social conditions improved wherever the social reforms related to female status, contraception, and abortion were enacted. Africa, Latin America, and Asia have not yet experienced such a change in conditions for women, nor change in attitudes about contraception and abortion. And even the USA and some European nations (such as Poland and Germany) may undo some of the prior reforms, a consequence of growing political strength of religious fundamentalism. As a result of these trends, the regions of the world where population growth continues at dramatic, explosive rates are the very same regions where social and environmental conditions are getting worse with each passing year.

The Doomsday Curve

*   If population on Earth continues growing at an exponential rate, then at some point in the future, the total number of people could theoretically approach an infinitely gargantuan number, in the billions of billions, or even a billion times that number, whatever it might be. This fact was postulated first in 1960, and a "Doomsday curve" was calculated, to determine when population would soar to infinity : "Doomsday" would take place on Friday the 13th, November, 2026 AD. Of course, the bit about Friday the 13th was thrown in for dramatic measure. But the idea that populations could continue to grow forever, to a point where bodies began to pile up on the Earth's surface so fast that the top of the pile would be propagating outward into space at the speed of light... well, it is all very ridiculous, except that the mathematical exercise forces one to recognize that there is, indeed, a natural limit to the growth of populations. Obviously, food shortages with political and economic contractions, and warfare over resources would occur; mass starvation and epidemics would set in long before this theoretical "Doomsday", particularly in regions where all available farmland was already put into production. Populations would then dramatically decline. In fact, most parts of the world have already reached their maximum agricultural potential, and the crisis implied in the Doomsday model has already started: parts of Africa, Latin America and Southern Asia are examples. In each of these locations, contraceptive availability and usage are low, as is the status of the female. Repetitive unplanned pregnancies are the typical norm. Population planners who had ridiculed the Doomsday curve in 1960 had themselves forecast a world population of 3 to 3.5 billion people in 1975. However, the Doomsday curve had more accurately forecast 3.65 billion people; the actual population in 1975 was a whopping 3.97 billion. And what of today? The Doomsday curve had forecast 3.969 billion people for 1980, but there actually were 4.414 billion, nearly half a billion more people than was forecast by the curve. By 1986 or 1987, we had passed the 5 billion mark, depending upon whose statistics one wishes to use, and still, the global population keeps climbing. So far, we are *ahead of schedule* for the 2026 AD Doomsday. Earth will achieve a Doomsday population, or rather, a *catastrophic series of events that will dramatically limit population growth*, sometime before the magical year of 2026! (*Science*, Letters, 25 Sept. 1987)

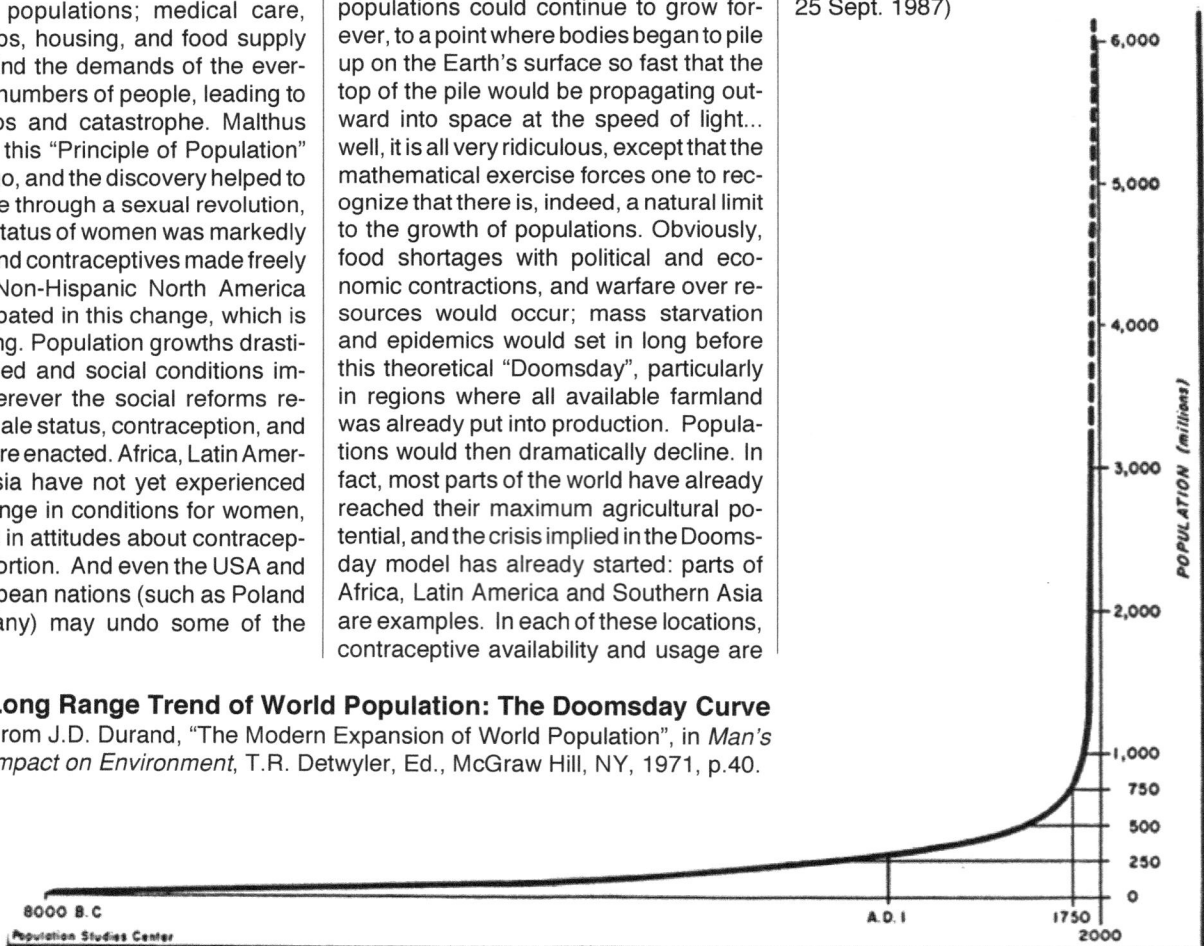

Long Range Trend of World Population: The Doomsday Curve
From J.D. Durand, "The Modern Expansion of World Population", in *Man's Impact on Environment*, T.R. Detwyler, Ed., McGraw Hill, NY, 1971, p.40.

More Population Woes

* A report by the United Nations Population Fund indicates that the Third World nations will account for 90% of global population growth in the next few decades. By the year 2025, they will produce four times as much carbon dioxide pollution as the developed world produces today.

* The population growth of Mexico fell from 2.9% in the 1970s to 2.3% in the 1980s. During this same time, their population more than doubled, and will reach 81 million in 1990 alone.

* In the United States, teenage birth rates increased sharply after nearly 18 years of decline. Back in 1972, some 230,000 babies were born to girls between the ages of 15-17. Since then, with increased availability of contraceptive and abortion services, the teen birth rate slowly declined. By 1986, however, the statistics indicate a slow reversal in the teen birth figures. In 1988, a sharp increase in the teen birth rate developed: one teenage girl in 30 gave birth, with roughly one in ten getting pregnant, roughly 2/3 of the girls choose to terminate their pregnancies. These shocking statistics are the end result of a planned assault upon family planning agnecies and services by conservative fundamentalist politicians. The present emphasis upon "abstinence education" and AIDS-scare talk has not reduced teen sexual activiity very much, but it certainly has muddied the waters regarding healthy, responsible, and loving sexual relations between teenagers -- this later approach is more widely used in Europe where, unlike the USA, sexual activity of teens is high, but unplanned pregnancies and abortions are much less frequent. (Nat. Center, Health Statistics)

Abortion

* Some 50 million legal and illegal abortions are performed each year, worldwide, most of which would be unnecessary if safe and effective contraceptive education and devices were available. Many nations still impose strict laws against birth control knowledge, and possession of a condom or birth control pills can land one in prison in some nations. Regarding abortions, about half of the 50 million performed are illegal, and 200,000 women die each year from such procedures. About 25 percent of the world's populations live in nations where abortion and contraception are greatly restricted or not available, mostly in Africa, Latin America, and Muslim Asia. These are the same nations experiencing extremely high population growth rates. (Worldwatch Institute)

* Abortion has been a top issue in the USA for years. It has also been a major issue slowing the unification of East and West Germany, but this fact has not emerged into the popular press, until most recently. East Germany provides free abortion on demand, while abortion is illegal in the West unless the woman can satisfy two reviewers that there are acceptable "social or medical reasons". As the review process can take months, many women travel from West Germany to the Netherlands or Yugoslavia for their abortions. With unification of the two Germanies, women could easily travel from West to East, and the Kohl government wanted to make this illegal also. Attempts were made, but mightly resisted, to impose the West German legal system on the East. A majority of East Germans, some 77%, want to retain current liberal abortion laws, which allow abortion up to the 12th week of pregnancy. At present, the pre-unification laws will be retained for two years, after which the Bundestag will decide the issue for the new unified Germany. (*Science*, 7 Sept. 1990)

* In the USA, the Federal Government recently ruled that doctors employed by agencies that use Federal funds are not allowed to tell pregnant women about abortion services in their local communities. Indeed, the new legislation banishes free speech between doctors and their female patients, by forbidding them from even mentioning the word "abortion". Meanwhile, Legislators in the State of Louisiana passed a law calling for stiff criminal penalties for doctors (or anyone else) who perform abortions, as well as for the desperate women who seek them.

* The first agenda of the anti-abortion movement will be to banish abortion in all or nearly all cases, followed by the banishment of birth control methods based upon prevention of implantation of the fertilized egg (such as the "morning after" pill, and the IUD). The second agenda will be the outright banishment of *all* forms of contraceptives, to make every sexual act capable of leading to a pregnancy. This agenda was made clear during the early days of the anti-abortion movement, by speakers who revealed their *hatred of* *sexual pleasure* as the primary motivation. Witness the following quotations from anti-abortion leaders, provided by Planned Parenthood; similar quotes by various preachers can be obtained without difficulty, from nearly every town and city across the USA, by watching the sunday religious programs on television.

"For those who say I can't impose my morality on others, I say just watch me."
- Joseph Scheidler,
Executive Director,
Pro-Life Action League.

"I don't think Christians should use birth control. You consumate your marriage as often as you like and if you have babies, you have babies."
- Randall Terry,
Executive Director,
Operation Rescue.

"Sex education classes in our public schools are promoting incest."
- Rev. Jimmy Swaggart

"We are totally opposed to abortion under any circumstances. We are also opposed to abortifacient drugs and chemicals like the Pill and the IUD, and we are also opposed to all forms of birth control with the exception of natural family planning."
- Judie Brown,
President,
American Life League.

It Can't Happen Here (?!)

* *"Nazareth, Israel -- About 15 young Arab women held an unusual demonstration yesterday to protest the killings of women by male relatives for having disgraced their family's honor. The women marched thorugh the center of this Israeli Arab town and distributed a leaflet saying that about 40 murders occur yearly among Arab families in Israel because daughters become pregnant out of wedlock, have affairs, or express wishes to marry men not approved by parents. The march was rare in the conservative Arab community. Taman Fuhaileh, a nurse from the coastal town of Acre, said the women planned the action after a pregnant, single woman was burned to death near Nazareth last week."* (*San Francisco Chronicle*, 25 June 1991)

Modern-Day Slavery

* Slavery still exists in many parts of the world, and the *Anti-Slavery Society*, founded in 1839, is still in business. Lesley Roberts directs this Society, which is the oldest human-rights organization in the world. The conditions leading to slavery, and which define it, are nearly the same as they were when the organization was founded. Chattel slavery, where one person totally owns another, was officially outlawed in Arabia only 20 years ago, and today it is increasing in parts of Northern Africa, particularly the Sudan. More prevalent is debt bondage, especially the giving of children to repay parental debts. Such children command a price of from $20 to $200 in the poor areas of the world, where it thrives. The practices are interwoven with social structures where the status of women and children are quite low to begin with, and where the dominance of the husband or father over the family is absolute. Such children sold into slavery work for extremely low wages: a carpet that sells in London for $6,000 may have been made by a slave for as little as $20. Millions of such children, boys and girls, are also devoured by the Asian prostitution industry. Contact: *Anti-Slavery Society*, 180 Brizton Rd., London SW9 6AT England.

* **We're Number One!** With more than 1 million people behind bars, the USA imprisons a bigger share of its population than any other nation on Earth. The USA jails 426 out of every 100,000 persons. South Africa is second (333 per 100,000) and the Soviet Union is third (268 per 100,000). European rates are from 35 to 120 per 100,000. These depressing statistics parallel a similar "top" rating in the USA for crime, violence, murder, drug usage, degenerative illness, and so forth. In recent decades, government leaders have declared "war" on these very same factors -- drugs, cancer, poverty, crime -- without being able to show any significant decreases. The reasons for the abysmally high imprisonment rates? The authors of the *Sentencing Project* study point to mandatory sentencing laws with stiffer penalties, reduction in use of parole, a reluctance to seriously condsider alternatives to imprisonment, and increased use of policemen to resolve social problems. The statistics suggest a near total failure in the USA to address the root causes of social problems. (*Sentencing Project*, Washington, D.C. 1990)

Native Peoples Under Assault

* Instances of outrageous genocide continue to crop up regarding the treatment of South American natives by their more "civilized" countrymen. In Columbia, cowboys engaged in deforesting the jungle and making ranchlands massacred 16 Cuiva Indians, including women and children. The court let them go after they argued that killing the indians was not a crime. The indians were called "animals, like deer or iguanas", and so the court let the killers go. Hundreds of Yanomami have also been killed in recent clashes with some 40,000 miners who have illegally rushed into Yanomami lands. According to the organization *Cultural Survival*, this situation is not unique, and the treatment of native peoples all over the globe is of a similar character: "Around the world, these outrages are perpetrated by capitalists, communists, industrial nations, Third World nations, and multinational corporations..." (*Cultural Survival*, 11 Divinity Ave., Cambridge, MA 02138)

* The government of Ecuador recently "legalized" the right of the Huaorani people to obtain title to the lands upon which they historically lived for generations. The Huaorani had previously threatened to fight workers building roads into their lands, an activity mainly undertaken by oil companies and lumber companies. Ecuadorian President Borja attempted to use the event as a propaganda coup, emphasizing the "gift" of Ecuador of the lands that the Huaorani felt they already owned. When the fine print on the documents was read in the sunlight, however, the ceremony was spoiled: the documents asserted that the Huaorani *"are not allowed to interfere with mineral and oil exploration by the national government and authorized companies"*. While speaking at the ceremony, Huaorani representative Ayuma Tenko reminded President Borja: *"This ratification of our territory is not completely your will. It is the product of years of struggle of our people, of national organizations, and of national and international solidarity. Furthermore, Mr. President, we ask that you immediately stop the construction of roads in our territory, that you evict the colonizers that have invaded our lands and the oil companies that are destroying our forests. Someday we will have nothing. If you do not meet our demands, we will defend with our spears what belongs to us."* (*Earth Island Journal,* Summer 1990)

* Bruno Manser, the Swiss anthropologist, has emerged from the jungles of Sarawak, Borneo to fight for decent and fair treatment of the Pean peoples he was living with. Their food-gathering grounds are threatened by bulldozers, and all appeals to the Malaysian government and to timer companies had fallen on deaf ears. From his jungle home, Maser prepared a press release about the problems of the Pean, and sent it to 30 regional and international magazines. A petition campaign was initiated to create a preserve for several of the native peoples of the forested area, and 7000 signatures were gathered, with endorsements from 14 different environmental and human rights groups. The response of the Malaysian government was to increase the pace of tree-cutting. When Manser organized blockades of the timber company roads, Pean tribesmen were jailed, and offers to negotiate were rejected by the government. Manser was thereafter pursued by Malaysian police for organizing the resistance to the logging firms. According to Manser, *"The rights of the original inhabitants have not in the least been looked after by the Sarawak government up until now, because leading politicians often are involved themselves in the business with tropical timber as holders of licenses. Since 1984, one-third of the area has been destroyed and two-thirds opened up by roads. On top of this, military, police and special units stationed in the crisis area now threaten the natives that had before lived in peace and harmony."* A survey of 13 logging contracts in Sarawak found politicians involved in every single one of them. Some owned shares in two or three concessions. Even the Chief Minister of Sarawak, Tan Sri Taib, his family and political allies control about four million acres of lumber concessions, about one-third of all forest land under contract in the country, valued at around $4 billion dollars. (*Earth Island Journal*, Summer 1990; *Greenpeace*, July 1990)

* The "Conquest of Indian America" has been dramatically documented in a series of 84 separate maps and color charts. A magnificent body of work, now available from *Historic Indian Publishers*, PO Box 16074, Salt Lake City, UT 84116.

ENVIRONMENTAL NOTES

Deforestation

* Deforestation is occurring at a fast pace in various places around the world, usually at the dictates of lumber firms based in the USA, Japan, or Europe. Protests against destructive, non-sustainable logging practices may be successful in nations with elected leadership, or where democracy prevails over dictatorial tendencies, but not so in nations where despotic rulers in government stand to reap financial rewards from deforestation. In those latter cases, destructive logging practices cannot be challenged within the system, and protesters are met by armed police or military forces. Worse, this use of military and police to enforce "rights" to deforest often takes place at the quiet urging of the multi-national corporations and their home governments. The USA is not immune from such anti-democratic associations. In Southeast Asia, for example, Buddhist monks in Thailand are ordaining trees into their holy order, wrapping trees with the sacred orange robes which are usually reserved for holy men. In this manner, they hope that the trees will be spared. Other villagers have taken up both non-violent resistance tactics, and sometimes the gun, knowing that deforestation of their homelands will result in complete destruction of their environment and subsistence. Pean tribesmen in Indonesia have been trying to stop buldozers by falling down before them, while Burmese in Asia's largest rainforest are now fighting a pitched battle with the nations military forces. Little is said of these environmental factors at work in the political strife, but they are real and present factors. Worse, deals are now being struck between various nations for collective exploitation of the forests; for example, Thai logging firms recently hatched a deforestation agreement with Cambodian guerrilla factions (including the bloody Khmer Rouge), in order to get money to buy weapons to fight off the Vietnamese-backed regime controlling the capital city. Cambodian guerrillas now act as paid mercenaries for Thai logging firms. One general pattern is that the people who live in the cities, with economic and commercial ties to central governments and overseas corporations, see the forests as vast resources to be exploited for cash profit; people who live in the forests, however, know that when the forests are gone, their sources of food, clothing, and shelter will be destroyed, and that poverty and a wasteland, with inevitable appropriation of land by the same wealthy city folk will soon follow. About 12,500 acres of tropical Southeast Asian rainforest is lost each day, with only 10 percent replanted. The Southeast Asian rainforests will likely be gone by 2010, according to regional environmentalists. (San Francisco Examiner, 13 July 1990)

* Forest Service Chief F. Dale Robertson acknowledged before Congress that only 10% of the Northwest's old-growth forests remain standing, and that half of those are scheduled to be cut over the next 50 years. Environmentalists have maintained this 10% figure for some time, but this is the first time that government officials have acknowledged it. (San Francisco Chronicle, 25 July 1990)

* According to the Wilderness Society, the US Forest Service currently sells entire trees in the Targhee National Forest of Idaho for around $1 each, even though they spend nearly three times the amounts collected to survey the land, build logging roads, and draft legal documents necessary for the sale of the trees to the lumber companies. Logging in that forest alone actually cost the US Treasury some $5 million in 1989, in the manner of subsidizing the logging of trees. The Forest Service was mandated to administer all US Forest lands for "multiple uses", but their interest is primarily in doing the bidding of the lumber companies.

* Some 200 years ago, California possessed around 2,000,000 acres of redwood forest. Today, around 95% of those forests have been cut, with less than 18,000 acres remaining in private hands — even so, the rates of cutting are up dramatically. In 1988, more trees were logged in California than at any prior time since the 1950s. Existing forest protection measures are inadequate, as demonstrated by the fact that redwood cutting exceeds regrowth statewide by more than 50%. In Mendocino County, overcutting exceeds 300% of regrowth. The worst offender in this mess is PALCO, the Pacific Lumber company of Humboldt County, which owns more than 80% of all remaining private virgin redwood forests in California. The company was once admired for its conservative, sustainable yield practices, but no more. Pacific Lumber was bought out with junk bonds by corporate raider Charles Hurwitz of the Houston-based Maxxam Group. In order to pay off the loan-shark interest rates on the junk bonds, Hurwitz ordered a dramatic increase in the rate of cutting on Pacific Lumber lands. Other companies have followed Hurwitz' lead, and likewise increased their rates of cutting.

* "Redwood Summer" was the name of a direct-action protest sponsored by the Earth First! organization in Northern California. The purpose of the action was to halt the destruction of some of the last remaining stands of virgin old growth forest in North America. The stands of trees in question are being heavily logged and clear-cut, in some cases merely for the purpose of repaying interest on massive loans assumed by the lumber companies, who purchased the stands of trees with high interest junk-bonds. The non-violent actions included blocking logging roads, mills, and log shipping facilities, and were designed to focus national attention on the issue at a time when legislation to protect the trees was log-jammed in the California State and Federal Legislatures. Two leaders of the direct action, Judi Bari and Darrul Cherney were recently injured when their car was bombed, but the protests continued anyhow. The Oakland, California police claimed that the pair was engaged in transporting the bomb for use in the protests, but evidence for this assertion never emerged, and charges were not brought against the two environmentalists. Irregularities in the police investigation have surfaced, however, implicating the Oakland police and the FBI in a cover-up attempt, and failure to properly investigate. Local and national news media have failed to report much of the details, and indeed, appeared most zealous in branding the Earth First! leaders as "eco-terrorists". Almost all of the violent acts, however, have been assaults by loggers and lumbermen on the protestors. The headline story of the Mendocino Country Environmentalist (Issue 132, July 1-15, 1990) "Redwood Summer Bombing: Police Framing, Not Investigating" provided details that have yet to appear in any "regular" news outlet. Likewise, the

Earth First! Newsletter provided a completely different version of events, as compared to the regular press (see "The Bombing: What Happened?" and "Someone Tried to Kill Us; the Cops Tried to Frame Us"). From another source, there is evidence that *agents provocateur* were at work in a previous FBI bust of Earth First! members. A costly FBI "discrediting campaign" of the organization, involving forged inflammatory letters, has also been alleged. An *Animals Agenda* magazine article ("Earth First! Founder Busted in Possible Set-Up", Sept. 1989, p.20) focused upon similar FBI involvement in an alleged bombing of a medical research laboratory. The Earth First! members are being defended by Gerry Spence, the attorney who successfully sued the Kerr-McGee nuclear empire on behalf of the late Karen Silkwood, who died mysteriously after exposing safety hazards at a Kerr-McGee nuclear facility.

* Other environmental activists are being targeted by local law enforcement officials, and by the FBI, for no other apparent reason than for the legal work, research, or social programs they undertake. One non-violent demonstrator against a Du Pont facility was jailed with a $100,000 bail; the man who appeared before the same judge just before the demonstrator, who had beaten his wife with a baseball bat, was released on a $1000 bail. Georgia police acted as "enforcers" for the supporters of a proposed toxic incinerator, preventing the opponents of the facility from speaking at a public meeting on the issue. Opponents of the incinerator were later harassed at their workplaces regarding their social activism. Another public hearing on the construction of a toxic incinerator in Arizona collapsed into anarchy as police attempted to clear the room of opponents; the issue under contention was the collusion between the state supporters of the incinerator and private corporations, over the interests of people who would be forced to live with the incinerator. Non-violent protestors were dragged out of the hearing by police, who used high voltage "stun guns" on some of them. They were locked up in a sheriff's van until after the hearing ended. Non-violent protestors marched in front of the headquarters of American Cyanamid, in Bound Brook New Jersey, protesting toxic mercury contamination. Police in riot gear rushed the crowd, grabbed several demonstrators and clubbed them to the ground. Said one beaten demonstrator: *"We were peaceful and we announced our intention to be non-confrontational in advance... The media's treatment of incidents like this paints a picture of wild and unreasonable environmentalists marching in the streets, and it portrays the pursuit of healthy debate as dangerous. The first point is not true, and if people are intimidated into not speaking out, we lose the most vital part of our democracy."* (*Greenpeace*, Sept./Oct. 1990)

* Environmentalists and park officials of the Abruzzo Altipiana Delle Rocche in Italy have been assaulted, and threatened with death on many occasions, due to their efforts to halt the development of this wilderness area. Developers, industrialists, and politicians had essentially divided the area up, and initiated construction of large ski areas and condominiums. As development and poaching in the rare ecosystem was halted by citizen complaints and political action, the various assaults upon the environmental leadership began: tires were slashed, the child of one park protection leader was roughed up by three hoods, and death threats began. (*Omni*, Oct. 88)

* Senhor Amazonino Mendes, governor of Brazil's Amazonas province, recently distributed 2,000 free chain saws to his political supporters, stating that he did not want the peasants to "live in misery". Meanwhile, ex-contra supporter Elliot Abrams — the Reagan Administration official who was embroiled in the Iran-scam dealings and who was recently declared *persona non grata* by the government of Costa Rica for shady dealings in guns and cocaine — is now arranging to sell Brazilian hardwood timber to Japan. Abrams said: *"I'm making a lot of money. It's great"*. (*Greenpeace*, Nov/Dec. 1989)

* The World Resources Institute (WRI) has estimated that some 29.7 million acres (46,390 square miles) of tropical forest were lost in only 8 nations in 1987. This figure is four times higher than the rate calculated by the United Nation's Food and Agriculture Organization for 1980. The estimates of WRI, made from satellite images, also indicate that tropical forests are today vanishing at the rate of 40 to 50 million acres per year, which is nearly twice the UN estimate of 28 million acres. Satellite images from India, for example, show that large areas legally designated as protected forest are now barren of trees. WRI has also estimated the relative contribution towards "greenhouse warming" for each country. The US and USSR are the top offenders on the list, due to their high consumptions of fossil fuels. Brazil, China, India and Laos come next, given their high levels of deforestation and coal burning. Qatar and the United Arab Emirates are next, due to the practices of gas flaring at oil fields. (*World Resources 1990-91*, $17.50 + $3 shipping from WRI, ~~PO Box 4852, Hampden Station, Baltimore, MD 21211~~)

* A new book documents, in tragic fashion, the enormous fires and smokes that fill the atmosphere over the Amazon during the *queimada*, the time when dead grass and underbrush are burned by farmers to clear the way for planting. The queimada is often carried out during the dry months, so flames often spread into nearby forest regions. Smoke from these

Clearcut "Moonscapes" in Northern California, an increasingly common sight in the Pacific Northwest.

fires is so think and dense that the ground is obscured, as seen by satellite images from space. (*The Burning Season*, by Andrew Revkin, Houghton Mifflin, 1990)

* "*The Effects [of Amazon burning] are much worse than those from saturation bombing. You end up with massive smoke clouds spread over millions of square kilometers. In some places, visibility is so bad you have to wait for days for things to clear up if you want to travel by plane or car. There are literally thousands of fires, and I'm just talking about the larger ones, at least 50 meters across.*" So says Alberto Setzer, head of Amazon studies for Brazil's National Space Research Institute. Approximately 12% of the rain forest has been lost by fire and flooding alone since 1978. (*Science*, 13 Oct. 1989)

* In 1982, some 52% of Quebec's forests were considered healthy; by 1987, only 1% remained in the healthy category, according to the Quebec Ministry of Agriculture, Fisheries, and Food. (*Acres, USA*, Sept.89)

Last Rites for Fresh Water

* The history of the Western States of the USA is largely a history of water: buying, stealing, and killing for water. The worst offenders in these battles are the major cities, such as Phoenix and Los Angeles. Using combined corporate and State power, these cities buy the water from out of stream beds hundreds of miles distance from the place where it will be used. Oftentimes, the water is squandered for lawn watering, swimming pools, and other domestic needs, without concern for the ecological effects upon the drained regions, and without attempts to conserve the scarce resource. Even water under the ground in one area may be owned by people or corporations living hundreds of miles away. The water rights can be inherited, passed along from one family to the next, or sold in perpetuity to big corporations. In the Western drylands, those with access to water become rich, while those without it, poor. Small wars were fought between regions of the West over the control of water, such as when the City of Los Angeles arranged, through shady back-room deals, to drain dry the Owens River in the Sierras, destroying the forested grassland ecology and small ranches of the region. The situation is reminiscent of the Middle East, with its Sheiks, Caliphs, and other war-lords, who

maintain power by virtue of their monopoly over water. The city of Los Angeles already has huge aqueducts that drain water from the Colorado, Owens and Sacramento Rivers; and new plans have been set into motion to drain water from the Columbia River in Washington State, and the Snake River in Idaho, all the way south to Los Angeles. Residents of Washington, Oregon and Idaho are hopping mad about this, though little about the plans is being reported in California newspapers. In spite of a five-year long drought, few meaningful conservation programs have been initiated in Southern California, and many politicians are preparing to finance mega-buck expenditures to steal even more water from other regions. Los Angeles County water supervisor Kenneth Hahn, a "smiling good'ole boy who likes strangers to call him 'Kenny'," complained that the Columbia River was wasteful, "dumping" billions of gallons of water into the Pacific Ocean each day. Wouldn't it be better, he argued, if most of that wasted water were being used to water lawns, fill pools, and flush toilets in Los Angeles? "All were asking for is 3 billion gallons", he said. (*Seattle Intelligencer*, April 1990)

* Irrigated lands account for only 17% of the world's area under cultivation, but produce around one third of the global harvest. According to a recent *Worldwatch Institute* report "Water for Agriculture, Facing the Limits", the world is becoming increasingly reliant upon irrigated cropland, and irrigation water, for its agricultural needs. Meanwhile, the costs for construction and maintenance of irrigation systems are increasing as irrigation water supplies dwindle. As world populations increase and irrigation projects push into lands less suitable for its development, overpumping of groundwaters (beyond sustainable rates), siltation of reservoirs and canal systems, and land salinization are on the increase. Cities are also increasingly demanding "their share" of increasingly scarce water supplies. The demand for water is fast approaching the amount available in rivers and lakes, and many of the world's largest freshwater lakes are now nearly drained dry for agricultural uses, such as the Aral Sea and Lake Chad. One region to feel the hard crunch of this water crisis is Egypt, where water supplies are barely adequate for today's needs, but where the population grows by one million every eight months. (*Science News*, 16 Dec. 1989)

Pesticides

* Pesticides banned in the USA continue to be manufactured here, and sold abroad. Recently, however, the Texas Supreme Court ruled that Costa Rican workers who were made sterile by exposure to the US-banned dibromochloropropane (DBCP) on a Standard Fruit banana plantation can sue the fungicide's manufacturer, Dow Chemical and Shell Oil, in Texas courts. Justice Lloyd Doggett ruled that the courts were open to such foreign lawsuits, and that the "convenience doctrine" defense, used to dismiss the case by lower courts, could not be used to "immunize multinational corporations from accountability for...causing injury abroad." (*Greenpeace*, July/August 1990)

* Citizens of California continue to be sprayed from the air with toxic malathion pesticide. Two Mediterranean fruit flies were found in a single grove, prompting the use of some 20 tons of the 96% industrial-grade spray. Families and schools are regularly warned not to allow children to play outside, or to let pets out of doors following the spraying of this chemical, which has been banned by 12 countries (including Japan and Germany) due to its side effects. The killing effects upon honeybees, ladybugs, and fish are considered inconsequential, and parents who complain about their sick children are simply ignored. However, the natural habitat of the red ground squirrel, near Los Angeles, is safe from the spraying, as the squirrel is an endangered species. Not so, the children of California. (*Earth Island Journal*, Summer 1990)

* According to Professor William Jordan of the University of California at Davis: In 1948, American farmers used 15 million pounds of pesticides and lost 7% of their crops to insects. Today, after 42 years of "wars of extermination" we have altered the natural controls and use 125 million pounds of pesticides yearly, and lose 15% of our crops. (Los Angeles Times, 23 April 1990) This reminds the editor of the past "war on cancer", and the current "war on drugs", both of which have led to results exactly the opposite of the intention — or, at least the *stated* intention. The real folly is in the making "war" of any kind as a means of solving social, health, or environmental problems.

* Noticed a decline in the songbird population in your area? Before you blame the neighborhood cats, take a close look at all those pristine, spotless green lawns around town. In nature, such golf-course greenery is non-existent, and it is only possible through the massive use of pesticides and herbicides. The Wildlife Center of Virginia has specifically targeted various toxic lawn-care products as the cause of this loss of bird life. The moral is, if you want happy, healthy birds, you must have happy, healthy bugs and worms in your lawn, and additionally various wild herbaceous plants ("weeds") to supplement the diets of the birds. Sorry, but your backyard feeders won't provide an adequate substitute for what nature provides.

* The World Bank will soon loan Brazil $51 million to spread some 3,000 tons of DDT across the Amazon Basin, in an effort to reduce malaria. Prior loans from the World Bank helped to build roads, dams, timber and mining operations, cattle ranches and farms, adding significantly to populations living in deep jungle areas. The activities not only add to the malaria susceptible human populations in those areas, but also help to create the ideal habitat for the Anophelene mosquitoes that carry malaria. Roads and mining projects create many large and small pools of water in which the mosquitoes breed, and deforestation exposes breeding mosquito populations to sunlight, speeding their development. DDT use to combat malaria is opposed by environmentalists due to its persistence (a half-life of some 40 years) and ability to concentrate in the food chain. Many medical people are opposed to DDT also because the mosquitoes become resistant to it. India, Sri Lanka, Indonesia, parts of Central America and the Caribbean have all witnessed a resurgence of malaria after mosquitoes became resistant to DDT. Mosquito resistance to DDT and other related insecticides is considered a major impediment by the World Health Organization. Even so, research into alternatives to pesticides, such as biological controls of mosquitos have yet to receive the same serious attention and research funding as the chemical pesticides do. (*Greenpeace*, Nov./Dec. 1989)

Notes on Drought and Desert

* *"The 1988 North American drought had widespread impacts, including $40 billion in estimated direct economic losses and costs in the United States. The 1988 drought was the worst since 1936 in the midwestern United States and parts of the northern Plains. Overall, during the height of the drought of 1988 (July), 40% of the country was experiencing either severe or extreme drought as measured by the Palmer Drought Severity Index. ... In the summer of 1987, the summer monsoon was weak in India and Pakistan, with many areas receiving less than half the normal rainfall. In 1988, Italy suffered one of the worst droughts in the last 175 years. The last three months of 1988 were the driest in at least 114 years in some regions of eastern China. In some regions of Turkey, January through April 1989 was the driest on record since 1951 (when reliable records began). The drought in the western regions of the African Sahel persisted from 1970 well into the mid-1980s."* (Bulletin, American Meteorological Association, July 1990)

* Droughts do more economic damage and harm more people worldwide than any other natural hazard. In the present era, at any given time, up to one fifth of the world's land mass is gripped by severe drought. In Africa, some 55 million people were badly affected during the 1980s; 1 million Ethiopians perished during the 1984 - 1985 famine. (*U.S. News & World Report*, 23 July 1990)

Electromagnetic Perils

* The Environmental Protection Agency recently concluded a two-year study, with the final recommendation that extremely low frequency electromagnetic fields be classified as "probable human carcinogens". This designation is now used to classify such chemicals as formaldehyde, dioxin, and PCBs. However, the paragraph with this recommendation was deleted from the final draft report by reviewers at the White House Office of Policy Development, who were concerned about the costs of implementing new safety regulations. As this report was being debated in Washington, a study published in the May issue of the American Journal of Epidemiology, revealed that mothers who slept under electric blankets during the first trimester of pregnancy had a four-fold increase in the risk of brain tumors in their children; the study likewise suggested that children who used electric blankets showed a 50% increase in childhood cancers. (*Science News*, 30 June 1990; *Time*, 30 July 1990)

* Electromagnetic emissions from computer word processors have been linked with various health problems, particularly for female office workers who have to spend hours on end in front of them. Among the common complaints and health-related findings are: loss of eye focusing ability, headaches, miscarriages, and general immune system disorders. One new finding is "computer squeal", a high-frequency sound that many female office workers have complained

about for years, but which male doctors and bosses have not taken very seriously, because they were not able to hear it themselves. The sound is usually accompanied by stressful feelings, and can be considered as a precursor to more serious health problems. (*San Francisco Examiner*, 12 August 1990)

* The deleterious health effects of electromagnetic fields from powerlines, electric blankets, microwave ovens, and computer terminals are now being taken more seriously by the public and some health professionals, following publication of several national magazine articles and two new books on the subject: *Currents of Death: Power Lines, Computer Terminals, and the Attempt to Cover Up Their Threat to Your Health*, by Paul Brodeur (Simon & Schuster, NY 1989), and *Cross Currents: The Promise of Electromedicine, the Perils of Electro-Pollution*, by Robert O. Becker (J.P. Tarcher & Co., Los Angeles 1990). Most recently, this Laboratory obtained a sensitive power line field meter, and we surveyed the local neighborhood. Incredibly high levels of 60 cycle magnetic fields were found at various points on the sidewalk, and a few homes had extremely high electromagnetic fields just outside their front doors. The walls and roof of a home will decrease these fields only slightly, and so the potential health effects from them are quite significant. Both of the above books, including sensitive electromagnetic field meters, are available from Natural Energy Works (naturalenergyworks.net). Similar meters are also widely advertised in various health magazines, indicating a growing public interest in this issue.

Oceans
* A 6% reduction in the extent of polar sea ice has taken place from 1973 to 1987, according to satellite measurements made at NASA's Goddard Space Flight Center. Most of the ice loss has taken place within the ice pack, in the form of open, ice-free areas. The observation was believed to be due to increased polar air temperature, a by-product of general global warming. (*Science News*, 8 October 1988)

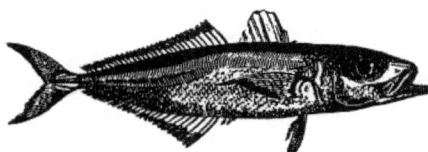

Missed Opportunities
* The US was the clear leader in photovoltaic (PV) technology until only a few years ago. During the past 8 years, the DOE budget for the PV program dropped from about $150 million to $35 million. The Japanese, German, and Italian governments now spend more on PV research and development that the US, and even Saudi Arabia outspends the US on solar energy research. US PV manufacturers held 80% of the world market in 1980, but our share today is less than 50%. The leading US PV manufacturer, ARCO Solar, was recently sold to a German investor. This erosion of American PV and other renewable energy production takes place at a time when energy from conventional sources is becoming harder and more costly to come by. Witness the end product of this social shortsightedness and irrationalism in the Persian Gulf war.

The Top 15 Contributors to Global Pollution:
In terms of total amounts of waste generated: 1) DuPont, 2) Royal Dutch Shell, 3) British Petroleum, 4) American Cyanamid, 5) Occidental Petroleum, 6) Agrico Chemical Co., 7)ASARCO, 8) EXXON, 9) Inland Steel, 10) Monsanto, 11) Eastman Kodak, 12) Vulcan Chemicals, 13) Dow Chemical, 14) Union Carbide, 15) Pfizer Pharmaceuticals. (*Earth Day Wall Street*, Box 1128, Old Chelsea St., New York, NY 10011) Note: the 1988 income figures for these 15 corporations ranged from $76.4 billion (EXXON) to a lowly $98 million (Agrico).

Pulse of the Planet not associated with DuPont Chemical Corp.
* As our regular readers are aware, the *Pulse* was announced in 1988, and first published in early 1989. In 1991, however, the DuPont company, identified above as the world's top polluter, began to broadcast a series of public service radio announcements, with disinformation and propaganda for various environmentally-toxic industrial schemes, under the banner "Pulse of the Planet". We just wanted our readers to know that neither this publication, nor the Orgone Biophysical Research Lab, are affiliated with or sponsored by the DuPont company. There is no connection between this publication and the announcements you may hear over the radio under the guise of environmental "information".

From Lester Brown, Worldwatch Institute
* "...if we don't create an environmentally sustainable economic system, there's no future. Economic indicators are flawed in a fundamental way: They don't distinguish between resource use that sustains progress and those that undermine it. The Gross National Product undervalues qualities like durability and environmental protection, while it overvalues planned obsolescence and waste. For example, if the oil spilled by the Exxon Valdez had reached a refinery and been processed, it would have modestly boosted the GNP. But spilling the oil generated far more economic activity in cleanup costs. According to the Alaska Department of Labor, the number of jobs statewide is rising. The March unemployment rate of 7.9% is the lowest for that month since 1976, when construction of the trans-Alaska pipeline was at its peak. So according to our system of accounts, you can boost the GNP by wrecking oil tankers... [To attack these problems, we] need to shift military expenditures into things like family planning, tree planting, energy efficiency and soil conservation. For example, armies in Africa might be mobilized to plant trees and reclaim the desert rather than engage in military maneuvers. Both Africa and Latin America ended the 1980s with lower living standards and food consumption than they had at the beginning of the decade. Social disintegration and political instability often follow. Take northern Ethiopia, for instance. Soil erosion has reached the point where there's not enough topsoil to support even subsistence-level farming. People have become environmental refugees. Water shortages will be another problem. If Ethiopia, which frequently suffers from drought, decides to develop irrigation from the headwaters of the Nile, it will be at the expense of Egypt. In years to come, water may replace religion and nationality as a principal source of conflict in this part of the world."

* "The frustration, pollution and death toll from automobiles are really quite high. In the US more than 45,000 people die in motor vehicle accidents each year, and hundreds of thousands more are seriously injured. One indication of how ludicrous our system has become is that thousands of people will get in a car and drive to a health club so they can sit on a stationary bike for half an hour and work out." (*Worldwatch Institute, ~~1776 Mass. Ave., NW, Washington, DC 20036~~*)

NUCLEAR HAZARDS

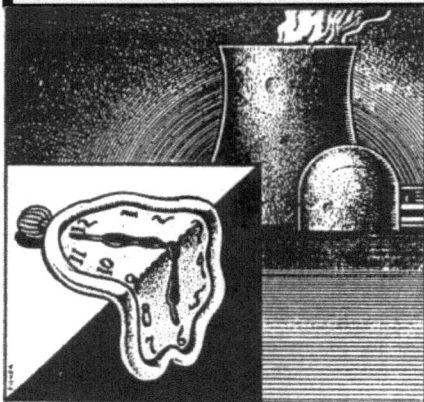

* Radiation leaks from the Hanford Waste "Repository" (dump) and the nuclear weapons fabricating facilities on the site are finally being discussed in the popular media, in spite of the fact that these abuses have been known and discussed by environmental organizations for years. Records dating back to the old Atomic Energy Commission (forerunner to the Nuclear Regulatory Commission and Department of Energy) have revealed that airborne emissions of radiation, more than 1000 times the levels considered to be "safe", were indeed released into the local atmosphere. The toxins emitted included radioactive iodine, plutonium, ruthenium and other very toxic products known to accumulate in the food chain. This was done, in spite of the fact that the administrators of the facility knew that the levels were extremely hazardous to both workers and local residents. One in 20 residents of the 10 counties surrounding Hanford facilities absorbed a "significant" radiation dose, particularly during the years 1945-1947. For example, so much radioactive iodine was dumped into the atmosphere that 1200 children living nearby received cumulative doses of from 15 to 650 rads each — one rad is roughly equal to a dozen chest x-rays. About 13,500 people, some 5% of the local populations, may have received 33 rads or more. More radiation was released over time at this facility than was released during the Three Mile Island accident of 1979. Milk from local cows was contaminated, but no warnings were issued from the Hanford plant managers. Water from the Columbia River was also contaminated, and this affected the health of people who drank the water and ate fish and shellfish. One administrator raised an objection to the secret emissions, in 1954, pointing out that workers at the plant would need medical assistance and help in later years. However, his objections and report on the problem were suppressed for "national security" reasons. Administrators heading the facility today say that the leaky plants at fault have been shut down, and that "the problem no longer exists". (*San Francisco Chronicle*, 13 and 14 July 1990, *Time*, 23 July 1990, *Science*, 3 August 1990)

* In a separate report, an advisory panel led by a former chairman of the Nuclear Regulatory Commission announced that large tanks of radioactive waste, currently in storage at the Hanford dump, were in danger of exploding. One or more of the 177 tanks may be "vulnerable to detonation". This fact has been known for nearly 20 years, as the editor of the *Pulse* recalls reading about the problem back in the early 1970s, when Nixon was president. Each of the large tanks holds thousands of gallons of highly radioactive waste sludge and waters. The radiogenetic heat that spontaneously develops in those tanks is so great that the liquid spontaneously boils. Some of the tanks are so radioactive that they must be cooled with refrigeration coils, and the mere walking by the tanks is enough to expose an individual to an extreme and significant dose of radiation. The tanks are constructed of stainless steel and other materials that were designed to last several hundred years, but in fact many of them began to rust and leak in less than 10 years. The original 300 year anticipated lifetime has turned out to be sheer fantasy, and in typical manner, the administrators at the dump have been routinely deceiving the public about the true nature of the hazard. Unlike reactors and other nuclear facilities, these storage tanks containing nuclear waste cannot be "shut down". An explosion similar to the one warned about in Hanford did occur in radioactive waste storage tanks in the Soviet Union, back in 1957. At that time, thousands of people were evacuated from hundreds of square miles in the Ural Mountains. Hundreds of people were killed by the radioactive fallout resulting from the explosions and fires. (*San Francisco Chronicle*, 31 July 1990; cf. Zhores Medvedev: *Nuclear Disaster in the Urals*, W.W. Norton, NY 1979)

* Congressional hearings on the Hanford nuclear dump continue to reveal more abuses of public trust by the various nuclear "regulators". In 1965, for example, a storage tank containing radioactive waste blew up, and was lifted some six feet off the ground, sending a geyser of radioactive steam 50 feet into the air. The "incident" was never reported, and was made public only because of secret communications between plant workers and the Washington State Department of Ecology. The Director of that Department indicated that she still relied upon secret communications from plant workers to find out what was really going on at the Hanford site. (*San Francisco Chronicle*, 1 August 1990)

* A "Letter to the Editor" recently highlighted the awful situation near the Hanford nuclear dump, and points to deliberate knowledge about nuclear dumping, and "use" of local populations to study the effects of radiation exposure:

"As 'downwinders', born and raised downwind of the Hanford Nuclear Reservation in Washington, we learned several years ago that the government decided to use us as guinea pigs by releasing radioactivity into our food, water, milk and air without our consent. Now, we've learned that we can expect continuing cancer cases from our exposure in their 'experiment'. The exposure began the same day our lives began. Several years ago, when the government admitted that releases had been made, we were assured there would be no observable health effects. Did the government really look? We had been seeing the effects for a long time. For us, the unusual was the usual! During my childhood I remember seeing men dressed in space suits walking in front of uniformed soldiers carrying shovels and sacks. We didn't know these were nuclear clean-up crews. What we also didn't know was that other kids didn't get 'neck massages' from the school nurse (looking for thyroid swelling) or have Geiger counters passed over them; that men didn't sample everyone's water and milk weekly. We thought everyone had deformed calves, sheep and kittens; that miscarriages — human and animal — were the norm. Common news was of neighbors and loved ones getting cancer. I was born a year after my stillborn brother. I struggled to breathe through underdeveloped lungs and suffered to overcome numerous birth defects. I underwent multiple surgeries, endured paralysis, endured thyroid medication, a stint in an iron lung, loss of hair, sores all over my body, fevers, dizziness, poor hearing, asthma, teeth rotting out and sterility. I am forced to question our government. Our

patriotism has been impugned, our credibility questioned. We have been redlined by the banking community since 1985. We have been put off by politicians — except for a brave few — until we become a popular issue. Who the hell do these people in the nuclear gang think they are? How can we citizens defend liars? We deserve fair and equitable treatment. Are we just so much nuclear waste? Will there be compensatory damages? There is a fine line of morality that none of us can cross and still claim membership in the human race. The government's nuclear gang deliberately crossed it. The price we had to pay, you say? We think we were worth more! " — Tom Bailie, Hanford Downwinders Coalition. (*San Francisco Chronicle*, 24 July 1990)

* A recent British study demonstrated a clear link between the low-level radiation exposure of men working at the Sellafield nuclear reprocessing facility and the increased levels of leukemia observed among their children. Radiation damage to the men's sperm is believed to have been responsible for the subsequent appearance of leukemia among the children. The finding is also believed to be related to the observed "clusters" of cancer in the vicinity of the facility. In one village near Sellafield, high levels of cancer, including leukemia and lymphoma, were observed in people under 25 years of age. Matched populations used as a control did not show the same high levels of disease, and the greatest correlation of risk was the employment of the fathers at the Sellafield reprocessing facility. (*British Medical Journal*, 16 Feb. 1990)

* The devastation of Navaho Indian uranium miners from lung cancer recently made national headlines, though the problem of miners dying from lung cancer at nearly 60% has been known for years. In fact, the Atomic Energy Commission once was presented a draft report on the urgency of placing high-volume air ventilation fans inside uranium mines, because the levels of radioactive radon gas were too high for worker safety. Indeed, the report predicted that a very high percentage of the Indian miners would die of lung cancer if the fans were not installed. The AEC suppressed this report for decades, however, because it would have been "too costly" to install the fans. Better to let the Indians die of cancer than to have the price of uranium ore go above $8 per ton. The same AEC, with the help of Senator Kerr of Oklahoma (the same personage

of Kerr-McGee Corp., which was implicated in the murder of reformer Karen Silkwood), and in collusion with other ignorant and contemptuous elected government officials, passed laws exempting the both the Federal Government and private uranium miners from responsibility for the vast heaps of radioactive sand — a by-product of uranium mining that is rich with radium and other isotopes. These sand heaps have since blown all over the territories where uranium mines were located, contaminating the air and streams, and creating lifeless landscapes that send Geiger-Counters singing off-scale. Radioactive sand was also used, with complete knowledge and approval of the Federal Government and mine owners, to make bricks and concrete for use in various homes, apartments, schools and hospitals. No records were kept of who used the radioactive sand, as it was sold to the unsuspecting public for only a few dollars per truck-load. At the time when this scandal was first raised in Congress, then retiring Senator Kerr stood on the Senate floor with his young daughter, claiming that the radium sands were "safe enough for children's sand-boxes", and pleaded with his fellow Senators to support nuclear power so that his kid would never be deprived of the benefits of electricity. Meanwhile, today, many of the homes and public buildings constructed from radioactive bricks and concrete have been condemned by State Public Health officials, due to excessively high radiation levels inside them. Some public schools were found to have radiation levels as high or higher than the uranium mines where Navaho miners are now dying in

large numbers. To date, not one single corporate executive or public official has been fined a single dollar, or served a day in jail, for these calculated and irresponsible, indeed murderous actions. (For more details, see *The Atomic Establishment*, by H. Peter Metzger, Simon & Schuster, NY, 1972)

* Given the massive problems associated with the storage and disposal of radioactive waste, which is piling up at a fast rate at nuclear reactor sites all over the country, the US Nuclear Regulatory Commission has come up with a marvelous solution: DEREGULATION. That's right. It worked for the Savings and Loan Industry, so why not for nuclear waste! Two marvelous terms, with suitably bureaucratic acronyms, have been developed to sooth public concerns: "Below Regulatory Concern" (BRC) wastes, which could be disposed of by incineration, or dumping into landfills or sewers, and "Exempt from Regulatory Control" (ERC) wastes, could be "recycled" into consumer products. These new policies would allow the unmonitored and unregulated disposal of nuclear wastes, and also allow unrestricted public use of radioactive lands and buildings at nuclear sites. The ERC-BRC formula, would reduce nuclear waste disposal costs by $82 million each year, and release between 30% to 60% of the projected volume of "low level" radioactive waste from the nations nuclear power plants. A report on the proposed measures is available from *Public Citizen Critical Mass Energy Project.*, 215 Penn. Ave., SE, Washington DC 20003.

* Over 2000 protestors converged on the Nevada Nuclear Test Site last March, in a "Global Action to End the Arms Race". Simultaneous protests organized by *American Peace Test* and *Parliamentarians for Global Action* took place in the USA, England, West and East Germany, the Soviet Union, and Japan, calling for a Comprehensive Test Ban Treaty. Representative of the USSR's "Nevada Movement" came to the US Nevada protest site, which was also attended by representatives of the Shoshone Nation. Some of the demonstrators were arrested for symbolically trespassing onto federal land (where trees were planted by the protestors), and for blocking Highway 95. The USSR, as of July 1990, has ceased all underground nuclear testing following demonstrations by citizens at its Semi-Palatinsk testing site; the USA has continued testing, though at a reduced rate in recent months. The goal of the American protests is to focus media attention on the issue. (What? Your local newspaper and news magazine didn't report on these mass protests?). (*Earth Island Journal*, Summer 1990)

Atomic Bomb Test Hotline!!
* For up to date information on the most recent nuclear tests, world-wide there is now a tape-recorded message on a telephone "Hotline" maintained by the group *American Peace Test*. For the Hotline, call 702/ 731-9646. To speak to a human being, call 702/ 731-9644.

Atomic Light Bulbs!!
* Certain brands of the new compact, screw-in fluorescent lights bulbs contain radioactive isotopes, much in the manner of radioactive smoke-detectors. The notation "15 nano Ci Kr-85" on the package indicates that the package of bulbs contains a small portion of radioactive krypton-85, a radioactive waste product of the nuclear industry that emits beta and gamma radiation. Other types of compact fluorescent bulbs contain promethium-147 and tritium. When the bulbs are discarded, the radioactive materials, as well as the toxic mercury inside them, eventually find their way into the atmosphere or groundwater. Another reason to keep away from fluorescent lights. (*Earth Island Journal*, Summer 1990)

* A new unit of radioactivity equal to 100,000,000 curies has been defined, and is now called the "Ch", for Chernobyl. More than 1000 Ch have been released by nuclear testing since the 1940s, and there are 27,000 Ch of radioactivity in the world's stockpile of spent nuclear fuel rods. This latter figure will dramatically increase over the next several decades, as several aging nuclear reactors are decommissioned, creating thousands of tons of high-level radioactive waste. (*Radiation Events Monitor*)

* Five thousand tons of radioactive milk, contaminated by fallout from the Chernobyl disaster over four years ago, is currently in dry storage in Germany. They are contemplating what to do with it, whether to sell it as milk, or as compost. Meanwhile, in the Netherlands, milk contaminated with dioxin, from toxic incinerators located too close to pasturelands, sits and waits for a decision. The Dutch Environmental Ministry wants to destroy the milk, while the Agriculture Ministry wants to mix the contaminated milk with uncontaminated milk, and sell it to the public. (*Earth Island Journal*, Summer 1990)

* Clean up and medical costs for the Chernobyl nuclear disaster will cost the Soviet Union some $320 billion over the next decade. Almost 4 million people are still living in regions with a higher than normal radiation level, and the call recently went out to evacuate 100,000

people from some of the worst areas in Byelorussia, Ukraine and the Russian Republic. A 10-fold increase in thyroid cancer is expected, due to radiation exposures of children. A new disorder, called "Chernobyl AIDS" has developed, affecting adults and some two million children, with the symptoms of leukemia, sarcoma, thyroid cancer and other related conditions of radiation poisoning. Toxemia of pregnancy, premature and still births, and birth defects, are also occurring at high rates. Mutations of plants and animals are also widespread. Official figures cite only 31 deaths from the reactor explosion, with another 250 reactor and clean-up workers dead by 1989 from various causes. (*Ukranian Echo*, 2 May 1990, *Newsletter, Women's Int. Coalition Against Rad Waste*, March 1990)

* Fred Anderson lives near the Seabrook nuclear power station, and often picks up stray radio broadcasts from the control room at Seabrook. He found the conversations to be so shocking, indicative of sloppy and dangerous work habits, that he decided to tape record the broadcasts. When the Nuclear Regulatory Commission found out about the tapes, they were not at all interested in investigating the safety situations at Seabrook. Instead, they subpoenaed Anderson and seized the tapes, in the manner to cover up the affair. (*Greenpeace*, July/August 1990)

* Of East Germany's four operating nuclear power plants, three were recently shut down due to shocking safety problems. The fourth remains open as it is currently the only source of electricity for the area it serves. All were built according to the same Soviet design. (*San Francisco Chronicle*, 30 August 1990)

* Structural engineer Paul Nestel, hired by the Department of Energy to survey the nuclear weapons plant at Oak Ridge, Tennessee, was fired after reporting that the unreinforced clay tile walls of the main Y-12 nuclear plant structure were not strong enough to withstand a moderate earthquake. Other engineers rewrote the study to say that the walls were strong enough. (*San Francisco Chronicle*, 9 Dec. 1989)

* For over 13 years, a minority of scientists and public health experts have asserted that sodium fluoride in the drinking water of American cities was a health risk that had not only failed to protect the teeth of children, as it was claimed, but which did lead to higher incidences of cancer in the fluoridated cities. A public outcry followed the publication of Dr. John Yiamouyannis' findings on this question, but it took a long time for the government to fund a new study addressing the problem. Finally, the results of the new study are in. The new study supports Yiamouyannis, and demonstrates that sodium fluoride might be causing bone cancer in experimental mice. Male rats given fluoride in doses of 79 parts per million also exhibited signs of osteoclerosis. The findings were cautiously endorsed by Dr. Michael A. Gallow, who led the research team, and a report is now winding its way through the federal bureaucracy, eventually to wind up at the Food and Drug Administration. Predictably, the study was attacked by the American Dental Association, which has staked its reputation to the previously untested assertion that fluoride was "safe". (*Acres, USA*, July 1990)

* Meanwhile, the same Food and Drug Administration that allows toxic fluoride and other questionable chemicals in drinking water and toothpaste has recently banished over the counter sales of 218 homeopathic remedies, to include 74 botanical preparations, including ginseng, asparagus, cucumber, and papain. To get these remedies now, you will have to obtain a prescription from an M.D. or O.D. physician. But surprise! The Catch-22 is that the vast majority of trained homeopathic healers are not M.D's. or O.D's! So the ruling is a de-facto banishment of those products, which by definition do not contain any *molecules* of the original substance, only the diluted energetic fields of the original substance. If this contradiction is not apparent, please realize that the FDA is currently prosecuting and trying to jail people who use such energetic methods for healing, on the claim that the life energy "does not exist".

* A recent study was published in the *British Journal of Clinical Pharmacology* (27:329-335, 1989), on a test for the effects of homeopathic flu medicines against placebo. The study was performed by non-homeopathic practitioners in a "double blind" setting, where neither the practitioners nor the patients knew the composition of the preparations. The result: homeopathic preparations showed a significant recovery rate for cold sufferers given the homeopathic preparations, as compared to the placebo. (*Health Freedom News*, September 1990; *Medical Self-Care*, March 1990)

* The *British Medical Journal* (302:316-323, 9 Feb. 1991) recently published a major review of 107 different clinical studies on homeopathy. Of the 107 studies examined, 81 showed positive results, significantly better than placebo controls. Of the best-constructed 25 studies, 15 showed clear positive results of homeopathic treatments for a variety of ailments. The authors of the review indicated they would be more inclined to accept the results of homeopathy if they could only understand the *mechanism of its action*.

* Americans presently spend approximately 12% of their total GNP on the "health care" industry, a truly large, Big Business which includes pharmaceuticals, hospitals, doctor bills, "health" insurance, and so on, from the expensive, unnecessary and Medieval rituals and surgeries performed at birth, to the equally expensive and unnecessary methods for prolongation of the death process.

* In 1987, there were 9.8 trillion cigarettes produced worldwide. If laid end to end, they would circle the planet 21,739 times.

* More evidence has surfaced to support Reich's priority on the unity of psyche and soma in the cancer biopathy. Some 13 years ago, Dr. David Spiegel of Stanford University began an evaluation of the short-term effects of group therapy on patients with advanced breast cancer. His study, recently published in the British medical journal *Lancet* (14 Oct. 89), indicated that the therapy lengthened the life of female patients with metastatic breast cancer by 1.5 years, and reduced anxiety and pain as well. (*Science*, 27 Oct. 89)

* New Medicare regulations, passed by executive decree and not through legislation, threaten to further undermine health freedoms in America. If a patient under treatment through Medicare self-treats an illness through use of some alternative strategy, such as vitamins or homeopathic treatment, or if they go to a chiropractor or some other practitioner not "approved" by Medicare bureaucrats, they may lose all rights to Medicare privileges -- this is so, even if they pay for such alternative treatments out of their own pocket! This is an attempt to keep the patient "obedient" and "in-line" with the normal regimins of orthodox medical treatments, which may or may not be effective against health problems, but which may have undesired side effects. The doctors are rendered "obedient" through a new requirement to tell the patient that any holistic, natural treatments provided are "experimental". If the doctor fails to tell the patient, they must return any payments made, and could be subject to an additional stiff fine. Of course, no similar mandate has been given to warn the patient about the potentially dangerous and experimental nature of many "accepted" orthodox medications and treatments, nor about the lack of controlled studies to demonstrate the efficacy of such treatments. These new regulations come at a time when the FDA is increasing its police actions against holistic health practitioners, health food stores, and nutritional supplement companies across the USA.

* A recent study indicates that as many as 50,000 deaths occur each year related to the poor quality of food in hospitals. Many hospitals serve up hot dogs, french fries, milk shakes (with fake milk), white flour and white sugar products, as if there were no problem with these foods. The practitioner directly responsible for this food-related sickness is the *Registered Dietician*, whom the hospitals routinely employ to "look after" the nutritional needs of patients. Registered Dieticians are also the primary supervisors of school lunch programs, which often serve up similar unhealthy junk-foods to children. Amazingly, the FDA and other "health care" institutions are engaged in a war against doctors and nutritional consultants who advocate a healthier diet, or who employ dietary methods for the treatment of illness. We propose a new public health slogan, to be taught to all school-children: *Just Say "No Way!" to Junk Food in Schools and Hospitals.* (Health Freedom News, ~~PO Box 688, Monrovia, CA 91016~~)

* According to the Congressional Budget Report, cited by the House Ways and Means Committee, over the last decade the US experienced the greatest "redistribution of wealth" in history; as you may have suspected, the rich got richer, while the poor got poorer. The tax rate for the poorest 20% of American taxpayers increased by 16.1%, while their real income only rose by 3.2%, which is much less than inflation. However, the richest 20% of Americans enjoyed a dramatic 31.7% increase in real income, while their tax rates actually *dropped*... by 5.5%! The richest 1% of Americans made out even better; their average yearly income increased from $280,000 to $550,000, while their tax rates declined by 25%. Americans in the upper 1% now pay an average of $40,000 *less* in income taxes today than in 1980. Wage earning potential also shows a similar disparity. Corporate executives showed a 149% inflation-adjusted increase in actual wages, from $116,000 to $289,000. However, hourly wages for workers in the largest corporations actually declined by 5% after inflation. (*Los Angeles Times*, 6 February 1990; *San Francisco Chronicle*, 31 August 1990)

* The wealthiest individual in Great Britain? Why, Queen Elizabeth, of course, with a royal fortune of $12.7 billion, including hundreds of royal masterpiece artworks, royal antiques which require a 75-volume catalog to summarize, a 330-volume royal stamp collection, royal jewels set in more than 20 royal tiaras, tens of thousands of royal acres of land, and a royal stock portfolio worth around $4.8 billion alone! Compared to this, Imelda Marcos and Leona Helmsley are rank amateurs. Of course, they only had one lifetime to accumulate their booty, whereas the royal monarchs of Great Britain have been raking it in for centuries! (*San Francisco Examiner*, 16 September 1980)

* *"One of the problems with understanding the dimensions of the thrift crisis is that the $500 billion cost now being predicted is just too large to comprehend. It doesn't relate to anything in our experience. Yesterday's newspapers and media contained the good news that U.S. agricultural officials predict a record breaking wheat crop this summer with Kansas expected to produce some 460 million bushels. The price of hard red winter wheat at the Beloit Kansas [grain] elevator was $3.83 a bushel. Consider that if Kansas farmers should harvest all of that record breaking crop, and if the price were to hold steady at $3.83 through harvest, and if we sent every last dollar from the sale of Kansas' wheat crop to Washington, D.C. to pay for the thrift bailout... IT WOULD TAKE THIS YEAR'S CROP, AND ANOTHER ONE JUST LIKE IT EVERY YEAR FOR 284 MORE YEARS TO PAY THE BILL OF $500 BILLION! That thought is stunning, yet the American taxpayer seems to be uncomprehending and stoic about the whole thing, and no one is worrying very much about who did what and why. Another way to look at it is to take the population of your own town times the $2000 per capita cost to every man, woman, and child in the U.S... Haven, Kansas, you owe $2,250,000. . . Great Bend, make your check out for $33,216,000 . . . Salina, Kansas, your bill comes to $83,686,000. . . Overland Park, get your checkbook out and send $163,568,000. That sure makes for a lot of bake sales and car washes, doesn't it! And think what we could do for economic development and quality of education and health care with those dollars. This isn't some "pie in the sky" scenario... The money is gone! Vanished. And the balance is due and owing to the federally insured depositors of failed thrifts. Right now! Of course, we may pass along the payment to our children and grandchildren, but sometime, Americans will pay, and pay dearly. And it could have been so easily avoided... Will the responsible people ever be held accountable? Yes, they surely will be, but the way things look now, it will be in some future history books."* Harold Stones, Kansas Banker's Association, as quoted in *Acres, USA.*

* According to Russell Mokhiber, editor of the *Corporate Crime Reporter* (PO Box 18384, Washington, DC 20035), white collar "crime in the suites" dwarfs the costs from more widely publicized street crime. *"The dollar cost of corporate crime in the United States is more than 10 times greater than the combined total from larcenies, robberies, burglaries, and auto thefts committed by individuals"*, amounting to some $100 billion or more each year. Corporate crime is so commonplace, says Mokhiber, that roughly two-thirds of this nations 500 largest companies were involved in some kind of illegal behavior during the past decade alone. In spite of this lawless behavior, there is no regular white-collar crime program at the FBI. *"The government can tell the public whether burglary is up or down in Los Angeles for any given month, but cannot say the same about insider trading, midnight dumping [of toxic wastes], consumer fraud, or illegal pollution."* Mokhiber's study of this issue was completed prior to the current Savings and Loan scandals, and does not reflect the fraud and over-charges paid out to Pentagon arms merchants, or the looting of the treasury which took place in the Federal Housing program over the last decade. These would considerably push up the total costs from corporate crime. When offending corporations or CEO's are caught in their wrong-doings, they usually are fined a sum which is far less than their ill-gotten gains, making corporate crime a profitable enterprise; jail sentences are rarely handed out.

Advice for New World Order

* Rumor has it that Reich once gave the following general advice regarding voting: *"Whomever is in power, vote for the opposing candidate. Keep recycling the politicians, such that no one individual or political party ever gains too much power".* We cannot verify if Reich actually said this or not, but it is sound advice. The "Great Compromise" and tripartate institutions worked out by the early colonists were designed primarily to accomplish just this: *to divide the powers up such that it would become very difficult for any one individual or group to gain too much political control and power.* One bit of advice we know for certain originated with Reich: *WORK, NOT POLITICS!"*

POSITIVE SIGNS AND CHANGES

* Several major companies selling canned tuna recently agreed to purchase from tuna fishermen who certifiably use fishing methods that do not harm or kill dolphin.

* Earth Day passed in April 1990, initiating a host of new environmental products into the marketplace, some of which are genuine, and some a sham. There's a new magazine that addresses many of these products, and focuses upon the problem of recycling. The magazine *Garbage*, is an excellent resource guide for recycled products, and carries important information on related environmental concerns. (*Garbage*, PO Box 51647, Boulder, CO 80321, $21 per year)

* Schoolchildren in Minneapolis recently protested the used of disposable plasticware in the school cafeteria, following a teacher's talk about recycling. The no-plastic movement led to a petition drive by the students, calling for a return of washable silverware and plates in the cafeteria. School officials responded by calling in an industry representative to lecture the kids about the "benefits of plastic", which provoked the kids to stage a sit-in protest. They were suspended from school for a day and a half, and we have no word as to what happened to the teacher who gave the talk on recycling. School officials apparently relented, however, and now say they will spend the bucks to equip the schools with dishwashers and reusable tableware. (*Earth Island Journal*, Summer 1990)

* West Germany has announced a complete ban on chlorofluorocarbons (CFCs) by the year 1995. Also, undertakers will no longer be allowed to use para-chlordibenzene in the cremation process, which releases dioxins into the ground-water near cemeteries.(*Earth Island Journal*, Summer 1990)

* Solar Electric scooters and automobiles, capable of highway speeds and with a range of 50 miles, are now being constructed and sold by the *Solar Electric Company* (175 Cascade Ct., Rohnert Park, CA 94928, 707/ 568-1987). The electric vehicles can be recharged from either household electrical sockets, or from the sunlight. In a recent test run, 15 different sun-powered cars competed in the 1990 American Tour de Sol, a 234-mile race from Vermont to Massachusetts. One of the solar vehicles made the trip in just over 6 hours (Contact: NESEA, 24 Ames St., Greenfield, MA 01301).

* There was a 17% growth in wind-generated electricity in 1989, according to the *American Wind Energy Association* (AWEA). California and Hawaii have more than 14,000 commercial wind turbines in place, generating some 1400 megawatts, at a cost of 7-9 cents per kilowatt hour. According to AWEA and the Department of Energy, the use of windpower could easily expand to 10-20 times its current usage by the year 2000. (Contact: AWEA, 1730 N. Lynn St., #610, Arlington, VA 22209; 703/ 276-8334)

*   The Report and Proceedings of the First International Ecological City Conference 1990, is available from *Urban Ecology* (~~Box 10144, Berkeley, CA 94709, $9~~). The 128 page illustrated report covers the work of more than 151 presenters from 14 nations, with insights on how the urban environment can be structure in an ecological, sustainable manner. Anyone listening?

*   While most loggers and lumber companies in the Pacific Northwest feel they have the "right" to clear-cut and deforest the landscape, there are a few lumber firms which have determined to harvest lumber according to sound, time-tested ecological practices. The Wild Iris Forestry, for example, harvests, mills, and cures hardwood and second-growth softwood in a manner that minimizes the damage to the forest ecology, and allows for a sustainable yield. They use existing roads to prune or fell trees that are carefully selected to reduce wildfire hazard and facilitate regeneration of new trees. They use all-age/all-species forest management techniques, in conjunction with restoration of natural drainage. "Reforestation forestry" is what the Iris' call the technique, which offers alternative employment to local workers who are uncomfortable with clearcutting. In short, there are ecologically sound alternatives for obtaining lumber, alternatives which guarantee the long-term viability of all creatures of the forest. Contact: *Wild Iris Forestry Catalog*, ~~PO Box 1423, Redway, CA 95560~~.

*   *The Power of the States: A Fifty State Survey of Renewable Energy*, by Nancy Rader, et al, is now available from the *Public Citizen Critical Mass Energy Project* (~~215 Penn. Ave. SE, Washington, DC 20003~~). Among the Survey's findings:
- California, Maine, Georgia and Washington are among the states leading the nation in developing solar, wind, hydroelectric, geothermal and biomass energy resources.
- Texas, Kansas, Indiana, Illinois, and New Jersey are among the states most lagging in renewable energy development.
-   Renewable energy resources now supply 13% of the nations electrical production, and nearly 10% of the US domestic energy supply. Nationwide, renewable energy provides 23% more energy than does nuclear power, and exceeds nuclear's contribution in 34 states.
- Renewable energy produced in the USA has reduced the CO_2 emissions by more than 550 million tons annually, an amount equal to the output of 138 typical coal-fired power plants.
- If every state had developed its renewable energy resource base as have Maine, Mississippi and California, the US would now derive about 23%, instead of only 10%, of its domestic energy from renewable resources.
-   Renewable energy use in California provided more electricity than the states five operating nuclear plants.

*   A landmark study on ecological, organic agricultural techniques, titled *Alternative Agriculture*, was recently published by the National Academy of Sciences; it has vindicated the long-held viewpoints of many small farm and environmental advocates, that farmers who use few or no chemicals can be as productive as those who use pesticides and synthetic fertilizers. (*Alternative Agriculture*, National Research Council, National Academy Press, Washington, DC 1989)

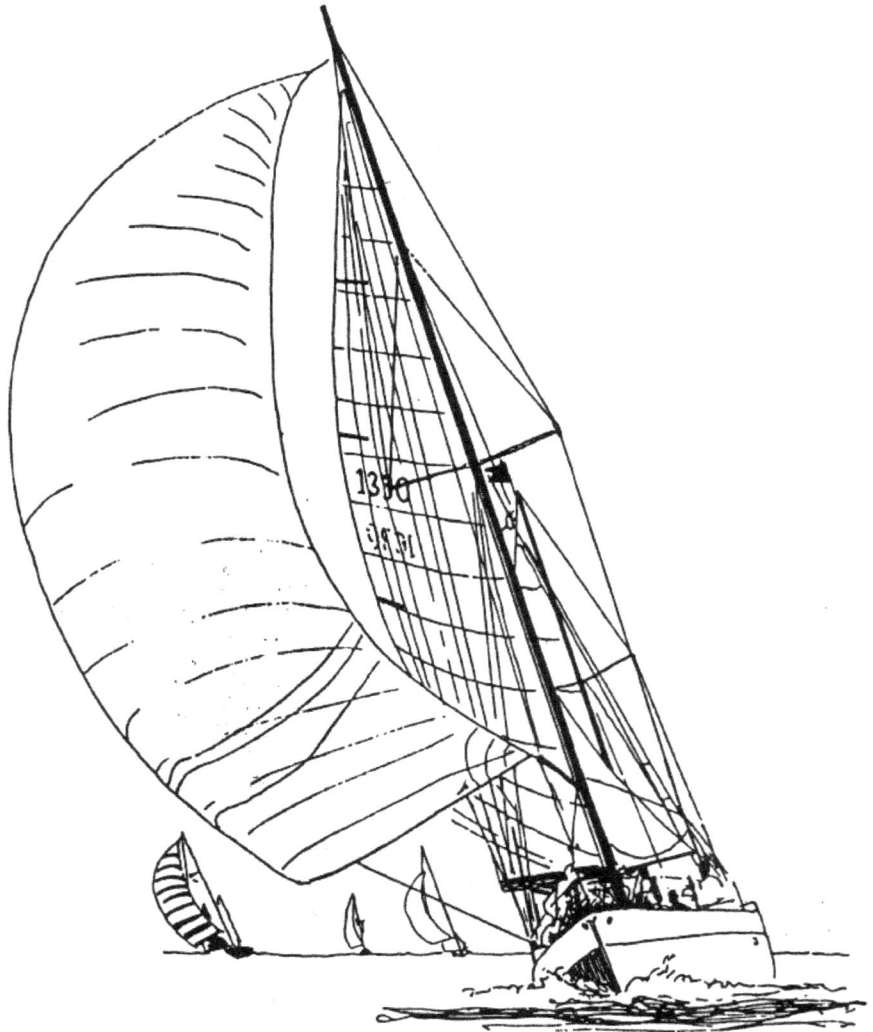

Appearing in *Orgonomic Functionalism* Volume 1, Spring 1990:
(Wilhelm Reich Museum, PO Box 687, Rangeley, ME 04970)
- The Developmental History of Orgonomic Functionalism, Part 1, by Wilhelm Reich, M.D.
- From *Homo Normalis* to the Child of the Future, by Wilhelm Reich, M.D.
- A note on "Sympathetic Understanding", by Wilhelm Reich, M.D.
- The Silent Observer, by Wilhelm Reich, M.D.
- Functional Thinking, A Discussion with Wilhelm Reich

Appearing in *Orgonomic Functionalism* Volume 2, Fall 1990:
- The Developmental History of Orgonomic Functionalism, Part 2, by Wilhelm Reich, M.D.
- The Silent Observer, by Wilhelm Reich, M.D.
- Wrong Thinking Kills, by Wilhelm Reich, M.D.
- On Using the Atom Bomb, by Wilhelm Reich, M.D.
- Man's Roots in Nature, Lecture and Discussion, by Wilhelm Reich, M.D.

Appearing in the *Journal of Orgonomy,* 23(2), November 1989:
(Orgonomic Publications, PO Box 490, Princeton, NJ 08542)
- Further Problems of Work Democracy (V), by Wilhelm Reich, M.D.
- In Seminar with Dr. Elsworth Baker
- Flight from the Essential, by Giuseppe Cammarella, M.D.
- Work Energy and the Character of Organizations (II) by Martin D. Goldberg, M.S.
- The Creation of Matter in Galaxies (III), by Charles Konia, M.D.
- SAPA Bions, by Mark R. Diamond, M.Psych. & Daniel D. Reidpath, Dip.Ed.Psych.
- Somatic Biopathies, by Charles Konia, M.D.
- Orgone Therapy (IX), by Charles Konia, M.D.
- Infant Cranial Deformation and Swaddling, by James DeMeo, Ph.D.
- The Orders of Function, by Jacob Meyerowitz, B.Arch.
- CORE Progress Report #21, by Richard Blasband, M.D. & James DeMeo, Ph.D.
- The F.D.A's. Evidence Against Wilhelm Reich: Postscript, by James DeMeo, Ph.D.

Appearing in the *Journal of Orgonomy,* 24(1), May 1990:
- Further Problems of Work Democracy (VI), by Wilhelm Reich, M.D.
- Radiation Victims and the Reich Blood Test, by Richard Blasband, M.D., et al.
- The Creation of Matter in Galaxies (IV), by Charles Konia, M.D.
- The Orgonotic Treatment of Transplanted Tumors, by Ernani Trotta, Ph.D. & Eugenio Marer, M.S.
- A Man on the Horns of a Dilemma, by Barbara G. Koopman, M.D., Ph.D.
- Acute Catatonic Withdrawal in a Three-Year-Old Child, by Charles Konia, M.D.
- Treatment of a Child with Elective Mutism, by Morris Osborn, M.D.
- Orgone Therapy (X), by Charles Konia, M.D.
- Work Energy and the Character of Organizations (III), by Martin D. Goldberg, M.S.
- Male Genital Mutilations, by James DeMeo, Ph.D.
- Core Progress Report #22: OROP Arizona 1989 (II), by James DeMeo, Ph.D.
- Bechamp's Microzymas and Reich's Bions: Similarities and Differences, by Bernard R. Grad, Ph.D.

Appearing in the *Journal of Orgonomy,* 24(2), November 1990:
- Further Problems of Work Democracy (VII), by Wilhelm Reich, M.D.
- In Seminar with Dr. Elsworth Baker
- The Creation of Matter in the Galaxy, by Charles Konia, M.D.
- Wilhelm Reich and UFO's, by Peter Robbins
- Somatic Biopathies (II), The Diaphragmatic Segment, by Charles Konia, M.D.
- Structural and Symptomatic Change and Early Therapeutic Termination, by Gary A. Karpf, M.D.
- An Adolescent in Orgone Therapy, by Jack Sands, M.D.
- A Case of Manic Depressive Character with Dissociation, by C. Andrews, M.D.
- Work Energy and the Character of Organizations (IV), by Martin D. Goldberg, M.S.
- Desertification and the Origins of Armoring (VI): Female Genital Mutilations, by James DeMeo, Ph.D.
- The Primordial Universe, by Jacob Myerowitz. B. Arch.
- CORE Progress Report #23: OROP Arizona 1989 (III), by James DeMeo, Ph.D.

Research and Publications Review (Continued)

Appearing in *Annals, Institute for Organomic Science,* 6(1), September 1989:
(PO Box 304, Gwynedd Valley, PA 19437)
- Reich's Experiment XX, by Robert Dew, M.D.
- Reports on Treatments with Orgone Energy, by Dorothea Opfermann-Fuckert, M.D.
- Human Armoring: An Introduction to Psychiatric Orgone Therapy, Chapter 6, by Morton Herskowitz, D.O.

Appearing in *Annals, Institute for Organomic Science,* 7(1), September 1990:
- Further Observations on the Air Germ Experiment, by Robert Dew, M.D.
- Orgone Treatment of Sprouting Mung Beans, by C.F. Baker, M.D. and P. Burlingame
- Human Armoring: An Introduction to Psychiatric Orgone Therapy, Chapter 7, by M. Herskowitz, D.O.

Appearing in *Pulse of the Planet* No.1, Fall 1989:
(Orgone Biophysical Research Lab, www.orgonelab.org)
- The Orgone Energy Continuum: Some Old and New Evidence, by James DeMeo, Ph.D.
- Earthquakes and Nuclear Testing: Dangerous Patterns and Trends, by Gary T. Whiteford, Ph.D.
- The Psycho-Physiological Effects of the Reich Orgone Accumulator, by Rainer Gebauer, D.Psych. & Stefan Muschenich, D. Psyche.
- Wilhelm Reich: Discoverer of Acupuncture Energy?, by Prof. Dr. Bernd Senf

Appearing in Other Publications:
- "Hidden Psychological Motives", by James DeMeo, Ph.D., *NOCIRC Newsletter*, 4(2):1, Summer 1990.
- "The Origins and Diffusion of Patrism in Saharasia, c.4000 BCE: Evidence for a Worldwide, Climate-Linked Geographical Pattern in Human Behavior", by James DeMeo,Ph.D., *Kyoto Review*, 23:19-38, Spring 1990, and *World Futures*, 30:247-271, 1991.
- "Patrism in Saharasia: Climate-Linked Human Behavior", by James DeMeo, Ph.D., *Women's International Network News*, 16(3):75-77, Summer 1990.
- "Orgone Energy for Everyone", by Ma Aneessha, *Osho Times International*, 3(22):18-19, 16 Nov. 1990.
- "Aquarian Manifesto", by Joseph Terrano, and "Bioelectric Energy Love", exerpts from various works by WilhelmReich, *New Frontier*, Philadelphia, PA, May 1990.
- "Wilhelm Reich's Legacy: The Orgone Cure", *Psychic Reader*, 13(4), San Rafael, CA, April 1990.
- "Cloud Nine: James DeMeo Does the Reich Thing", by Stephen Stocker, *Baltimore City Paper*, 2 Nov. 1990, p.8-10.

USA: Organomic Institutes, Research Centers, Sources of Information and Publications
- *American College of Orgonomy, Journal of Orgonomy*, PO Box 490, Princeton, NJ 08542
- *Eden Press,* Book Catalog, PO Box 399, Careywood, ID 83809
- *Institute for Orgonomic Science, Annals*, PO Box 304, Gwynedd Valley, PA 19437
- *Natural Energy Works*, Book Catalog, www.naturalenergyworks.net
- *Orgone Biophysical Research Lab, Pulse of the Planet,* www.orgonelab.org
- *Wilhelm Reich Museum, Orgonomic Functionalism*, www.wilhelmreichtrust.org

Foreign: Organomic Institutes, Research Centers, Sources of Information and Publications
- *Center for Study of Bio-Energy, Bioenergy,* Kodatsuno 5-9-7, Kanazawa 920, Japan
- *Zentrum für Orgonomie, Lebensenergie,* Memelstr. 3, Eberbach 6930, Germany
- *Hellenic Orgonomic Association, Greek J. of Orgonomy,* T.O. 10837, Thessaloniki 30136, Greece
- *Societe por la Diffusion d'la Orgonomie, Sciences Orgonomiques*, Allee du Chene Vert, Parc Liserb, Nice 06000, France
- *Summerhill School, Friends of Summerhill Trust Journal,* c/o Hylda Sims, 280 Lordship Lane, London SE22, England
- *Wilhelm Reich Society, Emotion,* Karlsbergallee 25-E, D-1000 Berlin 22, Germany

Calendar of Forthcoming Events, with Contact Information:

1991:

* **July 22-25: Rangeley, Maine**, Workshop on "The Orgone Energy Accumulator", various speakers, including James Strick & Chester Raphael, Wilhelm Reich Museum, Orgonon, Box 687, Rangeley, ME 04970. (207) 864-3443. Registration by mail or telephone. Brochure available.

* **August 4: Seattle, Washington**, Sunday Workshop on "Body-Mind Behavior: Introduction to the Work of Wilhelm Reich". Instructor: Daniel Schiff, Ph.D., 9:30 - 3:30 PM. Fee of $50. Call for details: (206) 523-0800.

* **August 31 & Sept. 1-2: Pasadena, California**, 19th Annual Cancer Convention, various speakers on non-toxic, holistic, & unorthodox treatments for degenerative disease. Sponsored by Cancer Control Society, 2043 N. Berendo, Los Angeles, CA 90027 Telephone: (213) 633-7801.

* **September 14: Seattle, Washington**, Saturday Workshop on "Wilhelm Reich's Discovery of Orgone Energy: An Introductory, 'Hands-On' Workshop", with C. Grier Sellers, C.A. For more Information and a descriptive flyer, write to PO Box 31604, Seattle, WA 98103, Telephone: (206) 526-5956.

* **September 20: San Francisco, California**, Friday Evening Lecture by James DeMeo, on "Wilhelm Reich's Discovery of the Orgone (Life) Energy", Fort Mason Center (between Fisherman's Wharf and the Golden Gate, entrance at the intersection of Buchanan St. and Marina Blvd.), 7:00 PM to 9:00 PM, Admission $5, or free to High School or College Students with ID. No preregistration required.

* **September 21: San Francisco, California,** Saturday Workshop with James DeMeo on "Sex-Economy and the 4000 BCE Origins of Armoring (Patriarchal Authoritarian Society) in Saharasia", Fort Mason Center (between Fisherman's Wharf and the Golden Gate, entrance at the intersection of Buchanan St. and Marina Blvd.), 10:00 AM to 6:00 PM, Admission $95; half-price to High School or College Students with ID. Informative brochure available from James DeMeo, Orgone Biophysical Research Lab, PO Box 1395, El Cerrito, CA 94530, (415) 526-5978. Preregistration suggested, but registration at the door is OK on a space-available basis, subject to Dr. DeMeo's approval.

* **September 22: San Francisco, California**, Sunday Workshop with James DeMeo on "The Role of Life Energy in Ecological Destruction, Droughts, and Global Desert Spreading", Fort Mason Center (between Fisherman's Wharf and the Golden Gate, entrance at the intersection of Buchanan St. and Marina Blvd.), 10:00 AM to 6:00 PM, Admission $95; half-price to High School or College Students with ID. Informative brochure available from James DeMeo, Orgone Biophysical Research Lab, PO Box 1395, El Cerrito, CA 94530, (415) 526-5978. Preregistration suggested, but registration at the door is OK on a space-available basis, subject to Dr. DeMeo's approval.

* **October-November, Seattle, Washington**, (Dates to be announced) Weekly evening course on "Sex-Economy and Orgonomy: An Introduction to the Work of Wilhelm Reich". Instructor: C. Grier Sellers, C.A. For more Information and a descriptive flyer, write to PO Box 31604, Seattle, WA 98103, Telephone: (206) 526-5956.

* **October 6: Princeton, New Jersey**, Annual Scientific Meeting, American College of Orgonomy, Henry Chauncey Conference Center, contact ACO, PO Box 490, Princeton, NJ 08542, Telephone: (908) 821-1144.

1992:

The Sixth International Orgonomic Conference will take place in Europe, at a city and location to be determined. To keep informed about this event, contact the American College of Orgonomy, PO Box 490, Princeton, NJ 08542, Telephone: (908) 821-1144.

LETTERS TO THE EDITOR

Dear Editor,

Did you really think *Skeptical Inquirer* would print your rebuttal to Martin Gardner? You must be unaware of their history. They never print Heresy (conflicting views). That's why Prof. Marcello Truzzi resigned as their editor way back in the 1970s.

Your comments on Reich renting the accumulators as experimental devices and avoiding extravagent claims, etc. matches my memory, but, alas... I don't know where to start looking for a reference that makes this clear. If you could give me the name of a book and a page number, that would be a big time-saver. In the latest *Skeptical Inquirer*, Robert Sheaffer says Reich was selling the accumulator as a cancer "cure".

Keep the lasagna flying,
Robert Anton Wilson,
Los Angeles, CA.

Dear Mr. Wilson,

Regarding *Skeptical Inquirer*, I learned of their policy of refusing rebuttal, and of Truzzi's experiences, only some months after being attacked in its pages. Do you know why Issac Asimov and Stephen Jay Gould maintain their association with CSICOP? Carl Sagan, I can understand, as he knows everything...

Regarding Reich's accumulators, there is a good summary of the financial aspects of the accumulator in an article "Report on Orgone Energy Accumulators in the USA" by Ilse Ollendorff, in *Orgone Energy Bulletin*, 3(1):53-58, 1951. This article states that there were, between 1943 and the early 1950s just over 300 accumulators constructed by Reich's organization. These were rented for $10 per month to people who could afford it, but a good number were provided free of charge or at reduced rates to those who could not afford the full rental fee. Proceeds from the rental fees went directly into Reich's research foundation, and he saw none of the money himself. All of it was used to support his scientific research. They were clearly identified as experimental devices, requiring the prescription of an attending physician, and a signed affidavit, before the rental would take place. The income from accumulators ranged from $4,594 in 1946 to $23,000 in 1950, when Ollendorff's report was published. A few years after this date, Reich's work was attacked by the FDA. By court order the accumulators were then recalled from patients, though a few were also sold to those same patients at that time.

According to Professor Jerome Greenfield (*Wilhelm Reich Versus the USA*, W.W. Norton, NY, 1974, p.93), the FDA pursued Reich for over 10 years, spending around $2 million, or 4% of its total 10 year budget "...to stop the interstate shipment of, by 1954, some 300 orgone accumulators, an enterprise that — making allowances for the accumulators that were loaned out free or at reduced charges — in its maximal year earned only some $30,000." J.D.

Dear Editor,

In *The Cancer Biopathy*, Reich states "...the development of living plasma on our planet preceded the organization of coal substance and carbohydrates... biochemical molecules did not exist prior ...to the development of plasmatic substance, but appeared as one of the mechanical constituents in the process of plasmatic organization." Krishna Bahadur and his colleagues have, since the mid 1950s, studied the origin of life from the point of view of living function. The central idea is that function precedes structure. He starts with simple mixtures of minerals, a simple organic compound such as formaldehyde, and an energy source such as light. Interestingly, the reactions he observers proceed even in the dark -- a lead box is required to stop them -- an orgone accumulator has not been tried. Bahadur obtains structures of a few micron in size, which bud and divide, show internal structure, and chemical analyses show not only minerals, but also amino acids, nucleic acid bases, sugars, peptides, etc. There is evidence that these structures, called *Jeewanu* (sanscrit for "particles of life") are photosynthetic bions -- they are blue when they form, and the solution becomes blue, too. They fix atmospheric carbon and nitrogen, and split water under some conditions, and they can evolve hydrogen gas. Most other origin-of-life researchers think that structure precedes function. That "somehow, by chance" (or, by God) a DNA molecule developed eons of years ago in the "primordial soup"...

Bahadur is also one of the few researchers who takes seriously the possibility of "neobiogenesis" -- that is, contemporary biogenesis. In an early publication he cites Reich's *Cancer Biopathy*... Articles by Bahadur (see citations below) discuss research I witnessed at NASA where Bahadur's findings were confirmed by physicist Adolph Smith and others. An as yet unpublished paper describes additional findings: When a solution of Jeewanu is exposed to sunlight, the solution pH decreases; in the dark, it increases. Unfortunately, Bahadur was not able to fully develop his promising research on the photolysis of water by Jeewanu. In 1981, the work at NASA stopped, defunded like most other rational energy research in the USA. Bahadur has come under emotional plague attack in this country ...at NASA, "memorandums" were circulated criticizing his work, but his work was not openly criticized.

Grier Sellers, C.A., Dipl. Ac.
Seattle, Washington

Krishna Bahadur: *Synthesis of Jeewanu - The Protocell*, Ram Narain Lal Bani Prasad, Allahabad, India, 1966.
Krishna Bahadur: *Origin of Life: A Functional Approach*, Ram Nairain Lal Bani Prasad, Allahabad, India, 1981.
Krishna Bahadur, Adolph Smith, Et al.: "A Functional Approach to the Origin of Life Problem", National Academy of Sciences, India, Golden Jubilee Commemorative Volume, Allahabad, India, 1980.
Adolph Smith, C. Folsome, and Krishna Bahadur: "Carbon Dioxide Reduction and Nitrogenase Activity in Organo-Molybdenum Microstructures", *Experimentia*, 37:357-359, 1981.

Dear Editor,

I appreciate the tremendous amount of work that went into producing *Pulse of the Planet* #2. It is packed with valuable research, insights, hyptheses, and updates. The only point that I find to be pretty weak is on page 45, "What you can Do". Since you are attempting to put the work in your journal in the context of the human condition (armor, emotional plague) in a very bold, forthright, and, I feel, accurate way, why give the reader a tepid suggestion to "go to the library"? The reasons why we are in limited contact with nature and ourselves, many to a severe degree, and the ways we're destroying it, are imbedded in ourselves, our armor -- as per Reich's powerful discoveries. Why not stress the importance of orgonomic therapy, with a trained orgonomist, as the basic, real, available way for scientists and laymen alike to make greater contact with their anxiety as well as their natural urges to support the various life-important efforts (social, sexual, economic, scientific, environmental) which you call such valuable attention to. Anyone who takes the time to read your journal is trying to maintain or make greater contact in some real way. I would encourage you to avail them of the opportunity to understand therapy -- perhaps in the form of a short article by an orgonomist. Isn't that the most important individual context (for someone already expressing an interest in orgonomic research by subscribing to your journal) from which to begin "getting involved"?   John Bratnober,
Oakland, CA.

Dear Mr. Bratnober,

Your point is excellent, and it brings up a number of important issues. We would welcome an article on orgone therapy by any of the trained orgonomists, to introduce our readers to this important means for individuals to resolve personal conflicts and difficulties, and to feel more contactful and alive. Orgone therapy is always recommended for persons interested in pursuing orgonomic research and studies, for those involved in the handling of infants and children (parents included), or for those who just want to better their lives. I've observed that individuals who undergo the therapy will move their lives towards more pleasurable circumstances, and become more effective and productive in their work. Almost every researcher in the field of orgonomy has undergone such a program of therapy, myself included. The organizations which maintain a central focus upon Reich's findings, without diluting or distorting them, and who can refer individuals to prospective orgone therapists are: the *American College of Orgonomy* (PO Box 490, Princeton, NJ 08542 / telephone 908/ 821-1144) and the *Institute for Orgonomic Science* (~~PO Box 304, Gwynedd Valley, PA 19437~~). The *Wilhelm Reich Museum* (PO Box 687, Rangeley, ME 04970) is also in contact with a number of qualified orgone therapists. There are additionally a smaller number of qualified therapists without such organizational ties, but we do not maintain a listing for distribution.

Having said this, I must point to another important aspect of the problem of human armoring, namely the larger social institutions which often work to create and maintain the armor, and which generally discourage or actively suppress normal and healthy life functions. Efforts must also be directed towards this larger social armor, to give the new generations a fighting chance. Therapy can help individuals move and act more freely, and to spontaneously be more assertive and caring, but the larger cultural armor, imbedded in the social institutions, must also be addressed.

In the above context, a major problem and dilemma arises: if we must rely upon therapy as the major means to disarmor and get the larger society moving towards health, to institute meaningful social change, then catastrophe cannot easily be avoided. Conditions in much of the USA and Europe are deteriorating socially, environmentally, and economically; and yet, by global comparison, these are "regions of plenty". Outside the USA and Europe, the situation is often quite stark. In Africa, Asia, and Latin America, the population crisis has steadily worsened living conditions; food supplies are often inadequate; wars over natural resources and water are taking place already; in some areas, increases in mortality and desert formation are already taking place in a rapid, widespread, and historically unprecedented manner. Individual therapy as a means to initiate social change will require hundreds of years, while the present rates of environmental deterioration and population expansion forecast major crisis conditions in most areas of the world within the next 25-50 years. By comparison: more human armoring is currently being created in a single day in the USA than all of the trained orgonomists (and neo-Reichians combined) can collectively remove in an entire decade! Additionally, if even 1% of the entire US population (2.5 million people) decided they wanted orgone therapy, there presently aren't enough trained people to provide the needed services. These sobering facts were probably evident to Reich also, as he increasingly emphasized the *prevention of armor* in infants and children as a more hopeful and effective strategy than only therapy for adults, after the armor had formed.

In the context of the prevention of armor, there currently are a lot of individuals and organizations engaged in life-affirmative work. My encouragement for persons to "go to the library" was a suggested *first step* towards obtaining the names and addresses of national and local citizen's organizations which could be helpful for individuals facing a problem within their own home, community or workplace. There are, for example, dozens of environmental organizations, child-abuse and family violence organizations, midwives groups, and government accountability and civil rights organizations, each of which publishes informative newsletters, and holds regular local and national meetings. These groups mostly work on concrete problems, and usually have good solid answers and help to offer to others facing similar problems. Just by joining such groups, supporting them, and reading their publications, one will become far better educated about the issues and possibilities for hopeful change than if one only relies upon typical newspapers, popular magazines, or television, which today provide highly *processed information*, much like processed cheese or baloney. What one does after joining or contacting such organizations is, of course, a matter of individual character and free time. My encouragement has been for people to take responsibility for, and initiate meaningful change within their own spheres of personal influence: home, family, work, school, community. By making contact and work-democratic alliances with like-minded people on issues of mutual interest, one can make small or even large steps towards meaningful personal, family, and social change.

Social structures which create armoring and destroy life do not spontaneous cease doing so; they are usually infested with the emotional plague, and must be courageously confronted and changed. Towards this end, there are a lot of decent, hard-working people already engaged in hard work and struggle towards life-positive ends; but the tasks involved are enormous, and additional help is always welcome.   J.D.

Watercolor by Deborah Carrino